Proceedings in Life Sciences

R. Austin, E. Buhks, B. Chance, D. De Vault,
P.L. Dutton, H. Frauenfelder, and
V.I. Gol'danskii EDITORS

Protein Structure
Molecular and Electronic Reactivity

With contributions by
G. ALEGRIA, T. BARRETT, R. BECHTOLD, W. BIALEK,
M. BIXON, E. BUHKS, B. CAMPBELL, B. CHANCE, L. DISSADO,
P.L. DUTTON, S. ENGLANDER, G. FEHER, V.N. FLEUROV,
H. FRAUENFELDER, J. FRIEDMAN, V.I. GOL'DANSKII,
R. GOLDSTEIN, D. GUST, H. HARTMANN, J.J. HOPFIELD,
S. ISIED, J. JORTNER, D. KLEINFELD, G. KOVACS, J. KRUMHANSL,
Y.F. KRUPYANSKII, A.M. KUZNETSOV, J. MILLER, T. MOORE,
M. OKAMURA, G. NIENHAUS, F. PARAK, M. POPE, Z. POPOVIC,
H. RODER, H. STAPLETON, D. STEIN, E. STERN,
J. ULSTRUP, A. VASSILIAN, P. VINCETT, A. WAND,
A. WARSHEL, AND P. WOLYNES

With 100 illustrations

Springer-Verlag
New York Berlin Heidelberg
London Paris Tokyo

Robert Austin
Princeton University
Princeton, New Jersey 08544, USA

Ephraim Buhks
B.F. Goodrich Co.
Brecksville, Ohio 44141, USA

Britton Chance
Paul Leslie Dutton
University of Pennsylvania
Philadelphia, Pennsylvania 19104, USA

Don De Vault
Hans Frauenfelder
University of Illinois
Urbana, Illinois 61807, USA

Vitalli I. Gol'danskii
Academy of Sciences of the USSR
Moscow, USSR

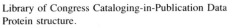

Library of Congress Cataloging-in-Publication Data
Protein structure.
    (Proceedings in life sciences)
    Proceedings of a conference held April 10–13, 1985
in Philadelphia, Pa.
    Bibliography: p.
    Includes index.
    1. Proteins—Structure—Congresses.   2. Proteins—
Reactivity—Congresses.   3. Proteins—Electric
properties—Congresses.   4. Biophysics—Congresses.
I. Austin, Robert H.   II. Series.
QP551.P69766   1987   574.19'245   87-16525

Text prepared by the editors using PC TₑX.
Printed and bound by R.R. Donnelley & Sons, Harrisonburg, Virginia.
Printed in the United States of America.

9 8 7 6 5 4 3 2 1

ISBN 0-387-96567-X Springer-Verlag New York Berlin Heidelberg
ISBN 3-540-96567-X Springer-Verlag Berlin Heidelberg New York

PREFACE

These are the edited proceedings of a conference held April 10-13 1985 in Philadelphia. The title of the conference was "Protein Structure: Molecular and Electronic Reactivity". This conference is the successor to an earlier conference held in Philadelphia on November 3-5, 1977, entitled "Tunneling in Biological Systems".

Britton Chance had proposed holding a similar conference to gauge our progress and had circulated a letter to that effect among interested parties. We are grateful to Professor John Hopfield for suggesting that since the 1977 conference had so carefully laid a solid foundation on tunneling it would be more interesting if the next conference would broaden its horizons and cover a wider perspective in biophysics, yet still remain firmly rooted in the PHYSICS of biomolecules. Professor Hans Frauenfelder made the suggestion that the number of talks be restricted and that time be set aside for frank and probing discussions.

The goals of this conference were to build from the initial ideas that arose in trying to understanding tunneling (electronic and nuclear) in macromoleucles, to explore new static/dynamic models of proteins and possible relation of these ideas to protein function (in the cell!!), and to explore a possible connection between protein physics and molecular machines, not necessarily in the cell.

Here then is the outgrowth of the three seminal suggestions and the assault on our lofty goals. Personally, I feel that we got about two-thirds of the way there towards the original goals. You will note that many of the talks in the conference are of a broad and reflective scope and summarize well our progress in comprehension of macromolecular tunneling since the last conference. A new and exciting perspective is given by the talks wrestling with the question of glass-like aspects of protein structures and dynamics. The physics connection remains intact, and in this period of tight budgets a new note has been injected: a lively debate will be found on the wisdom of doing biophysics for biophysics sake.

If I could fault anything, it would be in the over-reluctance of the participants to speculate on the "What if...." questions concerning molecular control and microengineering. Dr. Barrett's call for far-out ideas went unanswered. Perhaps the next time around we can prod the heavy hitters into going out onto a limb a bit more. In any event, in the course of editing these talks we have been informed, enlightened, entertained and made to think about the purpose of our life's labors. We hope you, the reader, will be similarly intrigued by the proceedings.

A note about the concluding talk by Brit Chance. We suspect that Brit got a bit concerned about the unremitting physical level of the talks and decided in his after-dinner speech at the banquet to give a "sweet dessert" about how physical techniques can lead to real clinical insight. We left the talk in the proceedings to remind the reader, as Professor Chance did the participants, of the very real problems that are calling for attention and the contributions we can make...a fresh wind into the ivory tower.

Two people made absolutely vital contributions to the success of the conference. Dr. Barrett of the Office of Naval Research provided critical funding and psychological support when things looked quite dark. Judy Flanagan, Associate Director of Communications at the University City Science Center was constantly in the thick of things searching for support, writing letters, organizing the actual proceedings, paying the bills, etc. etc. We suspect we aged Judy a bit in the process, but it was good for building her character. Further, Judy's coworkers Sarah, Suzanne, Rita, Chilton and Sandy helped make the actual proceedings a smoothly functioning machine. My valued colleague and friend Dr. Ephraim Buhks helped greatly in the editing of the charge transport section of the proceedings.

The creation of the actual manuscript is the result of two people: Dot somehow typed up preliminary drafts of the garbled tape recordings, and Lorraine Nelson, a TEX-nician of the first rank typed/typeset all the manuscripts into the computer. A note: Professor George Feher is the only author of equal TEX-nical abilities to Lorraine. Let us hope the talks from the next conference will be uniformly submitted in TeX.

Funding for the conference came from a number of sources:

Naval Air Systems Command

National Science Foundation (DMB 841-7085)

National Institutes of Health (GM-34592-01)

E.I. du Pont de Nemours, Inc.

Advanced Technology Center of Southeastern Pennsylvania

University Science Center

International Union of Pure and Applied Biophysics

We are most grateful for the generous support. Welcome to the written proceedings of a stimulating conference that we predict will over the years have some impact on how biophysics evolves.

Bob Austin

CONTENTS

Session IV: Charge Exchange in Proteins and Model Systems
Session Leader: E. Buhks

Session V: Charge Separation In Photosynthetic Reaction Centers
Session Leader: P.L. Dutton

Session VI: Applications

Welcome & Introduction
Dr. Terence Barrett, (NAVAIR)

Well, before getting down to the scientific presentations, I'd just like to take some time to very quickly explain the reason why the Naval Air Systems Command (NAVAIR) is interested in partially funding this conference. The reason is that we have problems in the field of biotechnoloy for materials applications, and those problems are in protein engineering.

We're considering using proteins in various ways which are sometimes unrelated to the *in vivo* functioning but derive from that functioning. We're also interested in using the results of the wild and wonderful field of biotechnology to obtain non-medical products.

Now, you can make a very good living from a small amount of insulin, for example, in the biomedical field. But small amounts are insufficient for economic survival in the commodity chemical field. There is no way, at the present time, to obtain bulk chemical production from biological processes. So we are interested in using protein engineering in some way to increase the yield from these biological processes.

We also see problems, which we hope you will address, if you are interested, in the separation of product from byproducts and microbes, and also in the ability of enzymes to function in various solvents.

We see problems particularly in the field of bioenergetics. For example, we don't understand how, if you have a high energy phosphate bond of only 0.46 e.v. usable for work purposes, how that energy can be transduced into an enzyme. Now there are various Ansätze for solving this vibrational energy transfer problem. And if you object to this approach, well, the reply is: If not these – what? Because one has to understand how you transduce that energy amount in order to obtain engineering control of enzymatic reactions. And you must also understand how you transduce that energy amount in order to have a rigorous science of bioenergetics.

So this endeavor comes under the general rubric of *molecular control** or the

---

* *Molecular Control,* a joint NAVAIR (Code AIR-310P) and NASA (Code EB) booklet.

obtainment of controlled energy transfer to molecular systems for useful purposes. We're looking forward to entropy engines and the understanding of how this energy transfer will enable us to *energy harvest*. Not only is NAVAIR interested in this, but so is NASA.

One would, as I say, hope for molecular control and e.m. field control of enzymes to speed up the reactions to get the yield we want. Now this is not easy to obtain, because this involves us in considering both classical and quantum mechanical resonance energy transfer for laser specific chemistry. The conceptual framework is there for classical energy transfer. It is not easy, and the conceptual framework has not been completed for quantum mechanical resonance energy transfer.

So, in order to solve our problems we are interested in developing the idea of molecular engines, that is: entropy and internal energy enzymes, which will bring in its wake a fundamental understanding of how proteins function *in vivo*. The transfer of energy in an entropy engine fashion is *not* represented in the case, say, of rhodopsin in which you have a one-shot transfer of energy. Unless you pump in energy using respiratory enzyme sources, the rhodopsin will not function again. But one might take bacteriorhodopsin, say, as a representative case of an entropy engine, which is able to regenerate itself for a second cycle and do useful work, using the external photonic energy source for both these processes.

All this boils down to some very fundamental problems and although an approach is beginning to be developed and understood in classical mechanics – namely, that one must, in order to achieve these goals, circumnavigate chaotic events to reach quasi-periodicity – we don't understand at the present time how you achieve resonance for energy storage in quantum mechanical systems.

So, in summary, the reason why we decided to partially fund this conference is that we have a number of problems and in this very fast review, I hope have indicated some of them. In the next three days, if any of these strike your interest, well, please come up and we'll have a discussion about it.

Debye-Waller Factor in Solid State and Biological Samples

E. Stern

Physics Dept., University of Washington, Seattle, WA 98195

Abstract

The Debye-Waller factor is discussed as it relates to X-Ray diffraction and to EXAFS.(Extended X-Ray Absorption Fine Structure). The two measure different quantities and are complementary. Biological metalloproteins in solution have a temperature dependence of the EXAFS Debye-Waller factor at the metal site which is similar to that of crystallized small molecules. Expressions for the free energy and entropy contributions of the vibrations that cause the Debye-Waller factor are presented. EXAFS measurements permit one to estimate these contributions. In the metalloprotein hemerythrin that serves to transport and store dioxygen, it is found that a large part of the entropy change as O_2 is released is localized at the active site and is accounted for by changes in the vibrational modes.

I. INTRODUCTION

There are two Debye-Waller factors (DWF) involved in structure determinations. In discussing these DWFs the assumption will be made initially that the harmonic approximation is appropriate. Anharmonic effects are important in biological macromolecules, and the same is also true in solid state samples. However, in the presentation of the physics of the DWF it is convenient to assume the harmonic approximation. The corrections to the harmonic approximations are known [1].

In diffraction from a primitive lattice the intensity of the diffraction line is decreased by a DWF given by [5]

$$I = I_0 \exp[-(\vec{G} \cdot \vec{u})^2],$$

where I_0 is the intensity if no displacement from the lattice point is present,

$$| \vec{G} | = \frac{4\pi}{\lambda} \sin \theta$$

is the reciprocal lattice vector which diffracts the incident wave of wavelength λ through an angle 2θ, and $\vec{u}$ is the rms displacement of the diffracting atom from the lattice point due to vibrations.

In EXAFS a DWF also occurs [1], but this factor has an essential difference from that in diffraction. It is usually written as

$$\exp[-2k^2\sigma^2].$$

Here $k = 2\pi/\lambda_e$ is the wave number of the excited photoelectron, while σ is the rms deviation about the average distance between the center atom and a neighboring atom. It is important to note that σ is the *relative* displacement between two atoms, while $| \vec{u} |$ is the displacement of a *single* atom from the lattice site.

In comparing EXAFS and diffraction, it should be noted that besides the difference in the DWF, EXAFS is a short-range probe within $\sim 5\text{Å}$ of the center atom [1], while diffraction is a long-range probe requiring long-range order [5].

II. COMPARISON OF σ AND u

To illustrate the difference between *sigma* and u, consider (as shown in Fig. 1) an imidazole ring which vibrates by rotating about the bottom N atom fixed at a lattice point. The imidazole ring remains rigid in this vibration. In this case u is nonzero for the four other atoms in the ring and is proportional to their distance to the N atom. However, σ remains between atoms to allow a molecule to pass through the ring, it is clear that σ is the pertinent quantity, not u.

If vibrations of two atoms are uncorrelated, then

$$\sigma^2 = u_1^2 + u_2^2, \tag{1}$$

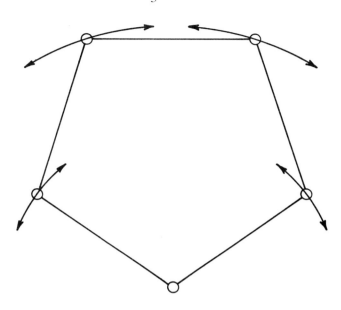

Figure 1. *Illustrating the difference between* u *and* σ *for an imidazole ring vibrating by pivoting rigidly about the atom at the lower vertex. The motion of the other atoms is* u *and is indicated by the arcs with arrows at each end. In this case* σ = 0.

where the subscripts 1 and 2 refer to the two atoms. However, if the motions of the two atoms are correlated, as in the case of the imidazole ring discussed above, then

$$\sigma^2 \neq u_1^2 + u_2^2. \tag{2}$$

For atoms far apart one expects no correlation and Eq. (1) applies. For neighboring atoms, correlation is expected and relation (2) applies. In general one expects significant correlation effects to extend out to second neighbors, or about 3-4 Å. Thus, one needs to exercise caution in employing (1) to interpret the u^2 from diffraction as a measure of the fluctuation in the relative distance between neighboring atoms.

In respect to the fluctuations in the relative distance between atoms, EXAFS and diffraction are complementary. Diffraction determines, the principle if not always in practice, the u^2 for all atoms. Use of (1) to determine the σ^2 for pairs of atoms is usually valid if the atoms are greater than $\approx 4\overset{\circ}{A}$ apart. EXAFS is able to measure σ^2 directly, but is usually limited to doing so for the nearest neighboring atoms less than about 4 $\overset{\circ}{A}$ apart.

III. EXAMPLES OF σ^2

Table I gives values of σ^2 as a function of temperature for solid state samples and biological macromolecules. As can be noted, there are not any large differences in the values of σ^2 in these two cases. The behavior of σ^2 as a function of temperature in biological macromolecules follows behavior similar to that in small molecules.

IV. IMPORTANCE OF σ IN UNDERSTANDING BIOLOGICAL FUNCTION

It has recently been emphasized that the kinetics of macromolecules and the diffusion of molecules to the active site are dynamic processes, and vibrations are an important contribution to these processes [6]. As discussed in Sect. II the quantity σ is the pertinent vibration variable.

It has been suggested that different configurations or conformations can be frozen out as the macromolecule is cooled, and these different conformations can change the barrier heights and thus tunneling rates by significant amounts [6]. Such a significant change in the potential distribution around the active site would be expected to also affect the vibrational frequencies. Such changes may be detected by measuring σ^2 as a function of temperature and comparing the low temperature disorder with the expected zero point disorder. The freezing of configurations may also introduce visible hysteresis effects in the temperature dependence of σ^2.

As shown in the discussion below, EXAFS measurements of Table I indicate that the free energy changes in biological reactions may have a significant con-

Table I. Einstein temperatures and $\Delta\sigma^2$ at room temperature of pairs of atoms for some crystallized small molecules and some forms of hemerythrin (Hr) in solution as obtained by EXAFS. The center atom in all cases except for Ge is Fe. The other atom in the pair is indicated in parentheses. The azidomet- and oxy-Hr have a short μ-oxo bond in addition to more standard O and N bonds in the first shell. These are distinguished in the Table. Listed are the values of $\Delta\sigma^2$ for the first and second coordination shells of atoms and their corresponding Einstein temperature θ_E.

Sample	$\Delta\sigma^2$ (10^{-3} $\mathring{A}^2$)		θ_E (K)	
	First	Second	First	Second
A. *Solid State Crystals of Small Molecules*				
Ge	1.6 (Ge)	6.3 (Ge)	353	207
Fe-gly.[a]	2.7 (O)	2.5 (Fe)	435	331
Fe-TIM[b]	1.0 (N)	--	621	--
FeOFe[c]	-- (O,N)	1.5 (Fe)	--	400
FeOHFe[d]	2.0 (O,N)	2.5 (Fe)	490	335
B. *Macromolecules in Solution*				
Azimomet-Hr	2.0 (O,N) 0.6 (μ-O)	3.1	490 720[e]	305
Oxy-Hr	2.6 (O,N) 0.6 (μ-O)	1.0	448 720[e]	463
Deoxy-Hr	4.3 (O,N)	8.4	375	217

[a] $[Fe_3O \cdot (glycinato)_6(H_2O)_3](ClO_4)_7$.

[b] bis(acetonitrile)(2,3,9,10-tetramethyl-1,4,8,11-tetrazacyclotetradeca-1,3,8,10-tetraene)iron(II)hexafluorophosphate.

[c] $Fe_2O(O_2CCH_3)_2(HB(pz)_3)_2$.

[d] $Fe_2(OH)(O_2CCH_3)_2(HB(pz)_3)_2$.

[e] Obtained from Raman spectroscopy; see, e.g., R.M. Solbrig, L.L. Duff, D.F. Shriver, and I.M. Klotz, J. Inorg. Biochim. 17, 69 (1982).

tribution from changes in vibrational frequencies. The formula that relates free energy changes to vibrational changes between a pair of atoms is

$$F_2 - F_1 = \frac{3k}{2}(\theta_1 - \theta_2) + 3kTln\frac{\Delta\sigma_1^2\theta_1}{\Delta\sigma_2^2\theta_2}. \qquad (3)$$

Here subscripts 1 and 2 refer to the states before and after, respectively. The change in mean square amplitude of vibration due to thermal excitation is denoted by $\Delta\sigma^2$, i.e., the increase in mean square vibration from 0 K to temperature T. The Einstein temperature is θ, and k is Boltzmann's constant. EXAFS degrees of freedom have the same free energy change as the radial distance.

To illustrate the importance of thermal vibrations to the energetics of biological processes, the case of hemerythrin is considered. Hemerythrin is a respiratory protein found in the coelomic fluid of a number of marine invertebrates. The active site of the protein contains two nonheme iron atoms that reversibly bind one molecule of oxygen. The iron in deoxyhemerythrin is in the ferrous state. Formation of oxyhemerythrin results in the oxidation of both iron atoms to an antiferromagnetically coupled ferric state with concomitant reduction of dioxygen to peroxide. Replacement of the bound peroxide by anions such as azide or hydroxide leads to methemerythrins in which the two Fe(III) ions retain their antiferromagnetic coupling but are no longer functional in transporting oxygen.

The energetics of the binding and release of O_2 in hemerythrin obtained from various invertebrates has been measured. There is a wide spread in the values, which probably is an indication more of the uncertainty in the measurements than of variations among the invertebrates. However, in order to have values to compare with, the values of enthalpy and entropy change for *P. gouldii* hemerythrin are chosen, the same source of hemerythrin as used in our EXAFS measurements. The experimental enthalpy values ΔH are 9.2 ± 0.4 kcal mole^{-1}, obtained from calorimetry, and 12.4 kcal mole^{-1}, obtained from measurements of the kinetics of the reaction [8]. Only one experiment value of the entropy change of $\Delta S = 18$ e.u. was found in the literature [8]. The entropy increases when the O_2 is released, and it has been suggested that this indicates that deoxyhemerythrin has fewer ligands at the active site. The increase in entropy is thus largely attributed to the extra degrees of freedom of the freed ligand, namely $)_2$.

Recent EXAFS results [9] suggest that the number of ligands at the active site does not decrease as O_2 unbinds. The site vacated by O_2 is replaced by another oxygen, presumably from a water or OH molecule. The entropy increase can be largely accounted for by changes in vibrational excitations. The total free energy change per atom pair is given by (3), while the vibrational energy change is given by

$$U_2 - U_1 = \frac{3\mu k^2}{\hbar^2}(\theta_2^2 \Delta \sigma_2^2 - \theta_1^2 \Delta \sigma_1^2) + \frac{3}{2}k(\theta_2 - \theta_1), \tag{4}$$

measures just one of the three degrees of vibrational energy, namely, the radial distance between the pair of atoms. In (3) it is assumed that the other two where μ is the reduced mass given by $\frac{1}{\mu} = \frac{1}{M_1} + \frac{1}{M_2}$ and M_1 and M_2 are the masses of the atoms in the pair, and $\hbar$ is Planck's constant divided by 2π.

In estimating the entropy changes at the active site due to vibrations, account must be taken of the fact that there are 13 atoms at the active site with 39 degrees of freedom. The first neighbors about the Fe atoms account for 12 pairs of atoms and the Fe-Fe distance is considered the second-neighbor shell and is the thirteenth pair of atoms. The first-neighbor pairs in the oxy- form are divided into 10 pairs at about 2 Å and 2 pairs between a μ-oxo bridge and the two irons at an Fe-O distance of 1.8 Å. In the deoxy- form the main change is the μ-oxo bridge converts to a μ-OH bridge with a lengthening of the Fe-O bond to about 2 Å. Our estimate is quite crude, treating each pair of atoms as an independent harmonic Einstein oscillator. Giving each harmonic oscillator 3 degrees of freedom totals 39 degrees of freedom, as required. This crude estimate, however, is sufficient to illustrate the point that the vibrational contribution to the free energy is significant. Using (3) and (4) for each pair of atoms and the values in Table I, the results, as O_2 is released, are

$$F_2 - F_1 = -0.32\text{eV/molecule} = -7.3 \text{ kcal/mole}$$
$$S_2 - S_1 = 16 \text{ e.u.} \tag{5}$$
$$U_2 - U_1 = -0.10 \text{ eV/molecule} = -2.3 \text{ kcal/mole.}$$

About half of the total vibrational change is contributed by the Fe-O-Fe atoms as the μ-oxo bridge changes to a μ-OH bridge.

Comparing the vibrational contributions with the measured values of $\Delta S = 18$ e.u. and $\Delta H = 9.2$ and 12.4 kcal·mol⁻1, it is clear that vibrational changes contribute a significant amount to both the enthalpy and entropy changes, but particularly to the entropy changes. It should be emphasized that the experimental measurements are for the total changes in the reaction, while the estimate in (5) is only for the vibrational changes localized at the active site. What is striking is that most of the entropy change can be accounted for by this mechanism.

CONCLUSION AND SUMMARY

The EXAFS σ^2 and diffraction u^2 are different. The mean square variation of the relative radial motion between a pair of atoms is σ^2, while u^2 is the mean square variation of a single atom about the lattice point. In understanding the behavior in the vicinity of an active site, σ^2 is physically more important. The change in free energy of the vibrational degrees of freedom depends directly on changes in σ^2. The diffusion of atoms will depend on σ. Because of correlation in the motion of atoms, σ^2 and u^2 differ for neighboring atoms, and only when typically third or more distant neighbors are considered does the correlation disappear and (1) hold.

In some metalloproteins such a hemerythrin EXAFS finds that changes are localized around the active site as the protein reacts. In the case of hemerythrin it is shown by a crude calculation that the entropy changes in the vibrational modes around the active site as O_2 is released constitute most of the entropy change in the reaction.

Table I shows that the values of σ^2 in the metalloprotein hemerythrin at the active site are of the same order as in small molecules. In biological systems the local vibrational modes and their changes can be understood by the same physics as applies to small molecules. The harmonic approximation and standard small corrections to it appear to be sufficient to explain the temperature dependence of σ^2. However, more investigations should be made to determine whether any hysteresis effects or large deviations from harmonic behavior as a function of temperature occur that could be related to the freezing out of different protein conformations, as suggested by Professor Hopfield in the discussion.

ACKNOWLEDGEMENTS

It is a pleasure to acknowledge many informative and helpful conversations with Professor Joann Loehr and Professor Ronald Felton. The results presented here are based on analyses of the EXAFS data by Mr. Ke Zhang, and his help is gratefully acknowledged. The work described here was supported by the National

Science Founation, grant no. PCM-82-04234.

REFERENCES

1. E.A. Stern and S.M. Heald, in *Handbook on Synchrotron Radiation*, vol. 1, ed. E.E. Kock (North Holland, N.Y., 1983), chap. 10.

2. J.M. Tranquada and R. Ingalls, Phys. Rev. B **28**, 3520 (1983).

3. E.A. Stern, in *Physics of Disordered Materials*, ed. D. Adler, H. Fritzsche, and S.R. Ovshinsky (Plenum, N.Y., 1985), p. 143.

4. J.A. Krumhansl, presentation at this conference and references therein.

5. B.E. Warren, *X-ray Diffraction* (Addison-Wesley, Reading, Mass., 1969).

6. H. Frauenfelder, presentation at this conference and references therein.

7. N. Langerman and J.M. Sturtevant, *Biochemistry* **10**, 2809 (1971).

8. D.J.A. de Waal and R.G. Wilkens, *J. Biol. Chem.* **251**, 2339 (1976).

9. Ke Zhang, E.A. Stern, J. Loehr, A.K. Schiemke, and F. Ellis (unpublished).

Discussion: Stern

Karplus: Two comments, one at the beginning you said in the macromolecules, you might expect the correlations to die out at 5 A or something, or so. Simulations show for pancreatic trypsin and inhibitor, if you look at the displacement correlation function of distance for the alpha carbons that basically go through a minimum around 10 A and then comes up again at around 20 A so that there are at least, in this one case, simulated in solution, very long range correlations – maybe an artifact of the simulation, but it's a result. As far as your calculation is concerned of the entropy changes, I didn't really understand it. What you measure is distances from the iron to a number of different atoms. Is that correct? How do they vary, the fluctuations or ...?

Stern: We measured two things: we measured the average distance, and we also measured the RMS, root mean square, variation about that average.

Karplus: Right, but you measure ... So you measure these various RMS fluctuations. Again, it's known for most molecules that the fluctuations - large scale fluctuations - are dominated by a few low frequency vibrations. I mean, if you take the simple case of an angle bending, then all of the atoms that are further away will have their distance changed by a large amount, but it may be only one mode, and my guess is that your mode counting, to include all the degrees of freedom of these thirteen atoms, gives a much inflated view of the actual entropy change. In addition to the question of what the Einstein temperature is, and the things of this sort. But, I don't think that you can conclude the fact that A and B have fluctuation difference, but its a different mode involved from A and C and so forth. And, as I say, if you look at small molecules and if you look at proteins, the main result is that out of the total number of modes, again if you want to take a protein case out of 1,700 modes, the number that contribute to the mean square fluctuations are maybe 20, so that it is a very small number that are changing in frequency probably, and my guess is that the entropy change is much smaller than your estimate.

Stern: Yes, of course a detailed calculation has to be made to verify either one of our two statements, but you have to remember that in the measurements that are made here, what we're measuring is a *relative* change in distance between the center atom and its neighbors. What is done is to model that with just a simple Einstein temperature, one single frequency for that mode. Since the relative motion between the center atom and its neighbor is correlated, the low frequency or the long wavelength phonons don't contribute to the relative motion, just the high frequency phonons. In other words, it's the optic modes that are giving most of the contribution to the relative distance and in that case, the Einstein approximation is valid. Now I agree with what you say, that one would have to be a little more sophisticated than I was in estimating the change in entropy, but I think it's suggestive that there is a significant contribution here and I'd like to stimulate more thinking along this line.

Spiro: The increase in disorder in the oxy-hemerythrin is striking. The iron/iron disorder might, of course, be due to the fact that you don't have a hydroxide at all. It's conceivable that it's just loose. But, the first shell disorder does surprise me. However, it's not clear whether the reference compounds included iron II or not which is the valence state for deoxy. It has a 0.10 A or so larger radius and might be expected to have looser bonds. Has that been looked at carefully?

Stern: You're right. The deoxy form has iron II while the oxy form has iron III. The reference compounds are iron III and the disorder in the reference compounds is closer to the oxy value than the deoxy value. So, I think you're at least partially right. The difference is certainly related to the change in oxidation states, but, except for the μ-oxo distance, the other distances did not increase by $0.10\text{\AA}$. The μ-oxo bond lengthening seems to accommodate most of the valence change with accompanying further changes in the other bonds of much less than $0.1\text{\AA}$. Except for the μ-oxo bond, the other first neighbor distances didn't change by more than $0.03\text{\AA}$. Thus, the increase in disorder must be related to the charge rearrangement as the oxo-bridge changes to an hydroxo-bridge and the charge state of the Fe atoms changes from III to II. However, it is important for a calculation to be made to verify these speculations.

Warshel: A half an electron volt is a very large number for chemists and should really be taken seriously and one of the things that you should, or could have done, is to compare the estimate of entropy to measurement of entropy of binding. It's more complicated because you have to take into account the entropy of the separated case, but it's obviously, or it seems, a far too large a number.

One should take into account the frequencies, but it's very useful to compare it, especially to model compounds of known entropy of binding which never exceed this value.

Stern: Yes, I agree, and I have made this comparison in the written version of my talk. A more careful estimate gives 0.32 eV for the free energy change and 0.22 eV for the entropy charge at room temperature in hemerythrin. The entropy change is somewhat less than the value measured for the reaction but the interesting thing is that most of the measured entropy change can be accounted for by the vibrational change!

Hopfield: From the point of view of x-ray analysis, it's very convenient of course to view things as harmonic, yet, one would ask if these do form glasses and so on, then things are really rather non-harmonic, one would ask whether there is anything actually in the experiments to say, Yes, it is harmonic rather than to say, there are actually say four structures at room temperature which have quite different bond lengths between which you average and then you go down to one of them at 77°. A radically non-harmonic view: Is there anything in the experiments which basically suggest a difference or is it simply a supposition in analysis?

Stern: No, in the measurements we actually can distinguish between harmonic vibrations causing disorder and anharmonic or static disorder.

The measurements are for a given structure where we always measure a reproducible behavior as a function of temperature. This behavior as a function of temperature fits what one would expect from an harmonic oscillator. If, as the temperature is lowered, some configuration freezes out or some hysterisis occurred, we would detect that difference. We don't see it. We see variations in disorder and absolute disorder which are consistent with just a simple harmonic oscillator. At low temperatures the absolute disorder is consistent with zero point motion of that harmonic oscillator.

Now there are some cases for atoms not directly bound to the center atom where the temperature dependence is strange. We have attributed the unusual behavior to interference between the atoms of interest and other atoms at about the same distance. Whether these effects are due to freezing out of different configurations as you suggest cannot be completely ruled out. The point deserves further investigations.

Krumhansl: I'd like to mention something that interested us a few years ago, calculations by Goldanskii of the potential for movement from an energy minimized of protein BBTI, you'll see that there are number of configurations that are of the sort that John was worried about. This is a normal mode that's associated with one of them; of approximately 240 normal modes, at least 30 by a classification here are highly anharmonic and because they are anharmonic and have that potential, then one asks the question of whether you can redo the Debye-Waller factor theory for that. In fact, the answer is Yes you can, and those of you who know the interest of Bob Schrieffer and I had several years ago know that this is quite close to home. Here is the kind of formula you get for a revised Debye-Waller factor for a double quadratic well which is quite soluable. That's for a single decoupled one, you can do the coupled chain of the double well systems, and you can find the correlation coefficients and they do extend some distance as Karplus indicated.

Then let me just show you one last transparancy.....

Editor's Note: Krumhansl's comments follow as the next paper.

Anharmonicity and Debye-Waller Factors in Biomolecules

J. Krumhansl

Department of Physics

Cornell University

Ithaca, NY 14853-2501

It is common practice to use Debye Waller factors $[\exp(-W)]$ in the interpretation of x-ray patterns or to infer atomic motions in biomolecules for comparison with molecular dynamics simulations. Usually W is taken to be proportional to $\langle u^2 \rangle$, which in turn is taken to be proportional to temperature T. For any real system both of these assumptions are wrong in principle, since they depend entirely on purely harmonic potentials. As a practical matter in many cases anharmonicity is weak, but these are cases where significant structural accommodations do not take place. However, when dealing with repuckering of the ribose in DNA or ground state multiplicity in biomolecular conformational motions the role of anharmonicity must be addressed properly; quasi-harmonic methods may be both quantitatively and qualitatively misleading. This is a short review of the theory of the Debye Waller factor, particularly in the classical limit.

I. Introduction

The purpose of this paper is to examine the special features of the physics of the thermal diffuse x-ray scattering of biopolymers which are caused by anharmonic effects. Considerable use is made of the Debye-Waller Factor (DWF) in the interpretation of biomolecular structure; we wish to call attention to the limited validity of this procedure when a significant anharmonicity is present e.g. conformational changes.

We summarize results by various authors who have addressed anharmonic effects in solids, particularly near structural transitions: Cowley [1], Mair et al. [3,4,5], Maradudin and Flinn [2], and Dawson et al. [6], and Dash, Johnson and

Visscher [7].

If, but *only* if, atomic vibrations are harmonic the intensity of scattered x-rays is proportional to $\exp(-2W)$ where W is the DWF, which in turn (harmonic) is proportional to $\langle (\underline{\kappa} \cdot \underline{u})^2 \rangle_T$; $\underline{\kappa}$ is the scattering vector, $\underline{u}$ the atomic displacement (a dynamic quantity) from equilibrium, and $\langle \ \rangle_T$ is the finite temperature average. In biopolymers, among other materials, this "rule" about the intensity is used to infer the mean square displacement (MSD) of various atoms, thence to check computer simulations. This is wrong in principle for any anharmonic system but may be particularly wrong for the significant conformational motions of biopolymers.

Since the regime of applications in biopolymers is different from some solid state situations, in passing we note several points. (i) For room temperature measurements it is not necessary to use the fully quantum mechanical treatment [1] of the DWF or MSD and the classical statistical mechanical averages suffice [2]. (ii) Many derivations of DWF are for phonons in periodic solids, for which the displacements are $u \propto e_q(\exp[iq \cdot x])$ and frequencies (and modes) are indexed by q, i.e. ω_q; in non-periodic polymers a general eigenvector form must be used $\underline{u}(x) \propto \underline{e}_\lambda(x), \omega_\lambda$ where λ is the eigenvalue index. In the harmonic regime the generalization to non-periodic polymers is straightforward; the anharmonic case is a different story.

2. General Formulae for the Debye-Waller Factor (DWF)

Below we will discuss a variety of regimes and conditions regarding the DWF; however, certain very general formula can be written. In complete generality the DWF is a "one-particle" attribute of moving atoms which multiplies positional or electronic structure factors in determining the x-ray scattering by an atom. We write ($\underline{\kappa}$ is the scattering vector)

$$\sqrt{f} = \sqrt{DWF} = e^{-w} = \langle \exp(-i\underline{\kappa} \cdot \underline{u}) \rangle \tag{2.1}$$

where $\langle \ \rangle$ denotes the appropriate classical or quantum finite-temperature average as determined by the system Hamiltonian H, i.e.

$$\sqrt{f} = \frac{Tr\left(e^{-\beta H}e^{-i\underline{\kappa}\cdot\underline{u}}\right)}{Tr\left(e^{-\beta H}\right)} \tag{2.2}$$

with $\beta = (KT)^{-1}$. Alternatively and formally sometimes more conveniently:

$$\sqrt{f} = \int d\underline{u}\, e^{i\underline{\kappa}\cdot\underline{u}}\rho(\underline{u}) \tag{2.3}$$

where $\rho(u)$ is the appropriate one-particle distribution function.

As noted in numerous condensed matter applications the expression $f = \exp(-\kappa^2\langle\underline{u}^2\rangle)$ is hardly ever correctly used for precision studies, being a purely harmonic idealization. The errors in its uses are not only quantitative but also may lead to phenomenologically incorrect interpretations. Simply put, anharmonicity makes the harmonic expression invalid; how serious the errors may be is the concern of this review.

3. Regimes

(i) At very low temperatures quantum effects must be considered; zero point motions prevent f from $\to 0$ as $T \to 0$. Most biomolecular studies are in the high temperature region.

(ii) In the high temperature region the classical Boltzmann averaging applies. Kinetic and potential terms separate, but since we average a function of coordinates only the kinetic terms cancel and

$$\sqrt{f} = \frac{\int \exp\left[-\beta V\{\underline{u}\}\right]\exp\left[-i(\underline{\kappa}\cdot\underline{u})\right]d\{\underline{u}\}}{\int \exp\left[-\beta V\{\underline{u}\}\right]d\{\underline{u}\}} \tag{3.1}$$

where $V(\{\underline{u}\})$ is the potential energy for the many-variable phase space $d\{u\}$. In general this potential is not quadratic.

(iii) If the anharmonicity is weak then perturbation series formalism may be used, and the methods of many body theory developed by Cowley [1] and applied by Mair [3], or the approach of Maradudin and Flinn [2] can be employed. However, if there are significant topological conformational changes these perturbation methods are not adequate [4].

(iv) Where molecules flop between quite distinct configurations or some regions show great flexibility perturbation methods break down. Special methods for discussing the statistical mechanics of this class of problem were developed by Krumhansl et al [8]. Examples of strong anharmonicity have been studied by Dash et al., and by Mair, and are discussed below.

(v) Finally, a large body of work on anharmonic systems has been done in the "effective one-particle potential" approximation [3,4,6]. However, there certainly are regimes where strong interparticle coupling is as important as the highly anharmonic local potentials, leading to 1-d polymers for example to "soliton" or kink excitations which are coherent over several atoms.

(vi) Finally, we mention the obvious difference that the many forms of condensed matter frequently studied by physicists are periodic, whereas biopolymers generally are not. This does not generally vitiate the various formal methods in principle, but it does make *detailed* comparison with the solid state literature difficult to the point of questionable utility. Generally a new start on biomolecules will be easier.

In summary, although the formal definition of the DWF is quite simply stated its evaluation is conditional on several physical and approximate considerations.

4. Relevance to Biopolymers: An Example, the Globular Protein BPT1

It is a priori clear that floppy biomolecules are highly anharmonic. We cite only one example here. An interesting computer simulation has been carried out by N. Gō et al. for bovine pancreatic trypsin inhibitor (BPTI) [9]. One minimal energy molecular configuration was obtained by a Monte Carlo technique, then a set of normal mode coordinates $\{\xi_i\}$ were determined from the generalized quadratic expansion of the potential about this equilibrium. Using these coordinates computations were then made of $V(\xi_i)$ for 241 modes (now no longer restricted to quadratic terms in ξ_i); a characterization of the potentials is shown in Table I.

This most revealing study points to the fact that the low energy (motions) are

Rank order of collective variables	Range of eigenvalues (kcal mol^1 deg^2)	No. of anharmonic collective variables					
		I	II	III	Type IV	V	Total
1- 53	$1.30 \times 10^4 - 1.00 \times 10^{-1}$						0
54-160	$0.99 \times 10^{-1} - 0.61 \times 10^{-2}$	3	1				4
161-180	$0.60 \times 10^{-2} - 0.45 \times 10^{-2}$				1		1
181-206	$0.44 \times 10^{-2} - 0.30 \times 10^{-2}$	2	6		1		9
207-229	$0.29 \times 10^{-2} - 0.10 \times 10^{-2}$	1	14	3			18
230-241	$0.99 \times 10^{-2} - 0.20 \times 10^{-3}$	1	3	5		2	11
Total		7	24	8	2	2	43

The energy curves for all the collective conformational changes are well fitted by fourth-order curves of the form $B\xi^2 + C\xi^3 + D\xi^4$ except for four curves classified as types IV and V. A curve is harmonic when the terms $C\xi^3$ and $D\xi^4$ are insignificant in the range of thermal fluctuations. As a measure of the range of thermal fluctuations, we take $[-\xi_0, \xi_0]$, where ξ_0 is defined by $\frac{1}{2} B\xi_0^2 = 0.3$ kcal mol^{-1}. A curve is anharmonic if $|C\xi_0^3/B\xi_0^2| \geq 0.1$ and/or $(D\xi_0^4/B\xi_0^2) \geq 0.1$, and may be one of five types according to its shape. Type I: curves where $|C\xi_0^3/B\xi_0^2| \geq 0.1$ and $(D\xi_0^4/B\xi_0^2) < 0.1$. In conformational changes caused by variations in these collective variables, large changes in dihedral angles and atom positions are confined to one or two side chains. II: curves where $1.0 > (D\xi_0^4/B\xi_0^2) \geq 0.1$. III: curves where $(D\xi_0^4/B\xi_0^2) \geq 1.0$. Conformational changes caused by variations in collective variables of types II and III are not localized in any part of the molecule. 4 of 24 type II curves and 5 of 8 entries in type III curves have non-negligible third-order terms, that is $|C\xi_0^3/B\xi_0^2| \geq 0.1$. IV: conformational changes are localized in one or two side chains with smaller changes elsewhere. A peak of the energy curve is due to torsional energy for dihedral angle(s) of the side chain(s). V: Almost free rotation of a side chain (lysine) dihedral angle. Peaks are due to the non-bonded energy.

Table I. Classification of anharmonic potentials from a simulation of BPTI. Reprinted by permission from *Nature*, Vol. 296, p. 776. Copyright 1982

highly anharmonic, indeed unstable in a few instances, so that the contribution of the significant conformational motions to x-ray scattering certainly cannot be handled by conventional Debye-Waller analysis to obtain $\langle u^2 \rangle$ from DWF. Lest it be assumed that these highly anharmonic motions are highly localized in the protein in Figure 1 we reproduce [from Figure 2c from Gō, et al. (1983)] for the quasi-harmonic approximation to the displacement pattern of the second lowest (out of 241) mode of BPTI; an evaluation of an anharmonicity index Table II by Gō shows this to be highly anharmonic. Quasi-harmonic analyses and their interpretation are of highly questionable validity in this light.

Similar behavior can be expected in many enzymes and polynucleotides. In addition, to x-ray structural determinations the n.m.r. experiments and m.d. simulations have frequently used harmonic model results for checking or interpretation, with equally questionable validity.

5. Summary of Theoretical Models

(i) Cumulant Methods

We follow the discussion by Lovesey [10]. Starting with $e^{-W} = \langle \exp(-i\underline{\kappa} \cdot \underline{u}) \rangle$,

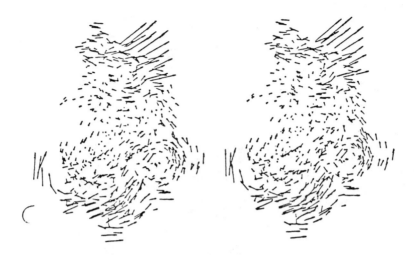

Figure 1. Stereo view of the second lowest mode (of 241 modes) in the simulation of BPTI by Gö et al. PNAS USA 80, 3699 (1983). Reprinted by permission of N. Gö.

taking logarithms and expanding in a Taylor series and denoting $A = -i\underline{\kappa} \cdot \underline{u}$ for brevity,

$$W = \frac{1}{2!}\langle \hat{A}^2 \rangle - \frac{1}{3!}\langle \hat{A}^3 \rangle - \frac{1}{4!}\left\{ \langle \hat{A}^4 \rangle - 3\langle \hat{A}^2 \rangle^2 \right\} + \dots . \tag{5.1}$$

The first term is simply $1/2\langle (\underline{\kappa} \cdot \underline{u})^2 \rangle$ the usual harmonic term; the second term being odd in the displacement $\underline{u}$ vanishes if the potential is even in $\underline{u}$ (to all orders of coupling and anharmonicity). The third term

$$-\frac{1}{24}\left\{ \langle (\underline{\kappa} \cdot \underline{u})^4 \rangle - 3\langle (\kappa \cdot u)^2 \rangle^2 \right\}$$

vanishes if the distribution of $\underline{u}$ is Gaussian, as just happens to be the case only for purely harmonic forces. However, in general these anharmonic corrections are not negligible, leading both to shifts from $\underline{u} = 0$ (therefore, displacements of Bragg spots in reciprocal space), and anomalous temperature dependence of the DWF. This series expansion is clearly in trouble for such an harmonic which occur when regions flop between conformational substates.

(ii) Harmonic Methods; Coupled Systems

Maradudin and Fein [2] (MF) provided both a review of the classical harmonic DWF including the coupled motions in periodic crystals, then extended the analysis in a particularly useful way to the anharmonic regime but again as a series expansion. To the same order as 5.1, and for a crystal with inversion symmetry, their anharmonic correction agrees with the cumulant expression.

As noted earlier biomolecules do not have the periodicity of crystals and the analysis of MF does not apply directly; however, in the classical *harmonic* regime it is straightforward to extend MF to use of a general set of the proper molecular normal modes with frequencies ω_λ and displacement $\underline{e}_\lambda(\ell)$ of the ℓth atom to be used as a basis and the Waller-MF procedure proceeds in a straightforward manner.

The result is

$$\ell n\left[\langle \exp -i\underline{\kappa}\cdot\underline{u}(\ell)\rangle\right] = -\frac{KT}{2}\sum_\lambda \frac{\left(\underline{\kappa}\cdot\underline{e}_\lambda(\ell)\right)^2}{M_\ell\omega_\ell^2} \tag{5.2}$$

which can be shown to be just $-1/2\langle(\underline{\kappa}\cdot\underline{u}(\ell))^2\rangle T$. Thus, as is not surprising, the harmonic result of coupled systems is not dependent on periodicity. In this approximation the DWF has no nodal structure in κ-space.

(iii) Strongly Anharmonic Systems: "One Particle Models"

There are many examples of structural transformations or multiple states of localized defects in condensed matter physics involving large displacements of atoms between different substates; ferroelectrics are examples of the former, and off-center impurities [12] in ionic crystals exemplify the latter. The potential functions (e.g. a Landau free energy expansion) representing these systems are highly anharmonic and may be of many approximate forms: spherical square well, flat bottom quadratic, double quadratic, and other multiminima potentials.

For one particle in an infinitely deep square well of radius R_0 Dash, Johnson, and Visscher (DJH) obtain in the classical limit

$$\sqrt{f} = \frac{3}{(\kappa R_o)^2} \left(\frac{\sin \kappa R_0}{k R_0} - \cos \kappa R_0 \right) \tag{5.3}$$

which is independent of temperature! (Moreover it has nodal structure as a function of scattering vector κ). DJH also treat a flat bottom of radius R_0 with harmonic elastic sides of stiffness σ and obtain (classical; high T)

$$\sqrt{f} = \cos(\kappa R_0) \exp\left[-\frac{KT}{2} \frac{\kappa^2}{\sigma} \right]. \tag{5.4}$$

The nodal positions are temperature independent, and at any scattering direction κ the temperature dependence of the DWF is similar to a harmonic system.

Mair [4] has examined a number of one-dimensional highly anharmonic one particle DWF examples, which would be appropriate to a strongly anisotropic linear motion. In the limit of an effective one particle potential, a possible regime in which the surrounding ion motion is effectively uncorrelated, Mair calculates the DWF for the double quadratic (DQ), the flat bottom quadratic results are similar to DJH 5.4, but for 1-d systems; the often used disorder model is found to have some serious limitations. The DQ model provides some useful lessons about DWF and strong anharmonicity so we quote Mair's results for this case (classical) limit:

$$\langle \exp(-i\kappa u) \rangle = \exp\left(-\frac{\kappa^2 b}{4} \right) \left\{ \exp(-i\kappa d) \; \mathrm{erf}\left(db^{-1/2} - \frac{\kappa b^{-1/2}}{2} \right) \right.$$
$$\left. + \text{ complex conjugate } + \cos \kappa d \right\} / \left[1 + \mathrm{erf}(db^{-1/2}) \right] \tag{5.5}$$

where 2d = well separation, $b = K_B T/\alpha$ where (2α) is the (quadratic) force constant. Clearly, neither the temperature nor scattering vector dependence are simply (except in the limit $d = 0$ which reduces to a single quadratic well and usual special result). For the same case the means square displacement (MSD) is given by

$$\langle u^2 \rangle = \left[b/2 + d^2 + d(b/\pi)^{1/2} \exp(-d^2/b) \right] / \left[1 + \mathrm{erf}(db^{-1/2}) \right]. \tag{5.6}$$

We note: (a) manifestly, $\ln(DWF) \neq -1/2 \left(\langle (\kappa u)^2 \rangle \right)$; indeed no simple relationship exists, (b) unlike any harmonic model $\langle u^2 \rangle$ does not extrapolate to zero at

$T \to 0$; the reason is not deep, at low temperature the particle spends most of its time in one well or another; (c) finally, one feature of the DQ model is found by Mair: the nodal structure of the DWF is temperature dependent. Indeed this last point leads to the anomalous feature that in some regions of κ-space the temperature dependence of the DWF is opposite (i.e. increasing with increasing temperature) to the conventional behavior.

(iv) Strongly Coupled-Strongly Anharmonic Systems

Having seen the anharmonic DWF anomalies to be expected for a "one particle" strongly anharmonic motion it is natural to ask what effects the interatomic coupling might have. It is not uncommon to find a situation with highly anharmonic local potentials but essentially harmonic coupling to neighboring regions.

A few simple models for a single anharmonic degree of freedom coupled to a harmonic reservoir were examined by Dash, Johnson, and Visscher [7]; they explained both the low temperature quantum limit and the high temperature classical limit of a spherical well whose sides are harmonically coupled to a lattice with many degrees of freedom. The results cannot be stated simply, and they eventually resort to perturbation expansions (which agree to comparable order with Maradudin and Fein) or to numerical computations. However, two general features appear again: (a) at high enough temperature the DWF factors into a harmonic-like term and one characteristic of the particular kind of anharmonicity, the latter having nodal structure in κ-space, (b) for a given scattering angle the high temperature DWF exponent (i.e. $-\kappa T/2\sigma$ in the harmonic approximation) does not extrapolate to zero at $T = 0$ (see Figure 5, DJV). We will return to those points below.

While DJV give a pedagogical insight into this anharmonic situation it is only for one special case. Recently Mair has provided a much more general and controlled analysis of the interplay between strong local anharmonicity and quadratic coupling to a surrounding lattice (molecule). This analysis is based on the facts that: (a) the DWF can be computed from the properly averaged one-particle distribution function $\langle \rho(\underline{u}) \rangle T$, (b) That one-particle DF can be computed exactly for several strongly coupled anharmonic 1-d models using methods developed for the statistical mechanics of structural phase transitions [8]. We refer the reader to Mair's papers [4] for the rather extensive details of the transfer calculations; here we limit to one particular set of her results which are probably instructive in the context of biopolymers.

In the second of two related papers (Mair, 1983b) the double quadratic chain with strong, medium and weak harmonic coupling is analyzed in detail. Results are obtained for the one-particle DF $\rho(u)$, the DWF, and the mean square displacement for various coupling regimes. These are illustrated in figures taken from that paper. Figure 2 shows the behavior of $\rho(u)$ vs. u for two temperatures and (a)-(c) strong to weak coupling. This shows one *very* important feature: even if there is strong local anharmonicity when harmonic coupling to the rest of the system is strong enough the one particle DF resembles a single peaked gaussian centered on $u = 0$, and the DWF will be like that of a harmonic system, at least for small κ.

In the extreme limit of weak coupling the two peaked "one particle" DF for a double well is found. For intermediate coupling the shape is quite temperature dependent. From the same calculations, for two different temperatures, the DWF vs. scattering vector are shown in Figure 3. We note (i) the curve (a) with strong intersite harmonic coupling is similar to that for a harmonic oscillator; nodal structure harmonic coupling is similar to that for a harmonic oscillator; nodal structure hardly develops; (ii) the curve (c) for weak coupling shows pronounced temperature dependent nodal structure. The model also allows computations of various cumulants from a knowledge of $\rho(\underline{u})$; the classical $\langle u^2 \rangle$ does not extrapolate to zero at $T \to 0$ and has other structure (*not* a quantum effect); the kurtosis $K = \langle y^4 \rangle - 3\langle y^2 \rangle^2) / \langle y^2 \rangle^2$ which would be zero for a harmonic system becomes very large as the intersite coupling is decreased and vice versa.

In summary, a system which has bimodal local anharmonicity but is coupled harmonically to its surrounding will have a one particle DF $\rho(\underline{u})$ which is doubly-peaked for small coupling but becomes singly packed for larger coupling and higher temperatures. Displaced gaussians may approximate this situation but

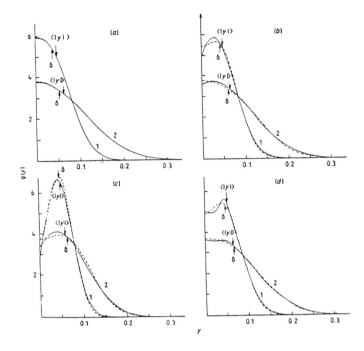

Figure 2. One-particle distribution function for coupled double quadratic (DQ) potentials with (a) strong, (b) medium, (c) weak coupling, (d) cusp smoothed DQ, intermediate coupling. S. Mair, J. Phys. C 16, 5591 (1983). Reprinted by permission of IOP Publishing Ltd.

their separation and parameters must be taken to vary with temperature, which makes their use heuristic at best.

6. Discussion; Experimental Signatures

The purpose of this conference report was to present an overview and key to some literature which discusses the theory of the Debye Waller factor in some generality, especially the effects of strong anharmonicity (i.e. multiple conformation substates). An analysis of a specific biomolecule has not been attempted as yet to seek out deviations from harmonic DWF theory and compare them with

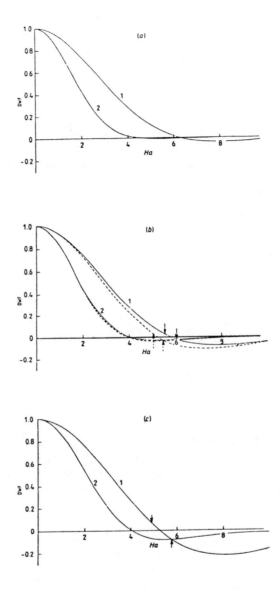

Figure 3. DQ-DWFs vs. scattering angle for two temperatures and (a) strong, (b) medium, (c) weak coupling. S. Mair, J. Phys. 16, 5591 (1983).) Reprinted by permission of IOP Publishing Ltd.

various cases of the models summarized here. That could and should be done, the choice of molecule is up to the biologist, through myoglobin or the Dickerson B-DNA dodecamer might be reasonable first choices.

However, we can point out certain obvious things to do based on the discussion above. Recall that the "B-factors" i.e. DWF in a structure refinement are determinable for many orders of reflection and in principle at different temperatures (though perhaps not much of a range for biomolecules in vivo).

(i) for a given order of reflection (scattering angle) the principal indicators of strong anharmonicity at an atom are that $ln(DWF)$ is not exactly linear in temperature and does not extrapolate to zero at $T = 0$.

(ii) For fixed temperature the indicator of fluctuations between substates is that the $ln(DWF)$ does not vary linearly with scattering angle $(\kappa)^2$; indeed the DWF itself is expected to have nodal structure.

It should be informative to reexamine x-ray structure refinements and Mössbauer data in this light to see whether some apparent anomalies are really manifestations of strong, localized anharmonicity.

Finally, it may be of some use to provide references to DWF anomalies seen in experiments on various materials of a non-biological nature. Jex, et al. [11] show diverging DWF near a structural phase transition at 168° K in KCN. Johnson and Dash [12] from Mössbauer measurements find clear evidence for anharmonic effects in $FeCl_2$. Mair [13] compares her anharmonic models with the observed DWF near a structural phase transition in $CsPbCl_3$. Reference 5 contains several recent experimental citations. Gilbert et al. [14] recognize the possibility of ring flips in proteins as a source of anomalous temperature dependence.

Acknowledgements

Support for this study was received from the Naval Air Systems Command.

Table 1 (on page 21) is reprinted from Nature, 296, 776 (1982), with the permission of Macmillan Journals Limited, New York, New York.

References

[1] R.A. Cowley, Adv. Phys. *12*, 421 (1963).

[2] A.M. Maradudin and P.A. Flinn, Phys. Rev. *129*, 2529 (1963).

[3] S.L. Mair, J. Phys. C. *13*, 2857 (1980).

[4] Sylvia L. Mair, J. Phys. C*15*, 25 (1982); ibid. *16*, 4811 (1983a); ibid. *16*, 5591 (1983).

[5] S.L. Mair, Phys. Rev. B*30*, 6560 (1984), C.H.J. Johnson and S.L. Mair, J. Phys. C*18*, 67 (1985).

[6] B. Dawson, A.C. Hurley, V.S. Maslen, Proc. R. Soc. A*298*, 289 (1967).

[7] J.G. Dash, D.P. Johnson, W.M. Visscher, Phys. Rev. *168*, 1087 (1968).

[8] J.A. Krumhansl and J.R. Schrieffer, Phys. Rev. B*11*, 3535 (1975); J.F. Currie, J.A. Krumhansl, A.R. Bishop and S.E. Trullinger, Phys. Rev. B*22*, 477 (1980).

[9] T. Noguti and N. Gö, Nature *296*, (1982); N. Gö, T. Noguti and T. Nishikawa, P.N.A.S. Biophysics *80*, 3699 (1983).

[10] S.W. Lovesey, *Theory of Neutron Scattering from Condensed Matter*, sec. 4.3, Clarendon Press Oxford.

[11] H. Jex, M. Mullner, R. Knoth and A. Loidl, Solid State Commun. *36*, 713 (1980).

[12] D.P. Johnson and J.G. Dash, Phys. Rev. *172*, 983 (1968).

[13] S.L. Mair, Acta. Cryst. A*38*, 790 (1982).

[14] W.A. Gilbert, J. Kuriyan, G.A. Petsko and D.R. Ponı, p. 405 in *Structure and Dynamics; Nucleic Acids and Proteins*, Ed. E. Clement and R.H. Sarma Adenine Press, N.Y. 1983, cf. discussion p. 409.

Proteins as Fractals

H. Stapleton

Department of Physics

University of Illinois at Urbana-Champaign

1110 W. Green Street

Urbana, IL 61801

Fig. 1 illustrates a fractal curve with properties relevant to a polypeptide. It is a fractal with geometric self-similarity. It has two elements: an initiator, which here is the largest N-sided structure that spans the interval 0 to R. In this example, N happens to be 8, and each leg of the structure, r, is 1/4 of the initiator length, R. The top equation in Fig. 1 is the most general relationship characterizing a fractal curve such as this. Here d is the fractal dimension. If, as in this example, each of the r_i is the same length, r, then the second expression in Fig. 1 is true.

At each stage of the construction, the figure would extend over the entire interval of the initiator. The fractal dimension, $\overline{d}$, is independent of the stage of the construction. If we were to attempt to measure the length of this curve, in its infinite stage of construction, we would obtain a length that grew as we used a smaller and smaller ruler. At higher and higher resolution, more detail is exposed, and it takes more ruler lengths to cover it. For an ordinary curve of topological dimension 1, the number of ruler lengths, $N(\ell_0)$ varies inversely with the length of the ruler, ℓ_0:

$$N(\ell_0) = L/\ell_0, \tag{1}$$

where L is a constant, the length. For a fractal curve:

$$N(\ell_0) = F/\ell_0^{\overline{d}} \quad (\overline{d} > 1). \tag{2}$$

If $\overline{d}$ is greater than the topological dimension of 1, then we have a fractal curve.

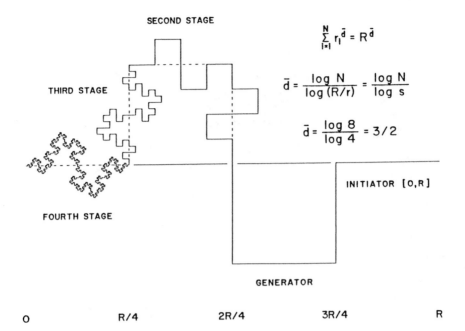

Figure 1. A fractal structure illustrating properties (as discussed in the text) relevant to a polypeptide.

Coastlines are an example of a fractal curve. Mandelbrot [1,2] has written two popular books on the subject of fractals.

There is a second type of fractal, a random fractal, in which the self similarity is statistical rather than geometric. An example would be an unrestricted random walk in 2 or more dimensions. N is the number of steps to cover, which in this case means we have the same RMS end-to-end length, R_{rms}, after scaling down the step size by a factor s. The fundamental relation for fractals is that N is proportional to $s^{\overline{d}}$ (see Eq. in Fig. 1), therefore:

$$R_{rms} = \text{(step size)} \sqrt{N} \propto (1/s)\sqrt{(s^{\overline{d}})}, \qquad (3)$$

In order for R_{rms} to remain constant d must equal 2. A self-avoiding random walk (SAW) in 3 dimensional space ($d = 3$) is characterized by $\overline{d} \approx 5/3$; in 2 dimensional space ($d = 2$), the $\overline{d}$ of a SAW is approximately $4/3$. You may recognize these values as being the inverse of the Flory constant from Polymer theory {Flory constant, $\nu = 3/(d+2)$} that relates the RMS lengths of a polymer, comprised of N monomers of length ℓ_0, to N as:

$$R_{rms} = \ell_0 N^\nu. \tag{4}$$

How do we compute the fractal dimension of a protein, which is neither self similar, random, nor strictly a polymer? We consider the alpha carbon coordinates from x-ray structural data as specifying the position of each monomer of the biopolymer and plot **Log** $\{R(N)/r\}$ vs **Log** N and determine from that the slope, $(1/\overline{d})$. Here N is greater than or equal to 2 and r is the RMS α-carbon/α-carbon separation. We use a weighted average of the statistical and geometrical R for each N. These are defined as [3]:

$$R_s^2(N) = \{1/(M-N)\} \sum_{i=1}^{M-N} R_{i,i+N}^2 \tag{5}$$

where $R_{i,i+N}$ is the through-space distance between residues i and $i+N$ separated by N sequential α-carbon segments, M is the number of residues, and

$$R_G^2(N,j) = (1/D) \sum_{i=1}^{D} R_{j+(i-1)N,j+iN}^2 \tag{6}$$

where j is a residue index, $j = 1, 2, 3, \ldots N - 1$, and D is limited solely by M. For $N = 10$ in Eq. (5), the terms on the right hand side would contain $R_{1,11}^2, R_{2,12}^2, R_{3,13}^2$, etc. For $N = 8, j = 1$, the right hand side of Eq. (6) would contain the terms $R_{1,9}^2, R_{9,17}^2, R_{17,25}^2$, etc. For each N, $R_G(N)$ is determined by the one value of j that yields a minimum in $\overline{d}$. This algorithm has been tested on the self-similar fractal of Fig. 1 and found to produce the proper fractal dimension.

The results of this type of computation for $\overline{d}$ for six proteins of particular interest are shown in Fig. 2. The fits are constrained to pass through the point 1,1 and the fit is stopped by considering changes in the correlation coefficient.

The best correlation of the fractal dimension with some other protein structural parameter seems to be with the amount of α-helix or β-turn conformation

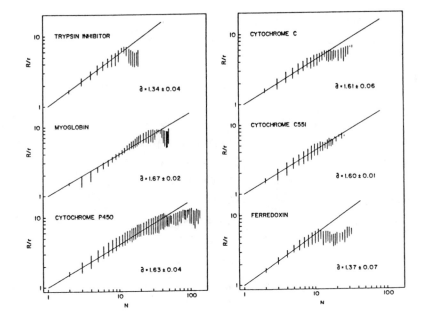

Figure 2. Graphic estimates of fractal dimensions in six protein structures. Protein coordinates are from data bank files. Bovine pancreatic trypsin inhibitor, 3PTI; sperm whale myoglobin, 2MBN; albacore cytochrome C, 3CYT, outer chain; P. aeruginosa cytochrome C551, 351C; 2. Plantinsis ferredoxin, 3FXC.

Coordinates of P. putida cytochrome P450 are from an unpublished preliminary data set (T.L. Poulos, B.C. Finzel, I.C. Gunsalus, G.C. Wagner, and J. Kraut). Vertical lines are twice the standard deviation in Log(R/r). For clarity, some data points are systematically deleted from all plots.

as measured by the fraction of tetrapeptide chain lengths under 7 Å. The most significant deviations from such a linear correlation correspond to the addition of repeating motifs at larger N values. Wheat germ agglutinin is a example of such a deviation. The unique structure of this protein ($M = 164$) is characterized by a conformation of 4 consecutive and apparently identical 41 residue domains.

We turn now to a brief discussion of the Raman electron spin relaxation rate. This relaxation process probes the vibrational density of states, $\rho(\nu)$, in the $10^{11} - 10^{13}$ Hz range at low temperatures. If the density of vibrational states varies with frequency as ν^{m-1} between the limits of 0 and some maximum frequency, μ_{max}, then we call m the spectral index. For ordinary 3 dimensional materials, $m = 3$; for 2 dimensional systems, $m = 2$; and for 1 dimensional systems, $m = 1$.

For this form of the vibrational density of states, the Raman relaxation rates varies with temperature as T^{3+2m} at low temperatures. The temperature dependence is thus T^9 in ordinary solids. At high temperatures, the rate varies as T^2, appropriate to a classical process involving two vibrational quanta. At temperatures in between, the rate has the following temperature dependence:

$$1/T_1 \text{ (Raman)} \propto T^{3+2m} J_{2+2m}(\Theta/T) \tag{7}$$

where $J_n(x)$ is a transport integral defined as

$$J_n(x) \equiv \int_0^x dz z^n e^z / (e^z - 1)^2, \tag{8}$$

and Θ is the Debye temperature, $h\nu_{max}/k$. Eqs. (7) and (8) have the property that:

$$1/T_1 \text{ (Raman)} \propto T^{3+2m} \quad \text{(for } T << \Theta) \tag{9}$$

and

$$1/T_1 \text{ (Raman)} \propto T^2 \quad \text{(for } T >> \Theta). \tag{10}$$

The assumptions that go into this are:

1. the paramagetic ion has an odd number of unpaired electrons;

2. an electrostatic interaction (or any other time even interaction) is involved.

Initially in this process, we have a spin up and a high frequency quantum of energy $h\nu_1$, which is absorbed; the spin flips a new quantum with energy $h\nu_2$ is created. There are many variations of the Raman rate. For an ion with an even number of electrons, the temperature dependence is T^{1+2m} at low temperatures, and thus T^7 in ordinary solids. All of this is well verified experimentally for ordinary solids.

In 1973 Mailer and Taylor [4] published relaxation data on single crystals of ferricytochrome C. They used a novel technique that was sensitive to the relaxation rates only at higher temperatures ($T > 8K$). They indicated in Fig. 4 of their paper that the temperature dependence of the relaxation rate followed a T^7 dependence as high as 20 K. There were two questionable features of these data. First, a T^7 dependence is not expected for Fe^{3+} since this ion possesses an odd number of unpaired electron spins. Secondly, the line through the data of their Fig. 4 actually follows a T^6, not a T^7 dependence. We did not realize this latter feature of the data until 1980.

In 1976 Herric and I published relaxation data on cytochrome P450 from P. putida [5]. We knew at the time that the relaxation rates fit a $T^{6.3}$ quite well, but we had no theoretical basis for such a dependence and we believed Mailer and Taylor's reported T^7 dependence for their hemoprotein. Therefore we attempted a fit with a spectral index of $m = 2$ and a low value of Θ. A low value of Θ has the effect of reducing the T^7 dependence, associated with $m = 2$, to a lower effective exponent of T, at least over a small temperature range.

In 1980, our group proposed a fractal model of spin relaxation in biopolymers [6]. Fig. 3 is from the publication. It shows relaxation data for four low spin hemoproteins: ferricytochrome C (cyt. c), myoglobin-OH (Mb), myoglobin azide (MbN$_3$), and cytochrome P450 from *P. putida* [5]. Each data set is separately fit to the sum of a direct process (varying with temperature as T or T^2) and a simple T^n power law. The best fitting values of n are indicated. Standard errors in n are typically 1%. The straight line in the figure represents a T^9 power law. For Mb and MbN$_3$ only every third datum is displayed between 5 and 11 K to improve clarity. An n value of 6.33 implies a spectral index of 5/3. All data corresponding to relaxation rates below $10^4 s^{-1}$ were measured at X-band microwave frequencies (9.5 GHz) on frozen protein solutions. The cytochrome c data above $T = 10$ K were taken on a single crystal using a microwave frequency of 24 GHz[4] and have been reduced by a factor of 0.106 from the values reported in the original paper. This has been done to aid in the determination of any common temperature dependence between the data of two experiments, one of which used real time measurements of the spin recovery after saturation, while the other did not.

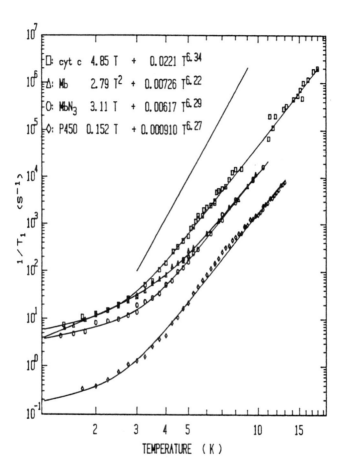

Figure 3. Relaxation rate [9] for four low spin hemoproteins: ferricytochrome C(cyt c), myoglobin.OH (MB), myoglobin azide (MbN3), and cytochrome P450 from P. putida [5]. Each data set is separately fit to the sum of a direct process (varying as T or T^2) and a simple T^n power law. The best fitting values of n are indicated. Standard error are typically 1%. The straight line illustrates a T^9 power law. Some data are systematically omitted for clarity.

The Raman rate temperature exponents, n, for the cyt c, Mb, MbN$_3$, and P450 data of Fig. 3 are 6.34 $\pm$ 0.06, 6.22 $\pm$ 0.09, 6.29 $\pm$ 0.08, and 6.27 $\pm$ 0.06, respectively, indicating spectral indices of 1.67 $\pm$ 0.03, 1.61 $\pm$ 0.05, 1.65 $\pm$ 0.04, and 1.64 $\pm$ 0.03, respectively. These values agree well with the fractal dimensions calculated from the data of Fig. 2.

If the relaxation rates under $10^4 s^{-1}$ in Fig. 3 are fit to a conventional spectral index of $m = 3$, i.e. $T^9 J_8(\Theta/T)$ from Eq. (7), a Θ value of 48.1 K is required to approximate the $T^{6.34}$ variation of the Raman rate. AT 20 K, the logarithmic slope of $T^9 J_8(48.1/T)$ is about 3, well below the value of 6.34, obtained from the relaxation data on a single crystal of ferricytochrome C.

Gayda et al. [7] published relaxation data on a 2-Fe/2-S ferredoxin in 1979. Fig. 4 is our reanalysis of their data assuming a fractal model. The exponent of a simple power law Raman rate that fits the data of Fig. 4 is 5.67 $\pm$ 0.11 from which a spectral index of 1.34 zpm 0.06 can be inferred. This should be compared with the value of $\overline{d} = 1.37 \pm 0.07$ computed for the ferredoxin S. *Platensis*, as shown in Fig. 2.

Solvent effects on the value of n obtained from relaxation data were first noted for putidaredoxin from *P. putida* [3]. These data are shown in Fig. 5 for two conditions of the solvent's ionic strength, l. In a solvent with 1 molar NaCl, the spectral index is 1.34 $\pm$ 0.03; a value in good agreement with the other iron-sulfur proteins studied. In low ionic strength, m appears to be 1.11 $\pm$ 0.03, a value well below that expected from the fractal dimension. It is worth noting that the larger spectral index is associated with a weaker direct process relaxation rate.

The relaxation data on cytochrome C551 [3] shows similar behavior. With a solvent of high ionic strength (1.1), the apparent value of m is 1.43 $\pm$ 0.09, while in low ionic strength (0.1) it is approximately 0.88 $\pm$ 0.14. The x-ray data on this protein yields a fractal dimension of 1.60 $\pm$ 0.01.

These results lead us to look systematically for solvent effects in myoglobin azide [8]. Eleven experiments were performed under different solvent conditions.

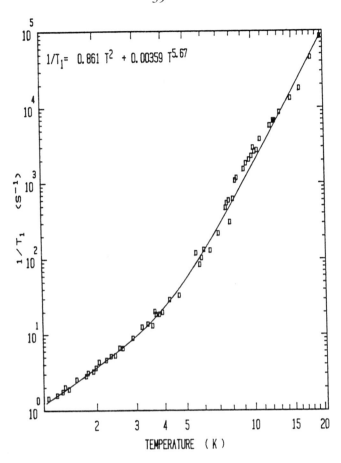

$$1/T_1 = 0.861 \ T^2 + 0.00359 \ T^{5.67}$$

Figure 4. Relaxation data on a 2-Fe/2-S ferredoxin (blue green algae Spirulina maxima) [7] up to a temperature of 20K. The curve is our reanalysis of these data to a fractal Raman relaxation process.

Low values of $m(\approx 1.2)$ were obtained in frozen aqueous solutions at high (1M) and low (mM) salt concentrations, and at high (21 mM) and low (mM) protein concentrations. Reasonable values of m were obtained on a sample concentrated under a partial vacuum and on samples dissolved in a 50% water-50% glycerol solvent. In all cases, slower direct process relaxation rates were associated with higher m values. Fig. 6 shows a feature present to a lesser degree in all the data with slow direct relaxation rates. These data refer to a dilute MbN_3 sample

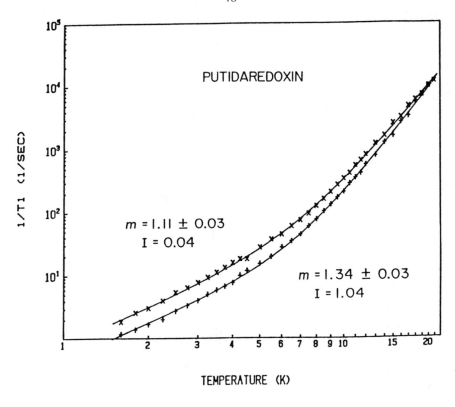

PUTIDAREDOXIN

$m = 1.11 \pm 0.03$
$I = 0.04$

$m = 1.34 \pm 0.03$
$I = 1.04$

Figure 5. X-band relaxation data ($\approx$ 0.1 GHz) on two frozen solutions of the 2-Fe/2-S protein, putidaredoxin from P. putida [3].

(1.5 mM protein in 50% glycerol and water). The effective temperature exponent fitting the data rises sharply at approximately 4 K, then drops to a value predicted by the fractal model at temperatures above 6 K. At 6 K, the integrand of the integral over vibrational mode frequencies (the Raman integral [8,9]) that leads to Eq. (7) peaks at a frequency of 627 GHz (20.9 cm^{-1}) for a spectral dimension of 1.54. This corresponds to an acoustic wavelength of 40 Å if a sound velocity of 2.5 x 10^5 cm/s is assumed. This characteristic length is in approximate agreement with myoglobin's molecular size ($44 \times 44 \times 25$ Å). Vibrations of wavelength greater than the molecular size become dominant in the Raman integral at temperatures below 6 K. The density of states, $\rho(\nu)$, for these excitations is not expected to reflect the protein structure, but rather that of the solvent. Below 6 K, the

Raman rate relaxation process utilizes vibrational modes that are characteristic of the solvent and anticipated to have an effective spectral index near 3.

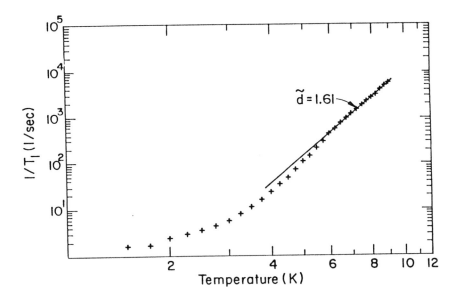

Figure 6. Relaxation data on a 1.5 mM myoglobin azide sample in a solvent of 50% glycerol and 50% water [8]. A kink in the slope of the data near a temperature of 6 K is discernable. Above that temperature a spectral index of d = 1.61 is indicated.

A model calculation has been made [8] utilizing a two component vibrational density of states with a transition frequency near 650 GHz. Below that frequency a spectral index of 3 was assumed, and above 650 GHz a value of 1.61 was used. In addition there can be a discontinuity in the mode density at the transition frequency. With this model two experimentally observed features are reproduced. The Raman relaxation data of Fig. 6 are nicely fit, but more surprising, the stronger the relative strength of the three dimensional solvent modes (as characterized by a discontinuous drop in the number of modes as the transition frequency is exceeded) the lower the exponent of the Raman rate above the transition temperature. With a strong contribution from the lower frequency solvent modes, the apparent spectral index falls below the assigned spectral dimension. This is consistent with our experimental observation that the fractal dimension is always observed in the Raman rate when the direct relaxation rate is weakest.

The relationship between the m, the spectral index, and $\overline{d}$, the fractal dimension, is [10-12]:

$$m = 2\overline{d}/d_w, \tag{11}$$

where d_w is the fractal dimension of a random walk on the fractal structure. It relates the number of steps to the end-to-end distance traveled. This is another way of considering connectivity. If you can move only along the backbone, then $d_w = 2\overline{d}$ and $m = 1$; i.e. if you have a one dimensional system, it remains a one dimensional system no matter how you wind it around in space. However, if you are able to move off the backbone via any type of bonds, then d_w approaches 2 (the fractal dimension of an unrestricted walk) and $m = \overline{d}$. This latter condition is what we seem to be observing in these relaxation data.

Go et al [13] have computed the low frequency normal modes of Bovine pancreatic trypsin inhibitor (BPTI) using a method appropriate to low frequency modes. There are 58 residues in BPTI and that implies 116 dihedral angles along the backbone. The authors also included 125 side chain angles. An incomplete histogram of the normal mode frequency distribution they reported is shown in Fig. 7. It corresponds to the lowest 201 modes out of a total of 241. The higher frequency modes were found to be quite localized, while vibrations below 3600 GHz (120 cm^{-1}) were predominantly delocalized, i.e. they correspond to concerted movements of the entire polypeptide backbone (including side chains). Since the lowest 117 vibrational modes are below 75 cm^{-1}, an estimate of 2250

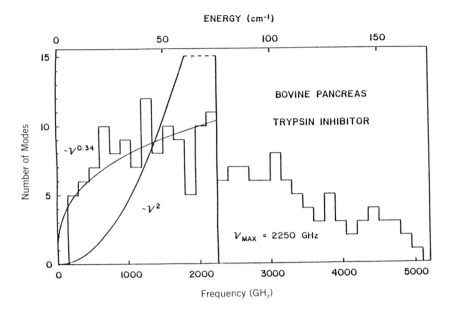

Figure 7. Histogram of the lowest 201 vibrational modes in bovine pancreatic trypsin inhibitor (BPTI). Data are reproduced from Go et al [13]. Superimposed are normalized densities if vibrational states varying as ν^2 (normal Debye model) and $\nu^{0.34}$ (fractal mode). The latter is derived from the fractal dimension of BPTI obtained from Figure 2.

GHz for $\nu_{\max}$ is not unreasonable.

Superimposed on the histogram of Fig. 7 are normalized densities of states predicted by the fractal model, $\rho(\nu) \propto \nu^{P0.34}$ (see Fig. 2) and by the classical Debye model for a regular three dimensional solid, $\rho(\nu) \propto \nu^2$. An inspection of Fig. 7 demonstrates convincingly that the fractal model provides a more accurate description of the low frequency vibrational mode distribution. An un-weighted, non-linear, least-squares analysis of the data in the frequency range 0 - $\nu_{\max}$ yields a best-fit normalized $\rho(\nu) \propto \nu^{0.24}$.

Our conclusions from all of the data discussed above are:

(1) spin-lattice relaxation rates fit a reduced density of states, not a reduced Debye temperature;

(2) the low frequency density of states $\rho(\nu)$, follows a $\nu^{\bar{d}-1}$ dependence, with $1 < \bar{d} < 2$;

(3) the theoretically computed low frequency normal modes fit a $\nu^{\bar{d}-1}$ dependence reasonably well; and

(4) the computed normal mode frequencies match those required for a Raman relaxation process in the 3-12 K temperature range.

This research was supported in part by the U.S. Public Health Service under NIH grant GM-24488.

References

1. B.B. Mandelbrot, "Fractals: Form, Chance, and Dimension"; W.H. Freeman & Co., San Francisco, **1977**.

2. B.B. Mandelbrot, "The Fractal Geometry of Nature"; W.H. Freeman & Co., San Francisco, **1982**.

3. G.C. Wagner, J.T. Colvin, J.P. Allen, & H. J. Stapleton, *J. Amer. Chem. Soc.* (to be published , Oct. **1985**).

4. C. Mailer, C.P.S. Taylor, *Biochim. Biophys. Acta. 322*, 195 (**1973**).

5. R.C. Herrick & H.J. Stapleton, *J. Chem. Phys. 65*, 4778 (**1976**).

6. H.J. Stapleton, J.P. Allen, C.P. Flynn, D.G. Stinson, and S.R. Kurtz, *Phys. Rev. Lett. 45*, 1456 (**1980**).

7. J.P. Gayda, P. Bertrand, A. Deville, C. More, G. Roger, J.F. Gibson, & R. Cammack, *Biochem. Biophys. Acta. 581*, 15 (**1979**).

8. J.T. Colvin & H.J. Stapleton, *J. Chem. Phys. 82*, 4699 (**1985**).

9. J.P. Allen, J.T. Colvin, D. G. Stinson, C.P. Flynn & H.J. Stapleton, *Biophys. J. 38*, 299 (**1982**).

10. S. Alexander & R. Orbach, *J. Physique LETTRES 43*, L625 (**1982**).

11. R. Rammal & G. Toulouse, *J. Physique LETTRES 44*, L13 (**1983**).

12. J.S. Helman, A. Coniglio & C. Tsallis, *Phys. Rev. Lett. 53*, 1195 (**1984**).

13. N. Go, T. Noguti & T. Nishikawa, *Proc. Nat. Acad. Sci. USA 80*. 3696 (**1983**).

Configurational Entropy and Dielectric Relaxations in Proteins

L. Dissado

The Dielectrics Group

Chelsea College

University of London

Pulton Place

London SW6 5PR, UK

I. Introduction

The morphology of proteins is complex [1] involving the embedding of one level of structure within grosser levels of organization. Thus, for example, the α-helices of the secondary structure are inter-woven in the tertiary structure. The ground state of proteins therefore does not have a unique structure as do ideal crystals but possesses a number of alternative configurations, a fact which is represented by a configurational entropy contribution to the Free Energy. It is the intention here to relate the configuration entropy to structural fluctuations and to show how these are involved in the dynamics of relaxation and transport. The paper will follow the outline given in figure 1, and the dielectric response technique will be used to illustrate the theoretical results.

2. Some Thermodynamic Considerations

When a field, E, is applied to a protein some configurations will be preferred to others and a displacement will take place in the positions and orientations of the components of the configuration which may for example be the segments of the tertiary structure. The set of displacements generated are defined as the difference in position between the field induced structure and the equilibrium structure and is the response of the system to the field [2]. The spatial symmetry of the response will be that of the applied field and can be regarded as being

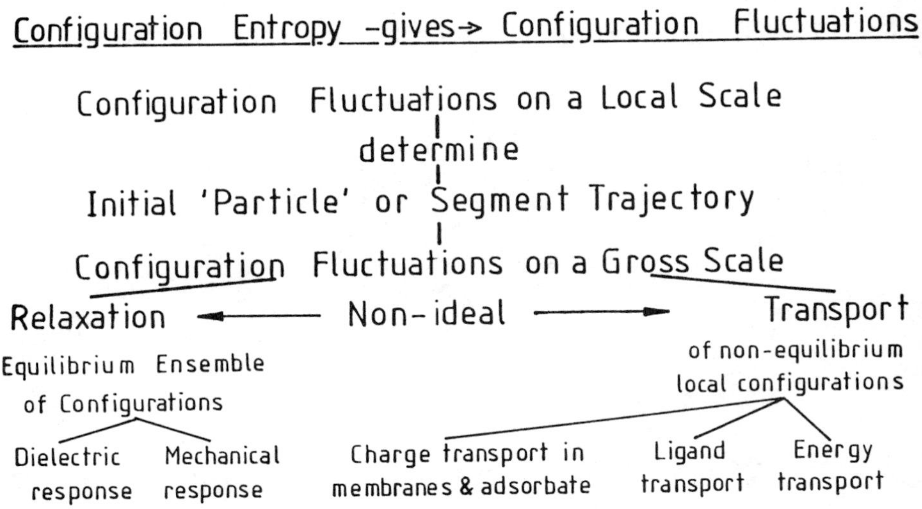

Configuration Entropy —gives→ Configuration Fluctuations

Configuration Fluctuations on a Local Scale
determine
Initial 'Particle' or Segment Trajectory
Configuration Fluctuations on a Gross Scale

Relaxation ←——— Non-ideal ———→ Transport

Equilibrium Ensemble
of Configurations

of non-equilibrium
local configurations

| Dielectric response | Mechanical response | Charge transport in membranes & adsorbate | Ligand transport | Energy transport |

Figure 1. A schematic outline of essential features.

composed of a number of identical structural fluctuations P, with that symmetry.

The average value of P in the presence of the field is given thermodynamically by

$$\langle\langle P(E)\rangle\rangle = Z^{-1}\sum_r P_r \exp(-P_r E/kT) \tag{1}$$

where P_r is the segment displacement and Z the partition function,

$$Z = \exp\{-G(P)/kT\}. \tag{2}$$

$G(P)$ is the Free Energy of the fluctuation which be formally defined in terms of an entropy, $S(P)$, and an internal energy $U(P)$ as

$$G(P) = U(P) - TS(P). \tag{3}$$

Relaxation of the system following removal of an applied field [3] is determined by the regression of P to zero, which involves the conversion of the specific spatial symmetry of the field to the alternative configurations of the equilibrium state, ie. the generation of configuration entropy. Thus

$$\langle\langle P(t)\rangle\rangle = \langle\langle P(E)\rangle\rangle \exp[-\Delta S(t)/k] \tag{4}$$

where $\Delta S(t)$ is the increase in configuration entropy as the equilibrium state evolves.

When a potential barrier must be overcome, which may be the case for chemical reactions at a protein site [4] as well as structural relaxation, the theory of Eyring [5] may be utilized to show that an increase in ground state configuration entropy prior to entering the activated state leads to a decaying rate constant. Thus

$$\omega_p(t) = \omega_p(t = 0) \exp[-\Delta S(t)/k] \tag{5}$$

where $\omega_p(t = 0)$ is the rate constant in the absence of any entropy evolution and is given by

$$\omega_p(t = 0) = \nu e^{-\Delta G^*/kT} \tag{6}$$

with ΔG^* the activation free energy given by

$$\begin{aligned} \Delta G^* &= (U_a - U_g) - T(S_a - S_g) \\ &= \Delta U^* - T\Delta S^* \end{aligned} \tag{7}$$

Here the a and g subscripts indicate the activated and ground states respectively.

In figure 2 the thermodynamic reaction path is shown, schematically, for two alternative situations; (a) where the components of a configuration relax by decoupling their motions and which corresponds to dissociation of a ligand and protein, say, in chemical kinetic terms, and case (b) where topological constraints only allow specific configurations to relax whose chemical kinetic equivalent is the necessity of particular protein configurations to allow access of a ligand to an active site.

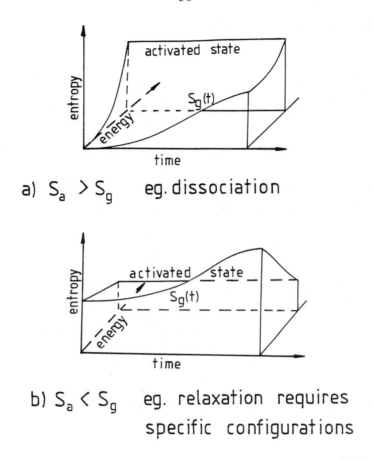

a) $S_a > S_g$ eg. dissociation

b) $S_a < S_g$ eg. relaxation requires specific configurations

Figure 2. Changes in configuration entropy during the relaxation over a barrier.

3. Fluctuating Potential Surface

The potential surface upon which any structural component moves will be determined substantially by the short range forces originating in its immediate surroundings. During the relaxation of a configurational fluctuation the environment of each component will alter as the other components also displace. The regression to equilibrium of any component will therefore by governed by a fluctuating potential surface with its motions at any time governed by the shape of the potential at that instant.

Figure 3 shows one such form of surface with curve (a) representing the average potential and curves (b) and (c) the behavior when displacements of surrounding components lead to stronger or weaker binding respectively. The shape of the potential shown has been chosen to allow for the possibility of displacements that remove a component from its original configuration and bind it to a neighboring one [6]. This construction can be related to that of a double minima potential [7] by extending the relaxation coordinate $R(\ell)$ through the transition state maxima to a similarly fluctuating alternative minima. The equation of motion of a component in such a potential is nonlinear in the displacement [6,8] and two classes of motion must be distinguished, namely those that retain the component within the original configuration and those that transfer it between configurations.

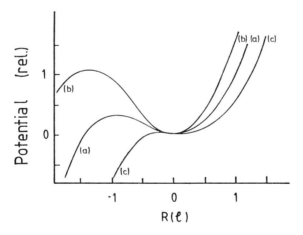

Figure 3. Fluctuating potential surface, (a) average potential; (b) configuration with tighter binding; (c) configuration with weaker binding.

Let us concentrate initially on motions that retain the binding of the component to its original configuration. Here displacement of the centers of motion along the relaxation coordinate alter the frequencies of the quantum vibrations about the center of motion and the time correlation function of the displacement is found to be given by the overlap of the quantum oscillators in the initial displaced configuration with those in all possible alternative equilibrium configurations [8]. The response function [3], $\phi(t)$, thus becomes

$$\phi(t) = Re\left(e^{-\Gamma t} \exp\left[- \int_{\Gamma}^{\varsigma} \{R(\Omega)/\beta\}^2 (1 - e^{-i\Omega t}) d\Omega/\varsigma\right]\right) \tag{8}$$

where the change in quantum frequencies, $\Omega(\Omega_{\max} = \varsigma)$ defines each possible displacement $R(\Omega)$ as an oscillator damped by the phenomenological damping constant Γ. The frequency Ω becomes smaller as the number of components displacing as a rigid body become larger, and integration over Ω gives the configuration entropy change during relaxation by counting the number of ways a single component can take part in joint displacements with other components up to a maximum $N(\Gamma)$.

Thus

$$\phi(t) = e^{-\Gamma t} e^{-\Delta S(t)/k} \tag{9}$$

and

$$S(t) = \begin{cases} n \ ln[\Gamma t N(\Gamma)] & \varsigma t < \Gamma t < 1 \\ n \ ln[N(\Gamma)] & \Gamma t > 1 \end{cases} \tag{10}$$

with

$$0 < n < 1. \tag{11}$$

This result shows that a time power law decay, t^{-n}, for $\Gamma t < 1$ arises from the evolution of configuration entropy in the relaxing system and is similar to the result of de Gennes [9] for reputation in polymeric systems. At times $\varsigma t < 1$, prior to the steady generation of entropy given in equation (10), the exact result for $\Delta S(t)$ oscillates as shown in figure 4, with a partially coherent exchange of energy between the quantum oscillators and the kinetic energy of displacement.

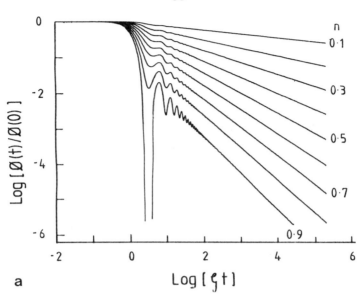

Figure 4. Exact response at short time plotted on (a) log-log scales to illustrate the transition to power law behavior with $\phi(t) \propto t^{-n}$. (b) Linear scales to emphasize the oscillatory behavior at short times.

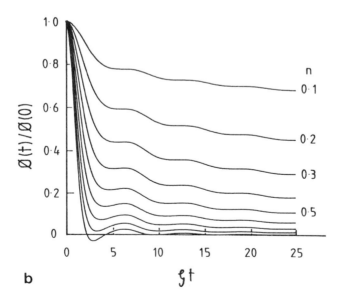

4. Initial Trajectory of Relaxation

Displacements with different values of Ω are related to each other by a change in scale, i.e., they are self-similar. Thus the model is equivalent to the fractal description [10] of relaxation in which a scaling ansatz is used to obtain a time dependent rate constant with power law form [11]. However the index n defines the degree of replication of displacement motion on changing scale and is not a universal value since it includes the structural irregularities typical of the system. The fractal concept can however be used to construct relaxation trajectories in model systems by combining the displacements appropriate to different scales.

Such a construct is shown, schematically, in figure 5. In (i) a single segment displaces rapidly in the rigid environment of its fellow segments, in (ii) double segments displace more slowly in the rigid environment of the rest of the chain, and so on through (iii) and (iv) doubling the scale each time to give the initial displacement trajectories shown in (b). Application of this theoretical formalism to particular model structures therefore allows the detailed computation of relaxation trajectories.

5. Relaxation to an Ensemble

We now consider the influence of the class of motions that transfer components from one configuration to another. In this way the number of components in both the donating and accepting configurations are changed, and an initial system of identical configurations is converted to an ensemble of different local configurations [6]. Inclusion of this feature in the theory gives the observable response function as an ensemble average [12,13],

$$\langle \phi(t) \rangle = \int_0^\infty e^{-\Gamma(y)t} \exp[-\Delta S(y,t)/k] g(y) dy \tag{12}$$

where

$$\Gamma(y) = \frac{\Gamma y^{1/m}}{1 + y^{1/m}} \tag{13}$$

and

$$\Delta S(y,t)/k = n \, ln[\Gamma(y)t \, N(y)] \tag{14}$$

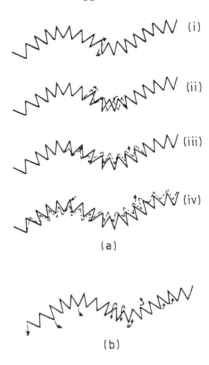

(a)

(b)

Figure 5. (a) Displacements related by a change of scale, (i) to (iv). (b) The relaxation trajectory derived from (a).

are the local relaxation rates and entropies respectively [6,13] for configurations defined by the number of components $N(y)$ that they contain, with

$$N(y) = N(\Gamma)/y^{1/n}. \tag{15}$$

Expression (12) shows that relaxation involves the generation of an ensemble with each configuration regressing towards its own instantaneous local configuration entropy at its own local rate $\Gamma(y)$.

The ensemble distribution density $g(y)$ can be obtained without arbitrary assumptions [14] and has the form shown in Figure 6, where the shape is determined by a single index $m(0 < m < 1)$ which defines the efficiency of transfer of components between local configurations. The extremes of behavior are given when (a) no transfer occurs, m is zero and $g(y)$ is a delta function and (b) when m is unity with each transfer identical and equally probable giving $g(y)$ an exponential form.

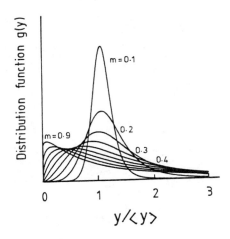

Figure 6. The theoretical distribution function of the ensemble $g(y)$ for $m = 0.1$ to 0.9 in steps of 0.1.

Fourier transformation of the response function gives the complex susceptibility of the system to the applied field as a function of its frequency [3] and figure 7 compares experimental data obtained for a synthetic polypeptide PMLG with the theoretical function [6,12]. The interpretation of this data in terms of the theoretical model [15] suggests that such response measurements can be used to obtain information both on the dynamic behavior of local structural fluctuations in proteins and also on their distribution in form.

6. Non-Ideal Transport

The transfer of components between neighboring configurations is clearly a transport process, but it can only be observed mechanically when the component

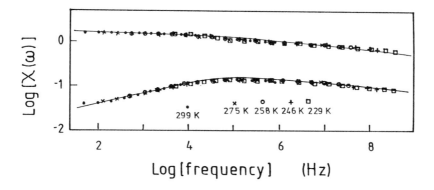

Figure 7. A plot of log $\chi^*(\omega)$ as a function of log ω. The experimental points have been obtained at different temperatures, as indicated by the datum symbols, and the experimental data have been temperature normalized. The theoretical curves through the points have been determined with n = 0.915 and m = 0.24.

is neutral. In a number of important situations, however, the component is charged and the transport can then be observed through a dielectric response.

The systems that will give this type of behavior are those in which ions are constrained to move between adsorption sites along partially occupied pathways, such as protons in a surface adsorbate or cations in 1-dimensional channels [16]. Because of the partial occupancy cluster configurations of charges will occur, and fluctuations of ion composition will be involved in addition to those of substrate conformation arising when the substrate is flexible such as in proteins. However when the ions are an intrinsic part of the molecular structure only molecular configuration fluctuations are involved.

In this form of transport actual movement of individual ions over small distances induces charge displacements over a much greater range through both ion-ion repulsions and local alterations in protein conformation. The "particle" transported is thus a charge-displacement which is illustrated, schematically, in figure 8.

When the system is homogeneous such that each transfer or charge leaves

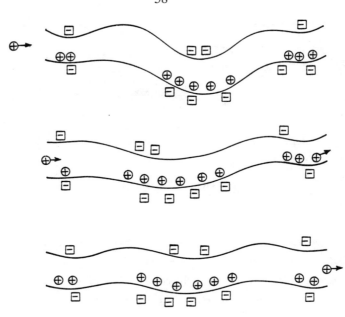

Figure 8. Approach of a cation induces charge displacements over a long range which is combined with changes of protein conformation.

behind a configuration with the same form as that to which it moves the transporting particle retains its integrity and the transport will be ideal. However, in general the configurations along the transport path will be irregular and thus the moving "particle" will lose structural integrity giving non-ideal transport [17]. As a result a system polarization will be produced by an electric field through the formation of a differential in the charge configuration differential, whether produced by chemical or chemi-physical means, will generate an electric potential. Theoretically [16,17] this mechanism gives a current density which steadily decays with time rather than the constant value expected of a d.c. transport. Figure 9 compares this situation with the relaxation behavior, d.c. conduction, and exponential relaxation. Fourier transformation again gives the complex susceptibility.

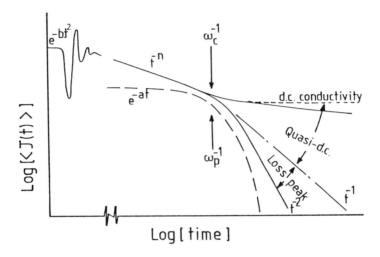

Figure 9. Relaxation current as a function of time for perfect d.c. transport (conductivity), non-ideal transport, non-ideal relaxation and exponential relaxation.

An example of a water adsorbate system in which charge transport has this form is the hydrated powders of the globular protein bovine serum albumin [18] whose response is shown in Figure 10. The data is presented as C' and imaginary C'' components of the capacitance, which correspond to real and imaginary susceptibilities (χ' and χ''), and the continuous lines are the theoretical curves. The most important feature of this data is the strong increase in polarisability at low frequencies due to the non-ideal nature of the charge transport.

Although the behavior of living systems is usually more complex, there is often a response region which can be identified with the transport of ions through channels in a membrane and therefore may be expected to be non-ideal in form. Figure 11 presents data from such a region which has been obtained from measurements on freshly picked leaves on the maize plant and shows that here also a non-ideal transport mechanism occurs in series with a weak d.c. transport.

The membranes in plants, however, are composed of cellulose fibers which do not contain proteins and are non-living, but Figure 12 shows that even in the case

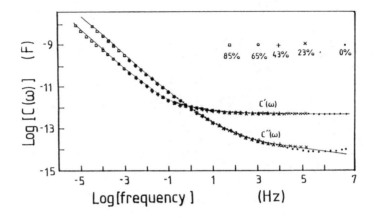

Figure 10. The dielectric response of hydrated bovine serum albumin. The fitted theoretical function for non-ideal transport is given by the continuous curves for m (denoted by p in ref. [16]) = 0.87 and n = 0.30. The plot is scaled at 43% relative humidity and has been obtained by normalizing data measured over the humidity range from zero to 85% R.H.

of flesh tissue the same mechanisms can be found, although there the range of frequency investigated may be insufficient to resolve this mechanism completely from other processes.

7. Final Comments

Since configuration fluctuations in protein structures can be related to relaxation experiments through changes in configuration entropy such experiments can be used to gain information on the details of local and ensemble dynamics in proteins, which may be particularly applicable to the functionally important chemical kinetic mechanisms. Another important area is the identification of those protein fluctuations that contribute to non-ideal charge transport through membranes, where the very non-ideality of the mechanism may be of functional significance, since the large polarisability allows transport of considerable amounts of information in the form of local configuration changes, by a small electric field.

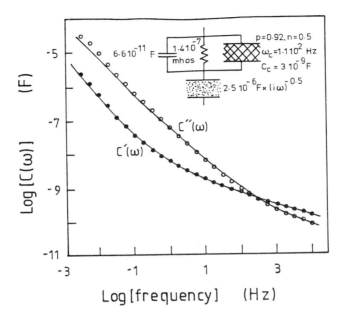

Figure 11. The dielectric response of maize leaves. The inset shows the equivalent electrical circuit which has been used to obtain the continuous curves through the data.

Acknowledgement

The author gratefully acknowledges financial support from the Conference Organizers for travel costs.

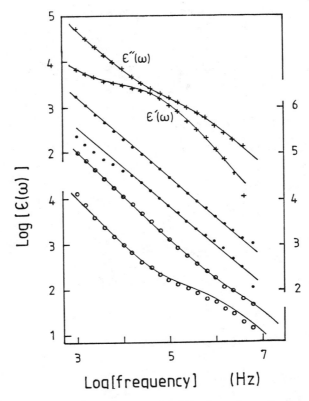

Figure 12. The dielectric response of rabbit tissue from the data reported in reference [19]. The theoretical non-ideal behavior, at low frequencies, is given by the power law divergence of polarization, $\varepsilon'(\omega)$, and loss $\varepsilon''(\omega)$, which obey the relationship

$$\varepsilon''(\omega) = \varepsilon' tan(m\pi/2)$$

with m = 0.95 in (a), 0.86 in (b) and 0.94 in (c). The detail in (a) and (c) is due to the interaction of series conductance elements and the curves through these data points have been derived from circuit models similar to that shown in figure 11. (a) Liver, (b) Kidney, and (v) Brain cortex.

References

[1] See, for example, R. Pethig, *Dielectric and Electronic Properties of Biological Materials*, (J. Wiley, Chichester), 1979.

[2] M. Lax, Rev. Mod. Phys. *32*, 25 (1960).

[3] R. Kubo, J. Phys. Soc. Jap. *12*, 570 (1957).

[4] A. Warshel, Proc. Natl. Acad. Sci. USA *81*, 444 (1984).

[5] S. Glasstone, K.J. Laidler and H. Eyring, *The Theory of Rate Processes*, McGraw-Hill, New York (1941).

[6] L.A. Dissado and R.M. Hill, Proc. Roy. Soc. (London) *A390*, 131 (1983).

[7] P. Hanggi, J. Stat. Phys. *30*, 401 (1983).

[8] L.A. Dissado, Chemical Physics *91*, 183 (1984).

[9] P-G. de Gennes, *Scaling Concepts in Polymer Physics*, (Cornell Univ. Press, Cornell), Chap. XI (1979).

[10] B. Mandelbrot, *The Fractal Geometry of Nature*, (Freeman Press, San Francisco) (1983).

[11] J.S. Newhouse, P. Argyrakis and R. Kopelman, Chem. Phys. Lett. *107*, 48 (1984).

[12] L.A. Dissado and R.M. Hill, Nature (London) *279*, 685 (1979).

[13] L.A. Dissado, R. Nigmatullin and R.M. Hill, in *Advances in Chemical Physics*, (J. Wiley, New York), Vol. *63*, Chap. 3 (1985).

[14] R.M. Hill, L.A. Dissado and R. Jackson, J. Phys. C. *14*, 3915 (1981).

[15] Details are contained in reference 6.

[16] L.A. Dissado and R.M. Hill, J. Chem. Soc. Far. Trans. 2 *80*, 291 (1984).

[17] B. Shapiro and E. Abrahams, Phys. Rev. B *24*, 4889 (1981).

[18] M. Shablakh, L.A. Dissado and R.M. Hill, J. Biol. Phys. *24* (1984).

[19] H. Ludt and H.D. Herrmann, Biophysik *10*, 337 (1973).

The Mössbauer Effect as a Probe of Protein Dynamics

F. Parak, H. Hartmann and G. Nienhaus

Institut für Physikalische Chemie der Universität Münster

Schloßplatz 4-7

4400 Münster BRD

1. Introduction

The flexibility of protein molecules reveals itself in several experiments. X-ray structure analysis of protein crystals shows that the unit cells are not identical. Different molecules have slightly different structures called conformational substates. The derivation of molecules from their average structure as determined by the x-ray method can be measured by the mean square displacement, $\langle x^2 \rangle^x$, which can be obtained for each atom of the molecule. Displacement of the atoms can be caused by thermal vibrations. However, since x-ray diffraction of single crystals is a coherent effect of all unit cells static disorder in the molecules contributes also to $\langle x^2 \rangle^x$. No information on the time scale of motions contributing to the $\langle x^2 \rangle^x$-values can be obtained from x-ray structure analysis.

For iron containing proteins Mössbauer absorption spectroscopy on ^{57}Fe can give complementary information to the x-ray structure analysis. The probability of the Mössbauer effect is determined by the Lamb Mössbauer factor f which depends on the mean square displacement, $\langle x^2 \rangle^\gamma$, of the iron. In contrast to x-ray structure determination Mössbauer absorption occurs incoherently at one individual iron nucleus. The lifetime $\tau_N = 1.41 x 10^{-7} s$ of the 14.4 keV level of the iron nucleus sets a time threshold. Only modes of motions with a characteristic time faster τ_N contribute to $\langle x^2 \rangle^\gamma$. Via the spectral shape Mössbauer spectroscopy sets also a time window which allows identification of the motions with a characteristic time between 10^{-7} and $10^{-9} s$.

This contribution deals with investigation on sperm whale myoglobin. Mössbauer spectroscopy is complemented by x-ray structure investigation at different temperatures.

2. Mössbauer absorption spectroscopy on deoxymyoglobin crystals

Figure 1 shows the Mössbauer spectrum of a sample consisting of a number of deoxymyoglobin crystals taken from [1]. The transmission of the 14.4 keV gamma radiation of a ^{57}Co source through the sample is given as a function of the Doppler velocity v of the source. To convert the x-axis into energy units one has to apply the relation: $E = E_\gamma \cdot v/c$ (c: velocity of light, $E_\gamma = 14.4$ keV). The experimental data are fitted by two Lorentzians (Figure 1a). The area $A^{(0)}$ of these Lorentzians measures the mean square displacement, $\langle x^2 \rangle^\gamma$, of the iron:

$$C A^{(0)} = f^{(0)} = \exp(-k^2 < x^2 >^\gamma) \tag{1}$$

C is a temperature independent constant and $k = 2\pi/\lambda$ with $\lambda = 0.86\text{Å}$. The Lorentzians have a width $\Gamma_{\exp}^{(0)}$ which is the sum of $\Gamma_s + \Gamma_a$. The indices s and a refer to the contributions of the source and the absorber, respectively. The Mössbauer spectrum contains two transmission minima at a distance ΔE_Q. This quadrupole splitting indicates that the iron is in a non cubic environment. The center of gravity of the spectrum is shifted by $S_{\exp}$ with respect to the velocity zero of the source.

It is obvious that two narrow Lorentzians with the width $\Gamma_{\exp}^{(0)}$ cannot explain the experimental results. A good fit is obtained by adding two Lorentzians with a line width $\Gamma_{\exp}^{(1)}$ as shown in Figure 1b.

The Mössbauer spectrum of deoxymyoglobin crystals can be measured as a function of temperature. Figure 2 shows the temperature dependence of the Mössbauer parameters. Coming from low temperatures the mean square displacement $\langle x^2 \rangle^\gamma$ of the iron increases linear with temperature suggesting a Debye behavior. At about 200 K a strong increase of $\langle x^2 \rangle^\gamma$ shows that new modes of motions are activated which can be attributed to structural fluctuations in the protein molecule. The total mean square displacement $\langle x^2 \rangle^\gamma$ contains, therefore, two contributions

$$\langle x^2 \rangle^\gamma = \langle x_v^2 \rangle + \langle x_t^2 \rangle \tag{2}$$

where the index v refers to solid state vibrations and the index t indicates protein

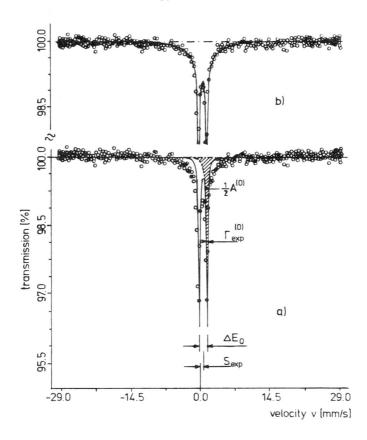

Figure 1. Mössbauer spectrum of deoxymyoglobin crystals at 269 K. The fit in a) (solid line) uses two narrow Lorentzians. One obtains as parameters: $A^{(0)}$: area of the Mössbauer spectrum; $\Gamma_{exp}^{(0)}$: width of the narrow lines; ΔE_Q: quadrupole splitting and S_{exp}: shift of the spectrum. Two narrow Lorentzians do not fit the experimental data. The addition of two broad lines with the width $\Gamma_{exp}^{(1)}$, however, gives good agreement with the experiments (Fig. 1b).

specific motions. It should be emphasized once more that $\langle x^2 \rangle^\gamma$ measures only motions occurring with a characteristic time faster $10^{-7}s$. The data shown in Figure 2 are taken from [1]. Investigation on the same subject described in [2] give essentially the same results.

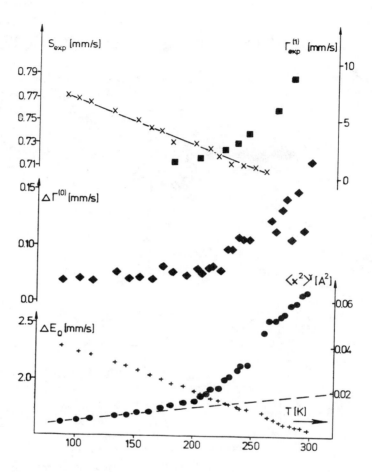

Figure 2. Temperature dependence of the Mössbauer parameters of deoxymyoglobin crystals. Lower part: mean square displacement, $\langle x^2 \rangle^\gamma$, of the iron and quadrupole splitting ΔE_Q center: line broadening $\Delta\Gamma^{(0)}$ of the narrow line; upper part: width of the broad Lorentzians $\Gamma_{exp}^{(1)}$ and shift, S_{exp}, of the center of gravity of the Mössbauer spectrum.

The quadrupole splitting ΔE_Q is caused by an average electronic field gradient at the position of the iron nucleus coming from the contributions of thermally populated low lying electronic levels of the iron [3]. The monotonic decrease of ΔE_Q with temperature reflects the Boltzmann factor weighing the contribution of the electronic levels. No change of this decrease occurs around 200 K. This shows that $\langle x^2 \rangle^\gamma$ labels correlated motions of the iron together with its neighbors.

Knowing from other experiments the widths Γ_s of the source the contribution of the absorber to $\Gamma_{\exp}^{(0)}$ can be determined as a function of temperature. With Γ_{nat} determined from τ_N by the Heisenberg uncertainty one gets: $\Gamma_{\exp}^{(0)} = \Gamma_s + \Gamma_a$ and $\Gamma_a = \Gamma_{\text{nat}} + \Delta\Gamma^{(0)}$. Here, Γ_a is the width of a "thin" absorber. In the ideal case $\Delta\Gamma^{(0)} = 0$. Figure 2 shows that even at low temperatures $\Delta\Gamma^{(0)}$ is not zero. This inhomogeneity broadening is a hint of the existence of conformational substates. The molecules of the sample are not absolutely identical with respect to the environment of the iron.

At about 200 K the $\Delta\Gamma^{(0)}$-value increases strongly with temperature, an indication for protein specific motions with a characteristic time of typically $10^{-7}s$ [1]. At temperatures above 180 K the shape of the Mössbauer spectrum can only be understood by the sum of broad and narrow Lorentzian as shown in Figure 1 [1,2]. The broad line is clearly absent at 160 K. From $\Gamma_{\exp}^{(1)}$ exceeding Γ_{nat} by more than a factor of 50 one can conclude that protein specific dynamics contain modes of motions with a characteristic time between 10^{-8} and $10^{-9}s$.

Recently the shift $S_{\exp}$ was investigated for myoglobin crystals [4]. $S_{\exp}$ contains contributions of the isomer shift, S_{iso} and of the second order Doppler shift S_{soD} which depends on the mean square velocity, $\langle v^2 \rangle$, of the iron. As seen in Figure 2 the value $S_{\exp}$ has a linear temperature dependence with no anomalies around 200 K. The mean square velocity of the iron is determined by the fast solid state vibrations which cause the mean square displacement $\langle x_v^2 \rangle$. Protein dynamics are essentially characterized by rather slow motions.

3. The calculation of Mössbauer absorption spectra

Neglecting for the sake of simplicity in the following the quadrupole splitting, the shape of a Mössbauer spectrum is determined by the spectral function [5]:

$$\hat{I}(E) = \frac{1}{\pi} \int\limits_0^\infty exp[i\,(E_A - E)t/\pi - \frac{\Gamma_{\text{nat}}}{2\hbar} \cdot t]\, S(k,t)\, dt \qquad (3)$$

E_A is the resonance energy of the absorber and $\hbar$ is the Planck constant. $\hat{I}(E)$ determines the energy dependence of the cross section $\sigma_a(E)$ for Mössbauer absorption. The intermediate scattering function $S(k,t)$ is the crucial parameter containing the dynamics at the absorbing ^{57}Fe. It depends on the wave vector k and the time. $S(k,t)$ can be calculated from models for the motion of the iron. Using Eq. (3) it is then possible to calculate the Mössbauer spectrum and to compare this calculated spectrum with the experimental results. Unfortunately, very different models give intermediate scattering functions yielding Mössbauer spectra similar to those experimentally obtained. Therefore, from the spectral shape alone it is impossible to come to an unambiguous decision which model describes the real physics. The situation is improved, however, if one takes into account the temperature dependence of the spectral area.

At present, two different types of models are used. The first type assumes that a molecule can exist in a small number of well defined substates. $S(k,t)$ is then determined by the jumps of the molecule between these conformational substates. The other limiting case is the assumption of a quasi continuum of conformational substates. In this picture segments of the molecule perform diffusion-like motions. Again, it is possible to calculate $S(k,t)$. It has to be emphasized that all models discussed in the following yield the correct spectral shape but fail in the description of the temperature dependence of $\langle x^2 \rangle^\gamma$. Additional assumptions are necessary in order to get the correct experimental results.

In the simplest picture one may assume that only two conformational substates of the molecule can be seen at the position of the iron [6]. The molecule has then two possible positions x_1 and x_2 of the iron. Jumping from one conformational substate to the other the molecule has to overcome the potential barrier ε which determines the transition rate $k_{12} = k_{21}$. The Mössbauer spectrum derived from this model is given by

$$T(V) = 1 - \frac{(1-A)\frac{\Gamma_a+\Gamma_s}{2}}{[E_a - E_s(1-\frac{v}{c}]^2 + \left[\frac{\Gamma_a+\Gamma_s}{2}\right]^2}$$
$$+ \frac{A\left[\frac{\Gamma_a+\Gamma_s}{2} + 2\hbar k_{12}\right]}{[E_a - E_s(1-\frac{v}{c})]^2 + \left[\frac{\Gamma_a+\Gamma_s}{2} + 2\hbar k_{12}\right]^2} \tag{4}$$

E_a, E_s, Γ_a and Γ_s are the resonance energy and the linewidth of the absorber (index a) and the source (index s), respectively. The transition rate k_{12} is given by:

$$k_{12} = k_o \exp[\varepsilon/(RT)] \tag{5}$$

where k_o is a constant normally taken to be $10^{13} s^{-1}$ and RT is the energy of the thermal bath. A is a parameter depending on k_{12} and the distance $|x_1 - x_2|$. According to Eq. (4) the Mössbauer spectrum consists of a narrow and a broad line with the widths $\Gamma_a + \Gamma_s$ and $\Gamma_a + \Gamma_s + 4\hbar k_{12}$, respectively. Adjusting k_{12} in order to fit the experimental data one obtains the height ε of the energy barrier. Eq. (4) shows that the decrease of the area of the narrow line is compensated by the increase of the area of the broad line. The sum of areas should be independent of the temperature. This is in contradiction to the experimental results. One could assume that there exists a substate where k_{12} is so large that the line broadens so much that it can no longer be resolved in the Mössbauer spectrometer. Simulations show that also this modification does not give the measured temperature dependence of the Mössbauer parameters.

It is easy to extend the two state model to more states. Analytical solutions from the Mössbauer spectrum are obtained if one assumes special geometries of the substates located i.e. in a tetrahedral (4 states) an octahedral (6 states) or a cubic (8 states) arrangement [7]. Also infinite states situated in equal distances on a sphere have been analyzed [8]. Although these geometries have no structural meaning they allow to study the generalization of the two state model. The Mössbauer spectrum is in all cases similar to that described by Eq. (4). Only A and k_{12} change in detail the meaning. All models do not explain the temperature dependence of $\langle x^2 \rangle^\gamma$.

Assuming a quasi continuum of conformational substates the Langevin equation can describe the diffusive like motion of the iron. However, free diffusion is impossible, a restoring force has to make sure that the average structure of the molecule is conserved. The motion of the iron is then described by the Brownian oscillator:

$$\ddot{x}(t) + 2\beta_t \dot{x}(t) + \omega_t^2 x(t) = F(t) \tag{6}$$

$F(t)$ is a stochastic force produced by the motion of the surrounding water molecules while β_t and ω_t characterize the friction and the backdriving force assumed to be harmonic in order to make the calculations easier [5,9,10]. In [5] an analytic solution is given in order to get the Mössbauer spectra from Eq. (3) and (6) via $S(k,t)$. As the result one obtains

$$T(v) = 1 - \exp -(k^2\langle x_t^2\rangle) \cdot C \cdot \sum_{N=0}^{\infty} \frac{[k^2\langle x_t^2\rangle]}{N!} \cdot \frac{\frac{\Gamma_a + \Gamma_s}{2}}{\left[E_a - E_s(1 - \frac{v}{c})\right]^2 + \left[\frac{\Gamma_a + \Gamma_s}{2} + N\hbar\alpha_t\right]^2}$$

(7)

with $\alpha_t = \omega_t/(2\beta_t)$. C is a constant containing trivial experimental parameters.

Eq. (7) is based on the solution of Eq. (6) for overdamped oscillations which is the case relevant for the present problem [5]. The spectrum consists of a sum of Lorentzians with increasing linewidth. Eq. (7) becomes formally similar to Eq. (4) if the lines for $N > 1$ are so broad that they can no longer be resolved experimentally. The parameter α_t gives the characteristic time for the motion and relates the Mössbauer spectrum to the parameters of the Brownian oscillation. It has to be emphasized that Eq. (7) by itself does not explain the temperature dependence of the Mössbauer parameters. However, it is possible to determine α_t and $\langle x_t\rangle$ as a function of temperature by comparison with the experiments and to derive additional properties of the Brownian oscillator. This will be discussed later. In discrete state models this freedom does not exist since the rates like k_{12} are already well done.

Recently, Frauenfelder emphasized that a distribution of the barrier heights between conformational substates has to be taken into account. He proved that such a model gives the right tendency of the temperature dependence of $\langle x_t\rangle$ [11]. Simulations have shown that the introduction of barrier distribution is not sufficient to get the correct temperature dependence of the Mössbauer parameter from discrete state models with a maximum of 8 states. The combination of a quasi continuum model containing a distribution of barrier heights is still under investigation.

4. X-ray structure investigation on myoglobin between 80 K and 300 K

In order to get a realistic picture of protein dynamics, the results of different experimental methods must be combined. Here, x-ray structure investigation on metmyoglobin crystals are reported which were obtained at 80 K, 115 K, 165 K, 180 K and 300 K. The 80 K structure is already published [12]. Details of the experiments and the data evaluation will be given elsewhere. Figure 3 shows some results. The overall B-value yields a mean square displacement, $\overline{\langle x^2 \rangle^x}$, which averages over all nonhydrogen atoms of the molecule. Within the limits of error a linear temperature dependence was found. The average mean square displacements of the backbone atoms of the proximal histidine HIS F8 and the distal histidine HIS E7 also show a linear temperature dependence. The same is true for nearly all residues in myoglobin. It is obvious that the individual $\overline{\langle x^2 \rangle^x}$-values show some deviations from the line obtained by linear regression. However, inspection of the results obtained on the 153 residues of myoglobin has given no justification for the assumption of a more complicated temperature dependence. In our opinion the deviations represent experimental errors and influences of the computer refinement.

The straight lines obtained from a least squares fit allows now to calculate $\langle x^2 \rangle^x$-values at 300 K and at 0 K. Both sets of $\langle x^2 \rangle^x$ take into account independent measurements at 5 temperatures. Figure 4 gives the results. The average mean square displacements of the backbone atoms of the residues are drawn as a function of the residue number in the sequence of the protein. The $\langle x^2 \rangle^x$-values at 300 K are rather similar to the values obtained by Frauenfelder et al. [13]. The linear extrapolation to 0 K neglects the fact that $\langle x^2 \rangle^x$ has to be temperature independent close to 0 K. Therefore, the extrapolated $\langle x^2 \rangle^x$-values in Figure 4 at 0 K are the lower limit of the real displacements. Considering an estimated value of 0.025 $\mathring{A}^2$ as zero point vibrations it becomes obvious that the molecules are not identical at 0 K. No common well defined energy minimum exists in which the molecules are driven. One has to assume that structural disorder is frozen in. In our opinion this result is a strong support of the picture of conformational substates obtained from CO flash experiments on myoglobin [14].

Figure 3 compares the mean square displacement of the iron obtained by x-ray structure analysis ($\langle x^2 \rangle^x$) and by Mössbauer absorptions spectroscopy ($\langle x^2 \rangle^\gamma$).

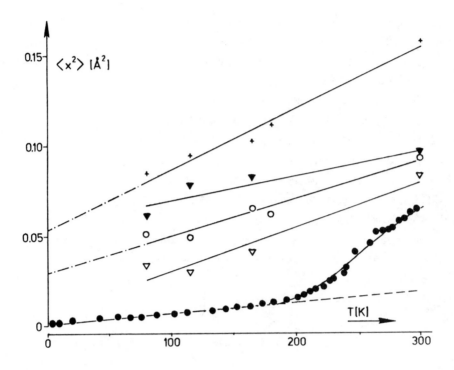

Figure 3. Mean square displacements, $\langle x^2 \rangle$ as a function of temperature. Full circles: $\langle x^2 \rangle^\gamma$ of the iron obtained from Mössbauer absorption spectroscopy. Open circles $\langle x^2 \rangle^x$ of the iron obtained from x-ray structure analysis. Triangles: $\langle x^2 \rangle^x$-values averaged over the backbone atoms of histidine, open symbols: HIS E7 (distal histidine), full symbols: HIS F8 (proximal histidine). Crosses: $\overline{\langle x^2 \rangle^x}$-values averaged over all non-hydrogen atoms of myoglobin. Straight lines give least square fits, dashed dotted lines broken lines give extrapolations.

Absolute values and temperature dependence differ remarkably. Nevertheless, these values are not independent of each other. As seen in Figure 3 the proximal histidineF8 to which the iron is linked at its 5th coordination place has larger $\langle x^2 \rangle^x$ values than the distal histidineE7 which is close to the 6th coordination of the iron. In a slightly denatured molecule the distal histidine binds to the iron forming hemochrome. Mössbauer absorption spectroscopy of deoxy hemoglobin

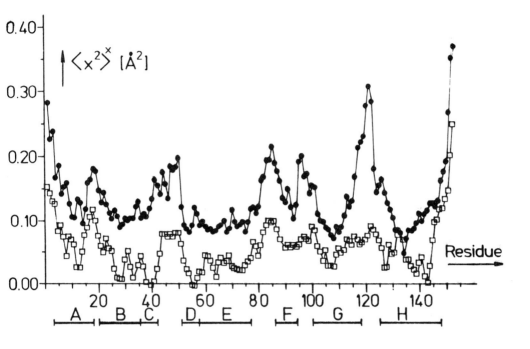

Figure 4. Mean square displacements, $\langle x^2 \rangle^x$ averaged over the backbone atoms of a residue as the function of the residue number. The values for 300 K (squares) and 0 K (circles) were obtained by linear regression of the experimental values between 80 K and 300 K.

and a hemochrome showed that binding of the iron to the more rigid HIS E8 decreases the $\langle x^2 \rangle^\gamma$-value [15].

5. The time scale of protein dynamics

Differences of $\langle x^2 \rangle^x$ and $\langle x^2 \rangle^\gamma$ reflect the fact that x-ray structure analysis has no time threshold while Mössbauer spectroscopy is sensitive only to motions with a characteristic time faster $10^{-7}s$. Comparison of the results allows to attribute a time scale to different contributions to $\langle x^2 \rangle$. At 300 K the total $\langle x^2 \rangle^x$ of the iron is 0.092 $\mathring{A}^2$. Extrapolation from the linear temperature dependence of

$\langle x^2 \rangle^\gamma$ between 50 K and 200 K to 300 K yields the contribution of the vibrations occurring with a characteristic time $\tau_c \ll 10^{-9}s$ and $\langle x_v^2 \rangle$ becomes about 0.018 $\AA^2$. The main contribution to $\langle x^2 \rangle^x$ comes from protein specific motions with $\langle x_f^2 \rangle \approx 0.046\AA^2$ at 300 K. The existence of the broad line indicates a characteristic time around $10^{-9}s$.

The remaining difference of $\langle x^2 \rangle^x - \langle x^2 \rangle^\gamma = 0.028\AA^2$ has to be attributed to processes with a characteristic time of $10^{-7}s$ of less. Partly it may stem from static lattice disorder or from static molecular disorder. The line broadening of the narrow line indicates that this part also contains slow quasi-diffusional motions, $\langle x_{qd}^2 \rangle$, occurring in a time of about $10^{-7}s$.

A problem arises if these results are compared with molecular dynamic simulations [16]. These calculations give $\langle x^2 \rangle^{\text{cal}}$-values which agree quite well with $\langle x^2 \rangle^x$-values as obtained from x-ray structure analysis. However, these calculations are sensitive only to motions with a characteristic time $\tau_c < 5 \times 10^{-11}s$. Comparing x-ray data and computer simulations means comparing an ensemble average (about 10^{15} molecules) and a time averaged on one molecule.

The equivalence of $\langle x^2 \rangle^x = \langle x^2 \rangle^{\text{cal}}$ proves equivalence of the ensemble average and the time average. This in turn means that the ensemble is large enough to occupy all possible coordinates of the atoms of the molecules in the crystal at the same time with the relevant weight factor while the averaging time is long enough to allow the atoms of the molecules in the crystal at the same time with the relevant weight factor while averaging time is long enough to allow the atoms of the molecule to pass through all possible coordinates. For the present example $\langle x^2 \rangle^x = \langle x^2 \rangle^{\text{cal}}$ demands that the mean square displacements occur in a characteristic time faster $5 \times 10^{-11}s$ while Mössbauer spectroscopy shows that the major contribution comes from motions with $\tau_c > 10^{-9}s$. The discrepancy can be solved if one assumes that the potentials used for protein dynamics simulation underestimate the friction. A relative smooth potential allows the molecule to fluctuate quite rapidly through all possible coordinates of the atoms yielding the correct $\langle x^2 \rangle^x$-value. However, if this potential is only the envelope of a more complicated potential containing a number of traps the actual motion is slowed down. The calculation gives the wrong time dependence although the amplitudes of the motions are correctly displayed. Strong friction may arise from the interaction of the molecule with the crystal water and the surrounding molecules. Such long range interactions are not taken into account in protein dynamics simulation.

6. A picture of protein dynamics explaining Mössbauer and X-ray measurements

The different temperature dependence of $\langle x^2 \rangle^x$ and $\langle x^2 \rangle^\gamma$ as shown in Figure 2 can be explained by a potential as drawn in Figure 5. This potential represents a modification of the Brownian oscillation, discussed before. Note that only a one dimensional picture of the real potential is drawn. The potential consists of a number of deep traps ($\varepsilon \gg RT$) and of shallow traps ($\varepsilon \approx RT$). The deep traps can be identified with conformational substates of the molecule influencing the iron coordinate. The occupied conformational substates determine the distribution $\langle x_c^2 \rangle$ which is essentially measured by x-ray structure analysis. The parabolic envelope attends to a linear temperature dependence of $\langle x_c^2 \rangle$ as found experimentally. Mössbauer spectroscopy measures only the small contribution $\langle x_v^2 \rangle$ as long as the molecule is frozen in one conformational substate. With increasing temperature the probability $p_t(T)$ rises to reach the transition state t and to move to another conformational substate. A motion in the potential of Figure 5 can be understood as an overdamped Brownian oscillation described by Eq. (6). The introduction of the deep traps (for simplicity only one type) yields the appropriate temperature dependence of β_t and explains at the same time the results of x-ray structure analysis. The protein specific motion, $\langle x_t^2 \rangle$, is correlated to the distribution of conformational substates by

$$\langle x_t^2 \rangle = p_t(T) \langle x_c^2 \rangle \tag{8}$$

$$p_t(T) = [1 + \exp \Delta G / (RT)]^{-1}. \tag{9}$$

The Gibbs free energy ΔG between the conformational substate and the transition state is determined by the large entropy contribution to $p_t(T)$ at higher temperatures. At room temperature the high entropy of the transition state makes a well defined structure of a molecule very improbable. The molecule segments perform a Brownian diffusion around their average positions.

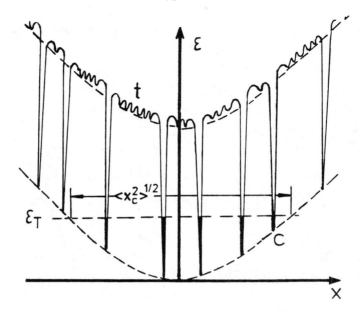

Figure 5. One dimensional picture of a potential which explains the different temperature dependence of $\langle x^2 \rangle^z$ and $\langle x^2 \rangle^\gamma$. The deep traps represent conformational substates.

7. The mobility of crystal water

Protein specific dynamics is only present if the molecule is surrounded by a water layer [17,18]. Drying myoglobin yields $\langle x^2 \rangle^\gamma \approx \langle x_v^2 \rangle$ even at room temperature. It is, therefore, important to correlate the mobility of the hydration shell and of the protein molecule.

A myoglobin crystal contains about 450 water molecules per myoglobin molecule. As shown by microwave absorption no free bulk water exists in the crystal [19]. In our x-ray structure investigation of myoglobin we were able to localize 123 water molecules showing an average mean square displacement: $\langle x^2 \rangle_{H_2 0} = 0.392 \mathring{A}^2$. What is the average mean square displacement of the other 327 water molecules?

The Rayleigh Scattering of Mössbauer Radiation (RSMR) is a good experimental tool to answer this question. In this technique the 14.4 keV [57]Fe radiation of a [57]Co-Mössbauer source is scattered by the electrons of the sample under investigation. Due to the dynamics part of the scattering occurs inelastically. A Mössbauer absorber introduced into the primary and into the scattered beam, respectively, is used in order to determine the elastic fraction of the scattering. This in turn yields an $\langle x^2 \rangle^R$-value which averages the mean square displacements of all atoms of the sample. In this respect it is similar to the value $\langle x^2 \rangle^x$ obtained from the overall B-factor of x-ray structure analysis. There are, however, two important differences. If polycrystalline samples are used the scattered radiation can be investigated at any angle. Consequently, coherent and incoherent scattering contributes. This means that the $\langle x^2 \rangle^R$-value contains also contributions of water molecules not found in x-ray structure analysis. Moreover, the determination of $\langle x^2 \rangle^R$ is performed with the help of a Mössbauer absorber. This way RSMR has the same time threshold as Mössbauer absorption spectroscopy.

The detection of the scattered radiation was performed with an area sensitive multi wire proportional counter of an area of 700 x 700 mm^2 and a spatial resolution of about 2 x 2 mm^2. This way, the radiation scattered into the angle $9° \leq 2\theta \leq 19°$ was measured at the same time. The angular dependence gives additional information concerning the contributions of the different components of the sample. Details will be given elsewhere.

One obtains from the RSMR measurements $\langle x^2 \rangle^R = 0.27 \text{Å}^2$ as the average value for one myoglobin molecule and 450 water molecules. From the overall B-factor, $\langle x^2 \rangle = 0.157 \text{Å}^2$ was obtained as the average value for one myoglobin molecule. For a comparison the $\langle x^2 \rangle^x$-value has to be corrected for contributions of very slow processes. An estimate of the reduction factor can be obtained by comparing with the situation at the iron position and reducing $\langle x^2 \rangle^x$ by the factor $\langle x^2 \rangle^\gamma_{Fe} / \rangle x^2 \rangle^x_{Fe}$. With these assumptions the average dynamic mean square displacement of the crystallographically non localized water molecules becomes 0.88 1Å^2. It should be emphasized that the angular dependence of the RSMR-data allows an estimate of the segments of the sample which move in phase and scatter collectively. While RSMR data on dry myoglobin could be best understood by assuming independent scattering of the atoms the investigation on myoglobin crystals indicates collective scattering in segments with a characteristic dimension of at least 6 Å. Detailed investigations are in progress.

This work was supported by the Deutsche Forschungsgemeinschaft and the Bundesministerium für Forschung und Technologie.

References

[1] Parak, F., Knapp, E.W. and Kucheida, D., J. Mol. Biol. *161*, 177-194 (1982).

[2] Bauminger, E.R., Cohen, S.G., Nowik, I., Ofer, S. and Yariv, J., Proc. Natl. Acad. Sci. USA *80*, 736-740 (1983).

[3] Eicher, H., Bade, D. and Parak, F., J. Chem. Phys. *64*, 1446-1455 (1976).

[4] Reinisch, L., Heidemeier, J. and Parak, F., Eur. Biophys. J. *12*, 167-172 (1985).

[5] Knapp, E.W., Fischer, S.F. and Parak, F., J. Chem. Phys. *78*, 4701-4711 (1983).

[6] Parak, F., Frolov, E.N., Mössbauer, R.L. and Goldanskii, V.I., J. Mol. Biol. *145*. 825-833 (1981).

[7] Bläsius, A., Preston, R.S. and Gonser, U., z. Physikal. Chem. Neue Folge *151*, 373-385 (1979).

[8] Nowik, I., Bauminger, E.R., Cohen, S.G. and Ofer, S., Phys. Rev. *31 A*, 2291-2299 (1985).

[9] Shaitan, K.V. and Rubin, A.R., Biofizika *25*, 796-802 (1980); Biophysics *25*, 809 (1981).

[10] Cohen, S.G., Bauminger, E.R., Nowik, I., Ofer, S. and Yariv, J., Phys. Rev. Lett. *46*, 1244 (1981).

[11] Frauenfelder, H., in *Structure and Motion: Membranes Nucleic Acids and Proteins*, ed.: Clementi, E., Corongiu, G., Sarma, M.H. and Sarma, R.H., Adenine Press, New York (1985) pp. 205-217.

[12] Hartmann, H., Parak, F., Steigemann, W., Petski, G.A., Ringe Ponzi, D. and Frauenfelder, H., Proc. Natl. Acad. Sci. USA *280*, 4967-4971 (1982).

[13] Frauenfelder, H., Petsko, G.A. and Tsernoglou, D., Nature (London) *280*, 558-563 (1979).

[14] Austin, R.H., Beeson, K.W., Eisenstein, L., Frauenfelder, H. and Gunsalus, I.C., Biochemistry *14*, 5355-5373 (1975).

[15] Mayo, K.H., Parak, F. and Mössbauer, R.L., Phys. Lett. *82 A*, 468-470 (1981).

[16] Swaminathan, S., Ichiye, T., van Gunsteren, W. and Karplus, M., Biochem. *21*, 5230-5241 (1982).

[17] Parak, F. and Formanek, H., Acta Cryst. *A27*, 573-578 (1971).

[18] Krupyanskii, Yu. F., Parak, F., Goldanskii, V.I., Mössbauer, R.L., Gaubmann, E., Engelman, H. and Suzdalev, I.P., z. Naturforsch. *37 c*, 57-62 (1982).

[19] Singh, G.P., Parak, F., Hunklinger, S. and Dransfeld, K., Phys. Rev. Lett. *47*, 685-688 (1981).

Discussion: Parak

Anderson: When I saw the view graph of your data I kept focussing on the points rather than the lines you drew through them. Can I see the graph of the temperature dependence of the mean square x-ray displacement?

Parak: You mean that there? (Figure 3 in the text).

Anderson: Yes. Looking from the back of the room, at least, it seemed to me that the line you drew did not describe the experimental points, which consisted of a group of three or four at low temperatures with little visible temperature dependency, in fact scattering both up and down from a horizontal line and one point at room temperature which was considerably higher. It would take a very careful analysis, which I still might not believe, to justify drawing a straight line with constant slope through these points. In fact, from the data as presented it seems to me you have no evidence that the T-dependence of x-ray and Mössbauer differ.

Parak: The experimental error is like this. (small)

Anderson: That makes my point stronger if anything.

Parak: I disagree completely. It's no argument.

Krumhansl: Following Phil's point, the other question you can ask is what about those inferred zero point mean square displacements. Could you make a calculation and find out about whether that's about what you would expect?

Parak: You mean is the extrapolation...

Krumhansl: No, up above, or are those shifted data?

Parak: You mean the extrapolation to $T = 0$?

Krumhansl: Yes.

Parak: Well, it's quite different and we have of course an analysis done for all atoms in the molecule and you cannot of course assume (intensive) heights of fitting these curves. But if you know that in all cases, we have an error bar like that (about) and if you then can fit the residuals, everything with a linear line which go through the error bars, then I can extrapolate, of course, to zero and then they are different types. 10% of the residues go to zero here, and the rest ends like the with larger values at $T = 0$ which indicate conformational substates in my opinion.

Krumhansl: Oh, OK quite. What I was asking was whether the displacements in those conformational substrates are reasonable.

Parak: Well, one can show this in a nice picture. Our extrapolation to $T = 0$ gives the following picture of myoglobin; dark is where it goes practically to zero, mean square displacement $= 0$ at $0°K$ and then blue and red is increasing order and so you see that this fragment is very well ordered if you extrapolate to $T = 0$ and then there is increase in disorder in some fragments and one sees, if one looks on the steric reaction coordinates really planes of good order and of disorder.

Karplus: Were the experiments done on different crystals.

Parak: They were done of different crystals.

Karplus: With different crystals at room temperature, you can have a variation in the absolute value of the temperature factors which is about half an angstrom something like that.

Parak: We have redone the whole experiment at 300 K which Frauenfelder et al did, and it fits excellently with our data. So it is two independent measurements with independent grown crystals. They used diffraction techniques and we used film techniques. Everything was different and it fits very nicely. The error bars are quite small.

Karplus: The myoglobin CO crystal done also in the same laboratory at MIT that just had much more relative fluctuations, behavior very similar to the absolute ones and much more than one finds in oxymtoglobin.

Parak: Yes, that I would believe. We compare now metmyoglobin crystals. I have done the same.

Dunbranner: We also see this in the Mössbauer experiments the mean square displacements are different from MbCO and MbO_2. So that makes it fairly disordered.

Lang: Did you say these were metmyoglobin?

Parak: The diffraction was done on metmyoglobin, the Mössbauer was done on metmyoglobin and on deoxymyoglobin. But I discussed here today deoxy data.

Lang: And did you ever compare crystalline myoglobin with frozen solution for the Mössbauer method? Because you were talking about effects on the outside and effects of the water.

Parak: We have not done in the case of myoglobin, but we did it in the case of hemoglobin where up to the limit where you can investigate both in solutions, they are quite the same. I did this together with Keven Meyer and in this case up to the temperature where could measure both in solutions; it was the same.

Lang: Can't tell the difference between a crystal and a solution?

Parak: That's the position of ions that goes up to 20K – no difference; but then we had to stop because diffusion processes set in we got no lines to investigate frozen solutions.

Lang: I see.

Hopfield: As I understand your basic position is that until Karplus' budget is increased by a factor of 10^4 he won't be able to compare with you. Now, it is true of course that if you were to start in different configurations and run his calculations, if you compare those different configurations as they essentially give the same answers, then he is already averaging sufficiently long and I would just wonder whether he's done it?

Karplus: In a sense, yes, it's always in the neighborhood of the x-ray structure, but essentially one does the Monte Carlo sampling of molecular dynamics and the results are essentially the same, obviously the individual residues has somewhat different fluctuations in each one of the trajectories, but the overall results, the magnitudes, are the same order.

Hopfield: The average result over all the configurations?

Karplus: It gets a little bit larger, but not much.

The Protein as a Glass

D. Stein

Joseph Henry Laboratories of Physics

Princeton University

Princeton, NJ 08544

Abstract

A spin glass-like model of protein conformational structure is described. Proteins such as myoglobin may be examples of "intrinsically complex" systems in the computational sense. Consequences for both thermodynamic and dynamic properties will be discussed.

I. Introduction

It has long been speculated that at sufficiently low temperatures certain proteins, such as myoglobin and hemoglobin, freeze into one of a large possible number of conformational substates about some given average tertiary structure [1]. Evidence for this freezing comes from X-ray [2], Mössbauer [3], and recombination studies [4]. The questions I concern myself with here include understanding why this should be so, to ask whether one can make a mathematical model of protein conformational substructure, and to explore the consequences of such a model [5]. The viewpoint I will adopt is that of condensed matter physics, which recognizes that this sort of freezing behavior is reminiscent of that of glasses and spin glasses, and that the existence of multiple metastable configurations arises from a universal feature of these systems commonly referred to as "frustration" [6]. Having developed a glass model of this aspect of protein behavior [7,8], we can then proceed to study its consequence for both the static and dynamic behavior of the relevant proteins.

II. The Problem of Complexity

Earlier work has recognized the importance of a glass transition in proteins [1,9]. One consequence of this which has so far received little attention is that as a consequence the protein may be an example of an "intrinsically complex system" [10], which implies that its behavior (at least at low temperatures) is describable by a random function of a very large number of variables. The spin glass (see Section III) is a prototype of such a system, and its property of freezing into one of a large number of metastable spin configurations has led various people to consider other systems as "spin glass-like" in the sense of having multiple ground states. For example, P.W. Anderson has proposed a spin glass model of molecular evolution [11], John Hopfield has developed a spin glass-like model of neural networks [12], and we certainly don't believe that all, or even most, of the possibilities have been exhausted.

These problems all fall under an important category in the theory of computational complexity in that they are all examples of NP-complete problems. Such problems often involve combinatorial optimization and include such examples as the travelling salesman problem, the satisfiability problem, and the graph partition problem, the last problem being equivalent to that of finding the lowest energy configuration of a spin glass. All of these problems are believed to be among the hardest computational problems in that no algorithm has ever been found for any of them that increases only polynomially in the length of the input problem size (e.g., number of points on a graph or spins in a spin glass). It is widely believed, in fact, that all such problems are exponentially hard, but this has never been proven; what *is* known is that all of these problems are equivalent in the sense that if a polynomial algorithm exists which can solve *one* of these problems, some such algorithm exists for *all* of them; and therefore if it can be proven that *any* of these problems is exponentially hard, then *all* are.

If the conformational structure of proteins belongs to this category, then at least certain aspects of protein behavior cannot be captured by modeling the protein with only a few variables; a large number will be required. In subsequent sections we shall explore a way to realize this; but first we will digress one last time to discuss some properties of spin glasses that will be useful for our purposes.

III. Spin Glasses

A (theorist's) spin glass is a magnetic system consisting of localized spins S_i at sites i, interacting via the Hamiltonian

$$H = -\sum_{i,j} J_{ij} S_i S_j \tag{1}$$

where the couplings J_{ij} are random functions of i and j and may take on both positive and negative values. Favored probability distributions $P(J_{ij})$ includes the Gaussian

$$P(J_{ij}) = \frac{1}{\sqrt{2\pi \tilde{J}^2}} e^{-(J_{ij} - J_0)^{2/2\tilde{J}^2}} \tag{2}$$

where $J_0 < \tilde{J}$ and the bimodal distribution

$$P(J_{ij}) = 1/2 \{ \delta(J_{ij} - J_0) + \delta(J_{ij} + J_0) \}. \tag{3}$$

The Hamiltonian given by Eq. (1) was first proposed by Edwards and Anderson [13] to describe the essential behavior of dilute magnetic alloys such as CuMn and insulators such as $Eu_{1-x} Sr_x S$. We will confine ourselves to the Ising model where the S_i equal ± 1.

An important property of spin glasses is *frustration*: if the product $J_{ij} J_{jk} \ldots J_{mi}$ around a loop is negative, then the loop is frustrated in the sense that not all of the interspin couplings can be simultaneously satisfied. If a system has many such frustrated cycles, then it will possess a large number of spin configurations which are local minima of the free energy and which cannot be transformed one to another via any simple symmetry transformation. The number of such local free energy minima diverges exponentially in the number of spins N in a mean-field spin glass, and the barriers between them diverge when $N \to \infty$. However, even a spin glass below the lower critical dimension can exhibit freezing into a particular configuration in that the time required for a thermally activated jump from a locally metastable state may exceed laboratory times, even though no true phase transition has occurred.

IV. Recombination Kinetics

Returning now to proteins, we recount the experiment of Austin et al. [1] which led to the idea of multiple conformations. In this experiment, MbCO was dissociated via flash photolysis and the number $N(t)$ of unrecombined protein vs. time was monitored at various temperatures T. At low enough T, recombination is geminate in that the CO does not leave the heme pocket, but still needs to surmount an activation energy barrier in order to recombine. At high $T (\gtrsim 220°K)$, $N(t)$ follows a simple exponential decay, but at lower temperatures $N(t)$ is better fitted by a power law:

$$N(t) \sim (1 + t/t_0)^{-n} \tag{4}$$

where both t_0 and n are T-dependent. Exponential recombination is indicative of a single average activation energy barrier, but recombination of the form eq. (4) indicates a *distribution* of barriers, in which each protein recombines exponentially but on a different timescale than its neighbor. (This corresponds to inhomogeneous broadening; the case of homogeneous non-exponential recombination was effectively ruled out by flash photolysis experiments [1].) The conclusion drawn in Ref. 1 is that for $T \lesssim 220°K$, the protein freezes into one of a large number of substates, each with barriers too high to surmount thermally during experimental times, and the activation barrier is related to the conformational substate. At higher temperatures, the protein is rapidly flipping among its substates, and only an average activation energy barrier is observed.

V. A Model for Conformational Substates

Why should the protein freeze into one of many conformations? The first clue comes from its compressibility, which lies between those of ice and lead. Myoglobin is rather incompressible, with atoms often close-packed to their van der Waals radii. If an atom should try to move, it often cannot do so without jostling its neighbor – there simply isn't much elbow room.

Nevertheless, we expect there to be many local degrees of freedom, since myoglobin is, in some sense, a three-dimensional disordered system. The work of Anderson, Halperin and Varma (AHV) [14] and Phillips [15] proposed that a glassy system has many degrees of freedom unavailable to a crystal, in the sense that, for example, atoms or even clusters of atoms might have two or more locally metastable configurations about some localized point in space. The two-level systems (TLS) are a universal feature of disordered systems, and give rise to a linear temperature dependence of the specific heat at low T. There is every reason to expect, therefore, that a protein such as myoglobin might possess many local degrees of freedom given a more or less unique tertiary structure. One might think of these as an atom or cluster of atoms in certain amino acid residues in the side chain as having two or more locally stable positions, which correspond to changes in bond angles. It is important to note, however, that the three dimensional covalent topology of the protein remains intact at all times.

Consider then the following simple picture. Hold all degrees of freedom fixed except for one TLS, which we may consider as an Ising spin $S_i = \pm 1$. Here, i is a spatial index and not the number of the amino acid along the polymer chain. Suppose that the energy difference between local minima is $2h$ and let the $s_i (= -1)$ state refer to the higher-energy configuration. In this case we may write

$$E_i = -h_i S_i \tag{5}$$

so that $\Delta E = 2h$ and the TLS acts like an isolated spin in an external magnetic field.

However, this cannot be all there is to it. Because the protein is so close-packed, it will usually be impossible to change the state of a TLS without rearranging neighboring TLS, which in turn will rearrange others, and so on. When a single TLS changes states, there will then exist many possible redistributions of its neighbor, some of which will raise the conformational strain energy and some of which will lower it. This blocking and interference among the shifting atoms is reminiscent of the frustration effect in glasses, and implies that we need to insert random interactions in the picture to make it realistic.

Let us denote a particular conformational substate by $\{S^{(\alpha)}\}$, and expand its conformational strain energy $E\{S^{(\alpha)}\}$ about the average energy of the tertiary structure E_0:

$$E\{S^{(\alpha)}\} - E_0 = -h\sum_{i=1}^{N} S_i - \sum_{\langle ij\rangle} J_{ij}S_iS_j \tag{6}$$

when the number of TLS in the spin system is N. We can ignore higher-order terms since higher-order correlations are increasingly unlikely, and including them won't change our conclusion in any case. The J_{ij} are given by

$$J_{ij} = \frac{\partial^2 E\{S^{(\alpha)}\}}{\partial S_i \partial S_j} \tag{7}$$

and are random, independent variables. Henceforth we shall assume the Gaussian distribution given by eq. (4). Because of the strongly interacting nature of the system, we further assert that $\tilde{J} \gg h$ [7].

VI. Conclusion

Because interactions are short-ranged and we believe [2] the number N of TLS to be of order 100, we do not expect a sharp phase transition in the protein to occur at some temperature T_f. Rather, as the temperature is lowered, more degrees of freedom will be frozen on experimental timescales while a few will remain active, or "melted", down to very low temperatures.

In this respect, the protein behaves like a glass slightly above the glass transition temperature or a spin glass below its lower critical dimension.

One easily calculable quantity [7] is the distribution of energies $P(E)$ among the 2^N possible states. For a Gaussian distribution of the J_{ij}'s

$$P(E) = \frac{1}{\sqrt{zN\pi\tilde{J}^2}}\exp\{-(E-E_0)^2/zN\tilde{J}^2\} \tag{8}$$

where z is the average coordination number. It is important to note that because we kept only two-spin interactions in eq. (6), the energies in the distribution eq. (8) are highly correlated. At a given temperature, the probability $D(E)$ of finding a protein with energy E is given by multiplying eq. (8) by a Boltzmann distribution $f(E)$:

$$D(E) = P(E)f(E) \sim \exp\left\{ -(E - E_0)^2/zN\tilde{J}^2 \right\} \times \exp\{-E/k_B T_0\} \qquad (9)$$

where

$$T_0 = \begin{cases} T & T \gtrsim T_f = z\tilde{J}/k_B \\ T_f & T \lesssim T_f \end{cases} \qquad (10)$$

Eq. (10) indicates that below an effective "freezing temperature" T_f the system goes out of equilibrium and is frozen in the distribution which pertained before freezing. In reality, the effective freezing temperature should to a first approximation be scaled by log time, that is, $T_f \to T_f \ell nt$, which indicates that longer observation times would indicate a lower "freezing" temperature. The distribution eq. (10) may be useful in studying recombination kinetics? If the distribution of activation energy barriers is strongly correlated with that of conformational structural energies, as we expect.

The frozen-in disorder at low T predicted by eq. (6) should be observable in x-ray and Mössbauer studies, each of which can measure the mean square displacement $\langle x^2 \rangle$ of each atom (the first through the Debye-Waller factor and the second through the recoilless fraction of nuclei). The difficulty here is in extracting the $\langle x^2 \rangle$ due to conformational substructures from that due to other causes, such as vibrations. The technique used to effect this subtraction is described in Ref. 2, where it is found that 95 side chain atoms have mean-square displacements greater than 0.025 Å^2 at low T.

The existence of TLS should manifest itself in the appearance of a linear term in the specific heat at low T. This has also been predicted, and observed, by Goldanskii and collaborators [8].

The most exciting possibilities for future study are in the dynamics of proteins during recombination, relaxation, and other processes. Glasses and spin glasses have a continuum of relevant timescales, stretching over several decades of time, in contrast to ordered systems which possess a few intrinsic timescales separated by large gaps. A continuum of timescales appears to be important to Mb recombination as well [16]. A recent paper by Palmer, Stein, Abrahams, and Anderson (PSAAP) [17] has postulated a hierarchy of constraints which inhibit relaxation in glassy systems, and which gives rise to a continuum of timescales and anomalous relaxation. This idea of a series, rather than a parallel, approach to equilibrium coincides in spirit with models for MbCO recombination put forward by the Illinois group [16,18].

It will be interesting to see how much of known protein dynamics can be understood with these ideas.

We conclude with a cautionary note: proteins are complex systems, and we don't expect that the simple model embodied in eq. (6) describes the full richness of protein structure. Nevertheless, it may serve as a first step toward a more complete description of the properties of certain proteins, and in the meantime may be helpful in understanding some of the observed freezing properties of proteins. Proteins are undoubtedly more complex in many ways than either glasses or spin glasses, which may serve as partial descriptions. The elucidation of those properties which are in some sense describable by glass or spin glass models looks to be a fruitful, and interesting, direction for future work.

References

1. R.H. Austin, K.W. Beeson, L. Eisenstein, H. Frauenfelder and I.C. Gunsalus, Biochemistry *14*, 5355 (1975).

2. H. Frauenfelder, G.A. Petsko and D. Tsernoglou, Nature (London) *280*, 558 (1979).

3. F. Parak, E.N. Frolov, R.L. Mössbauer and V.I. Goldanskii, J. Mol. Biol. *145*, 824 (1981).

4. H. Frauenfelder, in: *Structure and Dynamics of Nucleic Acids, Proteins, and Membranes.* Eds., E. Clementi and R.H. Sarma, Academic Press, NY (1984).

5. It is important to note that alternatives to the conformational substates model do exist, the most important perhaps being the quasiharmonic model. For recent work on this model, see W. Bialek and R.F. Goldstein, *Protein Vibrations can Markedly Affect Reaction Kinetics: Interpretation of*

Myoglobin-Co Recombination, Preprint; and references therein.

6. G. Toulouse, Comm. on Physics *2*, 115 (1977); P.W. Anderson, J. Less Common Metals *62*, 291 (1978).

7. D.L. Stein, Proc. Natl. Acad. Sci., in press.

8. Similar ideas have been proposed by V.I. Goldanskii and collaborators; see V.I. Goldanskii and Yu F. Krupyanskii, *Proc. of the Alma-Ata Conference on Mössbauer Spectroscopy*, Gordon and Breach (1985).

9. N. Agmon and J.J. Hopfield, J. Chem. Phys. *79*, 2042 (1983).

10. P.W. Anderson, Proc. Rio Grande Institute, 1984 (to be published).

11. P.W. Anderson, Proc. Natl. Acad. Sci. (USA) *80*, 3386 (1983); D.L. Stein and P.W. Anderson, Proc. Natl. Sci. (USA) *81*, 1751 (1984); D.L. Stein, Intl. J. Quant. Chem., Quantum Biology Symposium # 11 p. 73 (1984).

12. J.J. Hopfield, Proc. Natl. Acad. Sci. (USA) *79*, 2554 (1982).

13. S.F. Edwards and P.W. Anderson, J. Phys. F *5*, 965 (1975).

14. P.W. Anderson, B.I. Halperin and C. Varma, Phil. Mag. *25* (1), 1 (1972).

15. W.A. Phillips, J. Low Temp. Phys. *7*, 351 (1972).

16. A. Anjari, J. Berendzen, S.F. Bowno, H. Frauenfelder, I.E.T. Iben, T.E. Sauke, E. Shyamsunder and R.D. Young, *Protein States and Protein Quakes*, preprint.

17. R.G. Palmer, D.L. Stein, E. Abrahams and P.W. Anderson, Phys. Rev. Lett. *53*, 958 (1984).

18. S. Bowne, Ph.D. Thesis, University of Illinois (1984).

Rayleigh Scattering of Mössbauer Radiation (RSMR): Data, Hydration Effects and Glass-like Dynamical Model of Biopolymers

V.I. Gol'danskii, Y.F. Krupyanskii and V.N. Fleurov

Institute of Chemical Physics

USSR Academy of Sciences

117977 Ulitza Kosygina

4, Moscow, USSR

Abstract

Specific features of the Rayleigh Scattering of Mössbauer Radiation (RSMR) technique in the study of biological systems are described. Experimental data show that the temperature and hydration degree are the principal parameters which influence intramolecular mobility in biopolymers. Data in temperature dependencies of elastic fraction, f, and spectrum lineshape doesn't fit either Debye or Einstein models of solids or the free diffusion in liquids and demand for their explanation a multimode approximation (i.e. a wide spectrum of correlation times, at $T = 293K$ from $10^{-6}s$ to $10^{-12} - 10^{-13}s$). On the basis of RSMR, low temperature specific heat and x-ray dynamical analysis data and from the general considerations that macromolecule must be in a non-equilibrium state (an independent confirmation of this fact comes from the kinetic model of protein folding) a glass-like dynamical model of biopolymers is formulated. A possible interpretation of RSMR data shows that fluctuatively prepared tunneling between quasi-equilibrium positions (QEP) can prevail activated transitions up to a room temperature.

Introduction

During the last 10-15 years our understanding of the proteins has essentially changed mainly due to the studies of their dynamical properties.

Using as principal characteristics of the substance the presence $(+)$ or absence $(-)$ of: (i) order (i.e. definiteness of structural), (ii) periodicity, (iii) unambiguity of structure (here $+$ means the entropy tending to zero at $T \to 0$, while - means the inequality $S \neq 0$ at $T \to -$) and (iv) mobility of structure at $T \to 0$, (see Table 1), one can infer that the ideal crystal $(+ + + -)$ and the glass $(- - - +)$ are complete antipodes. At the same time, analyzing the dynamical properties a combination of the crystalline and glassy features is inherent to them.

Table I. Development of the description of dynamical properties of biopolymers

	Ordering	Periodicity	Unambiguity of structure at $T \to 0$	Mobility at $T \to 0$	Score glass:IC
ideal crystal (IC)	+	+	+	-	
glass	-	-	-	+ (since 1972)	
aperiodic crystal	+	-	+	-	1.3
biopolymer					2:2
(since 1973-79)	+	-	-	-	2:2
biopolymer					
(since 1983)	+	-	-	+	3:1

$+$ means $S_0 = 0$

$-$ means $S_0 \neq 0$

As long as forty years ago E. Schrödinger [1] characterized the proteins by the term, "aperiodic crystal", treating the ordering as the principal factor of their functional activity. Later on it became clear that the dynamical properties play no less an essential role in their reaction ability [2-4].

For a long period the conventional x-ray analysis made one treat the protein globules as having geometrically unambiguous atomic configurations. In other words the protein molecule was assumed to be in the absolute energy minimum (Figure 1a). However, Perutz and Mathews [5] in 1966 pointed out that the equilibrium Hb configuration left no room for an oxygen to reach the heme in spite of the fact that such a process was actually observed. To interpret this fact as well as the numerous data on the hydrogen exchange (which proceeded not only on the globule surface but also in its deepest bulk regions [6]) and also on the fluorescence quenching of the bulk tryptophane residues by oxygen [7] and acrylamide [8] short-lived excitations of the basic structure were introduced [9,4] (Figure 1a), which were assumed to be responsible for the conformational mobility of the macromolecule.

A principally new approach to the dynamical properties of the proteins is due to the assumption that the ground state of the macromolecule, i.e. the absolute minimum of its potential energy, is never achieved even at extremely low temperatures. Actually observed states are degenerate local energy minimum (Figure 1b).

Implicitly this assumption was used in 1975 by Lumri and Rosenberg [6] who had developed a mobile defects model in order to explain a high penetration rate of oxygen, water and acrylamide into the proteins as well as by Karplus, McCammon et al. [10] in their computer simulations of the protein dynamics.

A possible energy degeneracy of biopolymer molecules might have been inferred from the data of Frauenfelder et al [11,12] who studied the kinetics of the Fe-ligand $(CO,)_2$) rebinding after the flash photolysis of myoglobin and haemoglobin heme groups and found that at low enough temperatures ($T \sim< 160 - 200K$) this rebinding observes polychronic kinetics.

Most clear experimental evidences favoring these assumptions were presented

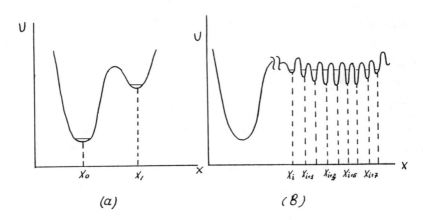

Figure 1. Versions of the potential energy curves for bio-macromolecules. (a) the absolute energy minimum corresponds to the ground state (which is unambiguously fixed at $T \to 0$), other minima (only one of them is shown) - to short lived and weakly populated excited state. (b) the absolute minimum is not realized at all; a system of local minima represents spatially delocalized and energetically close conformational substates (CS) or quasi-degenerate conformational states (QDCS).

by the x-ray dynamical analysis (XRDA) of proteins [13- 15]. This technique allows one to determine the mean square displacements $\langle x^2 \rangle$ of the backbone and side chain atoms in the protein globule, to reveal the "solid" ($\langle x^2 \rangle \leq 0.04 \text{Å}^2$) and "liquid-like" ($\langle x^2 \rangle$ up to 1Å^2) regions in these, i.e. a conformational heterogeneity in the biopolymers, and to study the temperature dependent $\langle x^2 \rangle$ within the range from 300 K [13] to 80 K [15]. The XRDA technique resulted in a considerable expansion of our knowledge of the protein dynamics and conformational substates (CS), or, better to say, quasi-degenerate conformational states (QDCS) were introduced [13], these corresponding to the designation of biopolymer molecule as $(+ - --)$. The last sign, $-$, in this notation implies an assumption, which held up to 1983, that the transitions between different CS's are exclusively thermally activated. That is why these transitions were assumed to be "frozen down" at low temperatures. The term CS means variations of the same gross-structure of a given protein molecule, which have equal or nearly equal energies and differ only slightly in their local configurations (reversible rotations around the σ-bonds, hydrogen bridge shifts, fluctuative ruptures and restorations of the hydrogen bonds and, hence, small shifts of large molecular groups with respect to the remaining parts of the globule). These are the transitions between the QDCS's which yield the large $\langle x^2 \rangle$ values observed by the XRDA technique. The QDCS concept and the transitions between them explained a sharp cusp in the temperature dependence of Mössbauer effect probability and its rapid fall at $T \geq 200K$ observed for the first time in 1973 [16]. This concept was quantitatively confirmed in later experiments on the Mössbauer absorption spectroscopy (MAS) [17- 20].

It is pointed out that both XRDA and MAS, contrary to the above mentioned hydrogen exchange and fluorescence quenching techniques, belong to a group of techniques describing the protein dynamics by using the Van Hove correlation functions [21,22].

This review presents data on a study of biopolymer dynamical properties by means of another "correlation" technique - Rayleigh scattering of Mössbauer radiation (RSMR). In order to interpret the RSMR data more precisely and to verify and develop the QDCS specific heat (LH) of the macromolecules is carried out as an additional technique.

Supplementing the earlier XRDA and MAS results with the RSMR and LH data allowed us to come in 1983 [23,24] to the conclusion that not only an ambiguity of structure (i.e. the presence of QDCS) but also a mobility, caused by the phonon assisted tunneling between these states, is inherent to the biopolymers even at $T \to 0$. This relates the biopolymers with the glasses and other amorphous phases. Therefore, the ref. [23,24] were the first to put forward a glass-like dynamical model of the proteins and other biopolymers. So we may consider a score-glass to ideal crystal (IC) using our notations in the Table 1. It was 1:3 in favor of IC for the Schrödinger aperiodic crystal, then after the introduction of CS's it became 2:2 and is now 3:1 in favor of glass.

The review formulates main reasons which have led us to the glass-like dynamical model for the biopolymers and discusses the tunneling in transitions between the QDCS on the basis of RSMR data.

1. Specific features of the RSMR technique in the study of biological systems [25]

The application of RSMR technique to the investigation of the dynamical properties of inorganic single crystals, organic and polymer glasses and super-cooled liquids has shown that the RSMR technique is one of the most effective methods of studying the dynamics of atoms in condensed phase [26-29].

Excellent energy resolution of the method ($\sim 10^{-9} ev$), achieved due to the combination of a Mössbauer detector and source is by several orders better than that provided by the most precise neutron spectrometers (up to $10^{-6} ev$) and, moreover, by the XRDA ($\sim 1 ev$). It is this fact which permits us to observe comparatively slow motions with correlation times $\sim 10^{-6} - 10^{-8} s$ from the broadening of energy spectrum lines or motions with correlation times $< 10^{-8} s$ from the decrease of the elastic scattering fraction. This method as compared to MAS possesses an essential advantage - it is universal since a scatterer may contain no Mössbauer nuclei. This advantage permits one to enlarge essentially the range of studied biological objects. The possibility to vary the scattering angle, and as a consequence, to choose comparatively small scattering angles allows one to carry out the measurements of the elastic scattering fraction at higher temperatures and hydration degrees than in MAS (for noncrystalline protein samples, for example, at room and higher temperatures).

At present a series of reviews on RSMR has been published which consider in detail experimental and theoretical fundamentals of the method [26-28]. We dwell here upon some specific features of the method application to the study of biological objects [25].

Figure 2 shows the principal scheme of the RSMR set up. Mössbauer radiation, emitted by Co^{57} source, mounted on a vibrator is Rayleigh scattered by all electrons of the biopolymer B. The radiation scattered at 2θ angle is counted by a detector D. To measure elastic fraction a resonant absorber is used, which in combination with the detector forms the so- called "Mössbauer" detector. The intensity of the incident beam I (0) is first measured with the help of the resonant absorber and then the intensity of the scattered radiation, $I(2\theta)$ is measured.

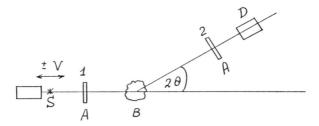

Figure 2. The scheme of RSMR set up. S-Mössbauer source, B-protein or an-other biopolymer under study, D- detector of radiation, A- resonance absorber, placed in front of B (position 1) or behind B (position 2) where corresponding intensities I(0) and I(2(θ)) are measured). To measure a spectrum shape g(ω) A is placed only in a position 2.

The elastic fraction is determined by the expression

$$f = \frac{I_\infty(2\theta) - I_0(2\theta)}{I_\infty(0) - I_0(0)} \tag{1}$$

where I_∞ and I_0 are measured at resonance (i.e. at zero source velocity, $V = 0$) and far from resonance ($V = \infty$). To measure the RSMR line shape, $g(w)$, the resonance absorber is placed in the position 2 only.

The intensity of γ-quanta Rayleigh scattered by an assembly of N equivalent atoms is [26-28,30]

$$I(\vec{Q}) = N \cdot h^2(\theta) \cdot S(\vec{Q}) \tag{2}$$

$Q = 4\pi \sin\theta / \lambda$-the transferred momentum, $h(\theta)$ is the atomic form-factor, $S(Q)$ is the so-called scattering law, determined only by the structure of a scatterer.

Even in protein crystals, consisting of small globular proteins of the myo-globin type (with the molecular weight 17800) a typical angular distance between the Bragg maxima is $\sim 30'$ [31]. Taking into account the available intensity of Mössbauer sources, this leads to a very small counting rate (~ 0.5 pulse/min) in the Bragg maximum, which gives rise to serious experimental difficulties when trying to realize a coherent version of RSMR for proteins and other biopoly-mers. At present only one paper has been published using the coherent version of the technique to determine the phase of $60\bar{3}$ reflection in myoglobin (Mb) single crystals [32].

Therefore the study of biopolymer dynamics for polycrystalline samples and in solutions is carried out in the incoherent version of RSMR with the use of a sufficiently large detector aperture. While designing the experimental set up, we have also to take into consideration that the biological objects are available only in small quantities. The experimental set up specially designed for RSMR study of biological objects is described in detail elsewhere [33]. For intensity reason an annular shape detector is used to collect all radiation scattered over azimuth at the average angle 2θ with respect to the symmetry axis. The method of 2θ determination is described in [34,31]. This angle turns out to be $12° \pm 1,5°$. The acceptance angle used is $5°$. In the case of a polycrystalline sample not less than 100 Bragg maxima are within the acceptance angle. Thus, in this case scattering can be treated as incoherent within rather high accuracy.

A specific feature of the biological object measurements lies in the necessity of measuring two or three component samples. Thus, in the study of a protein solution both protein and solvent molecules scatter the radiation. Therefore to get information on the protein, it is necessary to carry out two (and sometimes three) experiments - one with the protein solution and the other with the solvent (or its components). Since the size of the solvent molecule is much smaller than that of the protein it would seem that the solvent scattering would be coherent. Yet, because there is such a wide detector aperture, scattering by solvent can also be considered as incoherent in the first approximation [35], i.e., one can assume in all the experiments $S(\vec{Q}) = 1$.

Then, the intensity of γ-quanta Rayleigh scattered by the biopolymer solution equals the sum of intensities of quanta scattered by the biopolymer and solvent molecules [31]:

$$I_\Sigma = I_B + I_S \tag{3}$$

The elastic fraction, f_Σ, determined experimentally is

$$f_\Sigma = \frac{I_{\Sigma,el}}{I_{\Sigma,tot}} + I_c \tag{4}$$

where $I_{\Sigma,el}$ is the elastic fraction of the Rayleigh- scattering intensity, $I_{\Sigma,tot}$ is the intensity of Rayleigh scattering (including inelastic component), and I_C-is the intensity of Compton scattering.

Expression (3) can be rewritten in the following form [31].

$$f_\Sigma = \frac{m \cdot h^2(\theta)_B \cdot \exp\{-2W_B\} + (1-m)h^2(\theta)_S \cdot \exp\{-2W_S\}}{m(h^2(\theta)_B + F(\theta)_B)(1-m)(h^2(\theta)_S + F(\theta)_S)} \tag{5}$$

where m is the fraction of the biopolymer molecules in the total number of molecules in the solution, $(1-m)$ is the fraction of the solvent molecules, $h^2(\theta)_B$, $h^2(\theta)_S$, $F(\theta)_B$ and $F(\theta)_S$ – are the intensities of Rayleigh and Compton scattering by the biopolymer and solvent molecules, respectively, which can be expressed in terms of the corresponding atomic form factors and the intensities of Compton scattering from the atoms of given molecules, $\exp\{-2W_i\}$ are the overall Debye-Waller factors (the same for all atoms of the biopolymer ($i = B$) or solvent ($i = S$) molecule.

Since the fraction of elastic scattering for solvent can be measured independently, expression (5) can be reduced as follows

$$\frac{\phi f_B}{\gamma_B} = \exp\{-2W_B\} = C_\Sigma \cdot f_\Sigma - C_S \cdot f_S \tag{6}$$

where C_Σ and C_S are coefficients

$$C_\Sigma = 1 + \gamma_B' + \kappa F_{SB} \frac{\gamma_B'}{\gamma_S'}(1 + \gamma_s')$$

$$C_S = \kappa F_{SB} \frac{\gamma_B'}{\gamma_S'}(1 + \gamma_S') \tag{7}$$

$$\gamma_i' = \frac{1 - \gamma_i}{\gamma_i}; i = B, S; \kappa = \frac{1 - m}{m}; F_{SB} = \frac{F_s}{F_B}$$

the sign ϕ in front of f_B in (6) indicates that it is the calculated value of the elastic scattering fraction, γ_i equals the ratio of the Rayleigh scattering intensity to the total intensity of the Rayleigh and Compton scattering. The coefficients C_Σ and C_S can be calculated on the basis of the data from [36] for the given concentration m: f_Σ and f_S being taken from experiments. When scattering by a three-component system the third component can be taken into consideration in (6) and (7) in a similar way [37].

It follows from (6) that with an increase of the solvent concentration the error of the Debye-Waller factor determination for the studied biopolymer also increases as the C_Σ and C_S values become greater. Estimates of (6) and (7) show that the biopolymer concentration in the solution must be not less than 10-15%.

The RSMR technique due to the inherent high energy resolution reveals relatively slow motions with correlation times $\sim 10^{-6} - 10^{-8} s$ which can be obtained from the line broadening of the energy spectrum $g(\omega)$.

In the elastic, incoherent version of RSMR, in Gaussian approximation for the self-correlation function and in the case when atoms move classically, the lineshape of the spectrum $g(\omega)$ has the following form [38].

$$g(\omega) = \frac{1}{\pi} \int\limits_0^\infty dt \exp\left\{ -\frac{\Gamma}{2}t - \frac{1}{2}Q^l \langle [\Delta x(t)]^2 \cdot \cos(\omega - \omega_0)t \rangle \right\}. \tag{8}$$

Here the correlator $\langle [\Delta x(t)]^2 \rangle$ describes the motion of any atom of the system under investigation. It is provided, that all atoms move similarly.

Thus, the incoherent RSMR technique is used to study the biopolymer dynamics. This provides the determination of comparatively simple dynamical characteristics of the system under consideration, i.e. the overall Debye-Waller factor $e^{-2W} \equiv e^{-Q^2 \cdot \langle x^2 \rangle}$ of the "overall" means square displacement $\langle x^2 \rangle \cong 1/2\langle [\Delta x(\infty)]^2 \rangle$ of the biopolymer atoms and also allow characterization of the correlation times of motions, for which $\langle x^2 \rangle$ is measured. It is important that in the incoherent version we measure the amplitude of real motions i.e. of those which are not distorted by the contribution of the static disorder, which is essential in the case of XRDA. The averaging procedure, which really takes place in experiment (see (5)) leads to the fact, that the measured overall $\langle x^2 \rangle$ is always less than (or equal to) the arithmetical mean of $\langle x^2 \rangle$, since the large $\langle x^2 \rangle$ values give small contribution to the sum (5). Since the thermal-diffuse scattering results in a noticeable underestimation of the $\langle x^2 \rangle$ values obtained by the XRDA technique for proteins, the overall $\langle x^2 \rangle$ values measured by incoherent RSMR are closer to the true values than the overall values obtained by the XRDA [31]. It should be noted, that the possibility to obtain simultaneously an information about the mean square displacements and correlation times of the same motions in one experiment is a favorable property of Mössbauer spectroscopy (RSMR, in particular) which put this technique apart from another spectroscopic methods.

2. Experimental Temperature Dependencies of the Elastic RSMR Fraction, f, and RSMR spectra, $g(\omega)$

Figure 3 and 4 show the temperature dependencies of the elastic fraction f_Σ and overall mean-square displacement $\langle x^2 \rangle$ for different metmyoglobin (met-Mb) samples [39,40] ($M = 17800$). For met-Mb in the crystalline state and in solution the contribution of the solvent was approximately taken into account by using the Eq. (6). For met-Mb samples with $h = 0.05$, $h = 0.5$ and for met-Mb in a crystalline state RSMR spectra were measured within the temperature range from 80 K to 300 K, for met-Mb in the solution within the interval from 200 to 273 K. Preliminary treatment did not reveal any line broadening (see Figure 5). Similar dependencies - the drop of f with the temperature rise - were observed for the hydrated trypsin samples ($M = 23319$) [41,42], hydrated human serum albumen (HSA) samples and its concentrated solutions ($M = 66500$) [42,43], hydrated chromatophore samples [38,44,45] and chromatophores in solutions ($M \sim 10^7$) [37]. For hydrated samples of HSA and for HSA in solution, as well as for hydrated chromatophore samples RSMR spectra were studied within 215 to 300 K. For all investigated biopolymers with a relatively high water content ($h \geq 0.3$) - met-Mb, trypsin, HSA, chromatophores, DNA, as seen from Figures 3-5 a sharp decrease of f_Σ at $T > 230K$ is typical, that means a sharp increase of the overall $\langle x^2 \rangle$ whose values at room temperature and high hydration degree ($h \simeq 0.7$) are getting very large ($\sim 1\text{Å}^2$). For the dehydrated samples ($h \simeq 0.05$) $f(T)$ decreases linearly in the whole temperature

range from 100 K to 300 K, with a constant slope (at room temperature $\langle x^2 \rangle$ reach the value $\simeq 0.05\text{Å}^2$), Figure 5 shows that the drop of elastic fraction is not accompanied by the strong line broadening which is typical of the free unbounded diffusion. Small line broadening $\Delta\Gamma$, observed for example, in a HSA sample close to the room temperature (see Figure 8) can not explain such a strong decrease of $f(T)$ within this pattern.

Such a behavior of the elastic fraction, f, and linewidth, Γ, is typical neither

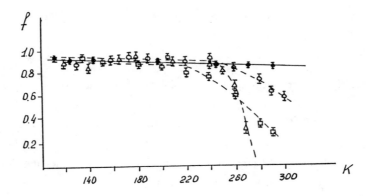

Figure 3. Elastic RSMR fraction versus temperature for various metmyoglobin samples, ($\phi - h = 0.05$; Φ-Mb crystalline; $\Phi - h = 0.5$; Φ-concentrated solution ($\sim 26,6\%$ Mb)).

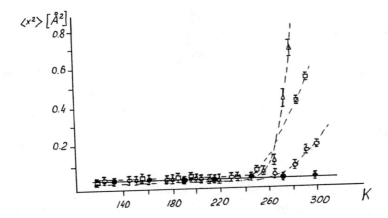

Figure 4. Overall mean-square displacement versus temperature for various metmyoglobin samples ($\Phi - h = 0.05$; Φ-crystalline; $\Phi - h = 0.5$; Φ-solution ($\sim 26,6\%$ Mb)).

of the Einstein or Debye solids, nor of liquids, where $f \simeq \frac{\Gamma}{\Gamma + \Delta \Gamma}$. Therefore, it is necessary to develop special models describing defreezing of the conformational mobility vs. temperature in the biopolymers. Since the behavior of $f(T)$ and

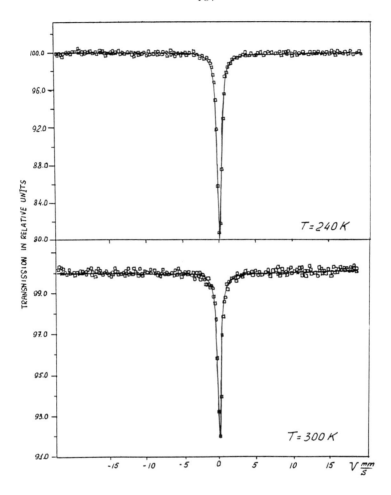

Figure 5. RSMR spectra for crystalline metmyoglobin at $T = 240K$ and $T = 300K$.

$g(\omega)$ is very similar to the analogous MAS data for met-Mb [19,20] and other biopolymers [16], the models used to interpret the RSMR data are analogous to models used in MAS [46- 48]. There are some reasons in favor of the suggestion that such a sharp increase of $\langle x^2 \rangle$ with temperature for sample hydrations, $0 < h < 1$ is associated, first and foremost, with the growth of the intramolecular conformational mobility but not with the different types of the motion of the molecule as a whole. These reasons will be analyzed in a separate paper [49]. Here we note one of them, namely, almost equal increase of $\langle x^2 \rangle$ with temperature (at equal hydration degree $h \simeq 0.5$) for the biopolymers with essentially different molecular weights - from 2.10^4 for met Mb up to $10^6 \sim 10^8$ for DNA and chromatophores.

In our earlier papers [38,39] the increase of the intramolecular mobility in biopolymers - transitions between CS - with the temperature increase was treated with the help of the model of single Brownian overdamped oscillator. In this model the main attention was drawn to the temperature dependence of the elastic fraction. However precise measurements of the temperature dependence of RSMR lineshape (see Figure 5,8) show only a weak line broadening (or no broadening at all). The appearance of the "wide" component is observed close to the room temperature, where the f_Σ values are already rather small ($f_\Sigma = 0.3 \sim 0.1$). This fact cannot be explained by a simple single oscillator model, in which the line broadening can be expected only in a region where f_Σ begins to drop, i.e. at 230-250 K, after this the line must narrow [38].

The detailed analysis of the models, able to describe specific changes of the RSMR line shape, accompanied by the strong decrease of the elastic fraction goes beyond the scope of this review. We note here only, that for a description of all available experimental RSMR and MAS data it is necessary to use multimode approximation, i.e. for a correct explanation of the conformational mobility increase with temperature one must use at least two or three overdamped Brownian oscillators [19] and onYC or two ordinary vibrational modes [50,51].

In other words, conformational mobility at the room temperature is characterized by a wide spectrum of correlation times – from 10^{-6}s (these times yield a broadening of the "narrow" component) and shorter, including times $\sim 10^{-8}$s (giving a contribution to the wide component) down to 10^{-12}s (times $< 10^{-8}$s determine the drop of the elastic fraction). One of the possible explanations of RSMR data is discussed in §5 of this review.

3. Experimental Hydration Dependence of the Elastic RSMR fraction, f, and RSMR spectra $g(\omega)$

Now to the hydration dependencies of the elastic RSMR fraction. The possibility of using RSMR for the investigation of the influence of biopolymer hydration on its dynamical properties arises due to the inherent high energy resolution of the technique and, consequently, the ability to separate with the high accuracy the elastic and inelastic scattering. Since the water bound to the protein is quite mobile, it contributes mainly to inelastic scattering. Therefore elastic RSMR gives information on changes of the dynamical properties of biopolymer as it is with the increase of the hydration degree h.

Figure 6 represents a typical hydration dependence of the elastic fraction f_Σ for HSA at the room temperature. Substitution in (6) of the numerical values of the coefficients C_Σ and C_S yields the following numerical expression for the HSA-water system

$$f_\Sigma = \frac{\phi f_B + 1.21 \cdot h \cdot f_W}{1 + 1.21 \cdot h}. \tag{9}$$

It is easy to conclude, by analyzing (9), that the dynamical properties of the water-protein system, HSA + water have no additivity in the whole hydration range [41, 52]. Really, if for f_B one takes the f_Σ value ($h \to 0$) ($f_B = f_\Sigma(0) = 0.75$), and for water $f_W \simeq 0$, (as for free water) then the calculated f_Σ curve (solid line in Figure 6) is in strong disagreement with the experimental data. In the whole range of the hydration degrees further addition of water not only gives no contribution to the elastic RSMR by itself (since $f_W = 0$) but moreover goes on to loosen the protein, to increase its mobility, to decrease values of f_B [41,52].

Thus, at all hydration degrees, $0.05 \leq h \leq 0.75$, water influences the dynamical properties of protein, which, in turn changes dynamical properties of water. This conclusion does not depend on the assumption that $f_W = 0$ and remains valid at any reasonable (positive) values of f_W [41].

The absence of the additivity of the dynamical properties in water-biopolymer systems is displayed in the dependencies of the calculated curves $\phi f(h)$. The curve calculated by the expression (9) at $f_W = 0$ (we always use below this simplest approximation) for the elastic fraction dependence for HSA is shown in Figure 7.

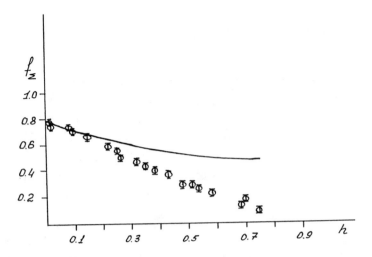

Figure 6. Hydration dependence of elastic fraction for HSA. Solid line - calculated on the basis of Eq. (9) curve.

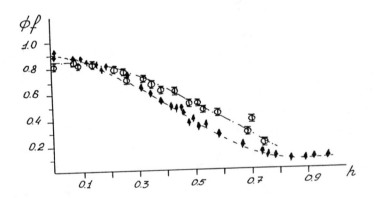

Figure 7. Calculated (on the basis of Eq. (9)) hydration dependencies of elastic fraction for HSA Φ and DNA $\square$.

We also have studied hydration dependencies for trypsin, lysozyme (see Figure 12), DNA (see Figure 7) and chromatophores. These investigations show that biopolymer hydration leads to a decrease of the elastic fraction $\phi f_B(h)$ and, consequently, to an increase of $\langle x^2 \rangle(h)$ within the range of hydration degrees $0.05 \le h \le 0.8$ (with the exception of lysozyme (see Figure 12) for which already at $h = 0.35$ further addition of water does not affect the dynamical properties of protein). Figure 8 shows RSMR spectra for HSA at $h = 0.05$ and $h = 0.73$.

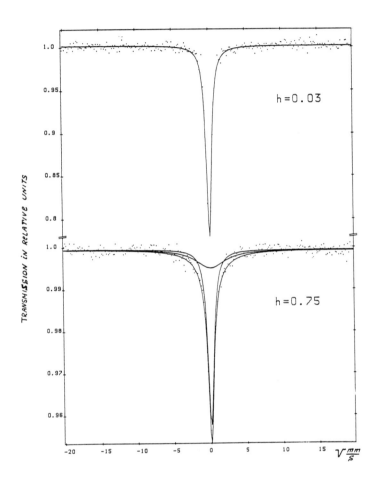

Figure 8. RSMR spectra of HSA at $T = 300K$ for different relative humidities: $h = 0.03$ *and* $h = 0.75$.

At $h = 0.73$ the broadening of the "narrow" line $\Delta\Gamma = 0.80$ mm/sec and the existence of the "wide" component are observed. However, the observed line broadening in this case is also too small to be responsible for such a strong drop of $\phi f_B(h)$.

Experimental data displayed show that the hydration degree as well as the temperature is the principal factor affecting the intramolecular conformational mobility of biopolymer. This statement is illustrated by Figure 9 with the help of three dimensional plot of the elastic fraction versus hydration and temperature.

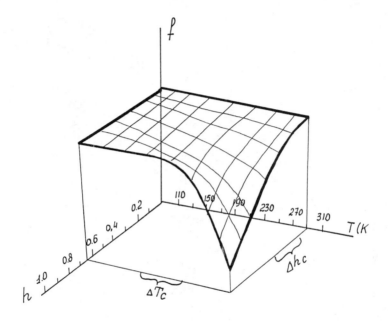

Figure 9. Three dimensional plot of elastic RSMR fraction versus temperature and hydration degree for the biopolymers.

4. Glass-like Dynamical Model of the Biopolymers [23,24]

Considering the RSMR studies of the organic and polymer glasses [27-29, 34,35] and supercooled liquids [27,34] one can be easily convinced that the pattern observed in these substances is quite similar to that described above. At temperatures higher than the vitrification temperature, Tg, the organic super-cooled liquids and polymers behave similarly to the biopolymers exhibiting an abrupt fall of the RSMR elastic fraction, the line broadening being inessential within the temperature range close to that where f falls down to 0.05-0.01.

Additional experimental data are presented favoring a phase transition in the biopolymers which is similar though not identical to the vitrification of the polymers.

The MAS data in Mb [20] are explained by assuming such a phase transition at $T \approx 220K$. The x-ray analysis of the trypsinogen [53,54] reveals the phase transition at temperatures close to 220K, the transition temperature being dependent on the solvent content (a methanol-water mixture). No phase transition is observed in the pure solvents (in the absence of the trypsinogen). The B-factor value $(B = 8\Pi^2 \langle x^2 \rangle)$ depends on the cooling rate of the trypsinogen crystals. These data are given in Figure 10.

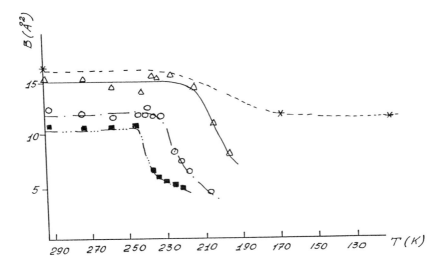

Figure 10. X-ray analysis data for trypsin in a different buffer solutions (Δ - 75% methanol + 25% water; 0-7% methanol + 20% water; $\square$ - 67% methanol + 33% water) at slow cooling rate [53] and at fast cooling rate (* - 70% methanol + 30% water) [54]. Data redrawn for this figure by permission of R. Huber.

The next Figure 11 presents calorimetric data [55] as compared to the temperature dependent elastic fraction for a DNA sample with hydration $h = 0.61$. A correlation is clearly seen between the start of the f_Σ fall and specific heat leap near 200 K. Such a leap usually implies a vitrification point in the polymers and is absent only in the polymers with the high crystallization degree, $W \geq 75\%$ [56].

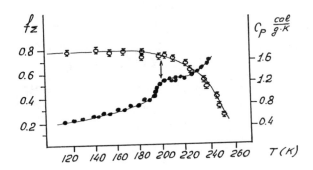

*Figure 11. Elastic RSMR fraction Φ and specific heat ● [55] for DNA versus temperature at $h = 0.6$. Reprinted by permission from Sov. Phys. Uspekhi **22**. Copyright 1979 American Institute of Physics. Copyright 1979*

A similar correlation between the specific heat leap [57] and the ϕf_B fall with the hydration rather than temperature change is found in our lysozyme study (Figure 12).

A clear example of how the myoglobin functional properties change within the temperature range from 200 K to 220 K is given in ref. [58] (see Figure 13). When the pressure rises from 1 bar to 2 kbar the Mb-CO rebinding markedly accelerates. However, if the sample has been first frozen down and only afterwards compressed (the way $a \to b \to c$) then this acceleration is considerably smaller than if the sample has been first compressed and only afterwards frozen down ($a \to b' \to c'$). The following experiment was carried out by the authors of the ref. [58]. The sample was transferred from a to c along the way $a \to b \to c$ and at the point (100 K, 2 kbar) the ligand rebinding kinetics $N(t)$ were studied.

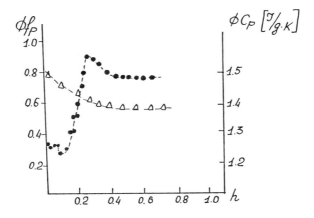

Figure 12. Elastic RSMR fraction Δ and specific heat • [57] for lysozyme versus hydration. Reprinted with permission from Biochemistry. Copyright 1979.

Keeping the 2 kbar pressure the sample temperature was increased from 100 K to T_m then the sample was cooled down back to 100 K and the curve $N(t)$ was measured again. The curve $N(t)$ turns out to remain the same as in the case $a \rightarrow b \rightarrow c$, if $T_m \sim< 200K$, whereas at $T_m \sim> 220K$ the CO rebinding kinetics corresponds to c', i.e. the transition $c \rightarrow c'$ occurs in Mb somewhere between 200 and 220 K.

Other similarities between the biopolymers and glasses are to be pointed out. As has been emphasized above the totality of the MAS and RSMR experimental data can be explained only when accounting for at least 4 or 5 modes with a broad spectrum of the correlation times (10^{-6}s to 10^{-12}S). Meanwhile such a broad spectrum of the correlation times is specific for the glassy state (contrary to the liquid and crystalline state) (see paper [59] and references therein).

An even broader spectrum of the correlation times from 1 second to 10^{-9}

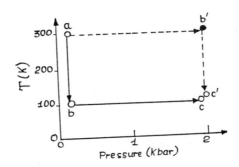

Figure 13. Pressure titration experiment on Mb-CO rebinding [58]. Data redrawn by permission of H. Frauenfelder.

seconds is experimentally observed in a number of the proteins and biopolymers using different physical labels [60]. A hierarchy of the amplitudes and correlation times of different internal motions is pointed out in proteins by means of the protein dynamics computer simulations [61].

Therefore, the above examples indicate a phase transition in the biopolymers, which is similar to the vitrification transition at the critical temperature $T_c \approx 200K$, as well as some other similarities between the biopolymer and glassy properties.

We emphasize that we are by no means going to prove the identity of the biopolymers and glasses. Such an assertion would have been misleading. The biopolymer is a complicated system capable of performing certain biological functions, and although it possesses some structural and dynamical features close to those of such systems as crystal, glass, supercooled liquid, liquid, it does not certainly mean an identity of the biopolymer and one of this entities.

Now, the origin of the glassy features of biopolymers is discussed. The biopolymer carries a certain information written in by rigid ordering of atomic groups which make a primary sequence. A relation between the information and entropy has been revealed during the last decades (see, e.g. [62]) and it is now clear that only a non-equilibrium open system is capable of carrying information. In particular the biopolymer molecule carrying information is necessarily non-equilibrium and, hence it has a non-zero entropy. This entropy relates only to a part of the globular degrees of freedom and any evolutions of the globule that affect the primary sequence. Also, according to the kinetic hypothesis of folding the protein globule may only in a topologically simple structure (e.g. without nodes) [63] which restricts additionally the degrees of freedom along which the globule evolution may proceed. Thus, two kinds of kinetic character are present. Thermodynamic equilibrium is achievable with respect to the remaining degrees of freedom which means that the globule is in a metastable state.

Usually (e.g., in ideal crystals) the equilibrium state of the system is determined by minimizing a thermodynamic potential,

$$\delta F(\{\xi_i\}) = 0 \qquad (10)$$

where $\{\xi_i\}$ is a set of generalized coordinates. For the globule, as pointed above, motion along a part of the generalized coordinates is forbidden, i.e. the minimization (10) is carried out only within a restricted region of the whole phase space. That is why the minimum found is local (see Fig. 1b) and the globule state should be most probably degenerate even at zero temperature.

A part of the degrees of freedom in the structural and spin glasses is also frozen down, i.e. some evolutions are kinetically banned. And though the physical mechanisms of these bans in proteins and glasses are at variance we believe it most instructive to emphasize a fundamental alliance between two entities which is exhibited the transition from the aperiodic crystal concept $(+ - +-)$ to the CS or QDCS states $(+ - --)$.

Above considerations lead to a conclusion that QDCS's must result in an alliance between not only the structural features of the biopolymer and glass but also between their dynamical features.

A striking feature, specific of glasses, is the retaining of the atomic (or atomic group) mobility even at extremely low temperatures, (i.e. at $T \to 0$) which is implied by the last plus in the notation $(- - -+)$ of the Table 1. The atomic (or atomic group) mobility at $T \to 0$ is caused in glasses by the phonon assisted tunneling (PAT) in the double-well potentials (DWP) [62-64] or quasi-equilibrium positions (QEP) of these atoms or atomic groups as we use to say when dealing with the biopolymers. This phonon assisted tunneling results in an additional term in the specific heat of the amorphous solids which is linear in temperature and dominates at low temperatures, this term being absent in crystalline substances.

Remarkable similarities of various aspects of the biopolymer and glass behavior have drawn us to a conclusion that the biopolymers at $T < T_c$ and $h < h_c$ (where T_c and h_c are similar, though not identical, to the vitrification point) are in a state which is more glassy-like rather than crystal-like. But then a pronounced contribution of the linear specific heat should be also exhibited by the biopolymers at $T \to 0$. To verify this hypothesis we have analyzed the experimental data on the low temperature specific heat of the biopolymers [23,24].

The specific heat of both inorganic and organic dielectric glasses at low temperatures is often described [66,56] as

$$C_V(T) = C_1 T + C_3 T^3 + C_E E(\Theta_E/T)$$

where C_1, C_3, C_E and Θ_E are experimentally fitted constants. Here the second Debye term is due to the longwave acoustic phonons, whereas the third Einstein term seems to be caused by the boundary frequency of the transverse phonons:

$$E\left(\frac{\Theta_E}{T}\right) = \left(\frac{\Theta_E}{T}\right)^2 \cdot \frac{\exp\left(\frac{\Theta_E}{T}\right)}{\left[\exp\left(\frac{\Theta_E}{T}\right) - 1\right]^2}$$

where Θ_E is the Einstein temperature.

The C_1 value depends on the QEP density $n(\Delta)$ at $\Delta \to 0$.

$$C_1 = \frac{\Pi^2}{6} \cdot K_B^2 \cdot n(0). \tag{12}$$

Here Δ is the energy difference between the states in the DWP (or QEP).

Now to the experimental data on the low temperature calorimetry of the biological macromolecules [55]. These data allow one to conclude that neither of the presently studied biopolymers has a specific heat observing the Debye law at $T < 10K$. When processing the experimental data on the low temperature specific heat taken from the refs. [53, 65-68] by means of the Eq. (11) a special program fitted the C_1, C_3, C_E and Θ_E values so as to minimize the sum of the squared deviations of the experimental and theoretical values. Values of two variable parameters, C_1 and C_3 producing the best fit of the experimental and theoretical points are presented in the Table 2.

The accuracy of the C_1 and C_3 values essentially depends on the temperature ranges in which the data are treated, hence these ranges are also listed in the Table 2. All the experimental data on the low temperature specific heat of biopolymers [55, 67-69] appeared to be, without exception, well described by the Eq. (11). The T_{min} value for DNA and collagen being markedly higher than for the other objects, the C_1 and C_3 values for them are found with a lower accuracy than for the polypeptides. This may result in overestimated C_1 values of DNA and collagen in the Table 2.

Table 2. Parameters C_1, C_3 and $n(o)$ for various compounds

Compound	C_1 $\frac{\mu T}{g \cdot \kappa^2}$	$n(0)$ $[1/T \cdot g]$ 10^{-40}	C_3 $\frac{\mu \cdot T}{g \cdot \kappa^4}$	Number of DWP in macromolecule in a temperature region $\Delta = 100K$	ΔT (K)	Reference
SiO_2	1.2	0.38	1.8		0.1-1.5	66
Polymethyl methylthacrylate	4.8	1.5	29		1.5-10	66
Melanin $(1\% H_2 0)$	5.0	1.6	12		2-4.5	67
Melanin (20% DEA)	9.5	3	9.9		2-4.5	67
Melanosome	22	7	9.9		2-4.5	67
Poly-L-alanine-α-form*	45	14	25		1.5-10	68*
Poly-L-Alanine-β-form*	36	11	20		1.5-10	68*
Polyglycine II*	18	5.9	8.8		1.5-22	69*
Collagen*	$\sim 10^3$	300	10	$2, 5.10^3$	5-20	55*

DNA*	$\sim 5x10^3$	1600	10	$1.1x10^5$	4-20	70*
HSA	39	12.5	8	19	1.5-25	71
Met Mb		3.3		1.4	0.1-3	72

*Asterisk denotes compounds for which literature data were treated by Eq. 11.

After the ref. [23] being published new experimental data appeared on the low temperature specific heat of the dehydrated human serum albumen (HSA) [71] and met-Mb in the crystalline state [72]. These data are also presented in the Table 2. A processing of the experimental data on HSA using the Eq. (11) are exemplified in the Figure 14.

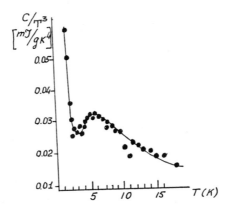

Figure 14. Recalculation of experimental data [71] for the LH of HSA - illustration of validity of Eq. 11.

The analysis of the low temperature specific heat of the biopolymers carried out in the ref. [23] as well as the special measurements of the refs. [71,72] confirm not only the structural similarities between the biopolymers and glasses (sign-in the third column of the Table 1) but also the obvious resemblance of their dynamical features at low temperatures. The mobility of the atomic groups in both biopolymers and glasses caused by the phonon assisted tunneling is still retained even at low temperatures. That is why the Table 1 denotes now the biopolymers as $(+ - -+)$ and their low temperature features exhibit a peculiar combination of both crystalline and glassy properties which is in a sense closer to glasses.

It should be especially emphasized that the biopolymer features as presented in the Table 1 always relate to a single macromolecule rather than to ensembles of such macromolecules.

Microscopically the QEP's for the atomic group of a globule result from the metastable non-equilibrium globule conformation of a native protein possessing many packing defects. An atom-to- atom distance may be more or less its equilibrium value, strains of bihedral angles may arise, etc.

Similarly to SiO_2 [66], QEPs arise near defects. Thus in SiO_2 [73] double-well potentials are shown to appear only at very strong bond deformation. The Si-to-Si distance being in crystal, $\rho = 3.26\text{Å}$ (see figure 15) the conversion of a single well potential for the O^* atom into a double-well potential occurs only when stretching the Si-to-Si distance up to $\rho > 4.32\text{Å}$. Now, the relative deformation is $\delta\rho/\rho \approx 1/3$ and the probability of such strong defects is small ($\sim 10^{-5} - 10^{-4}$). They are formed in the regions in which the interatomic distance is fluctuatively increased and the chemical bond is on the verge of rupture or in fact ruptured.

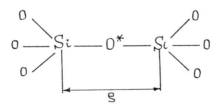

Figure 15. The scheme of the arrangement of Si and O atoms in a SiO_2. Stretching of the coordinates ρ yields a double-well potential.

The volume per atom certainly increases which relates this approach to the free volume model proposed in the ref. [74-76]. The double-well potentials in [74-76] are explained by assuming that these are due to a number of free volumes in the glasses each volume being of the atomic order. These free volumes are caused by the fluctuations at temperatures above the vitrification temperature and are then frozen in when rapidly cooling [73,74].

Since, however, the difference in the physical mechanisms of QEP creation in usual glasses and globules was already mentioned, a formula $T_g \propto n(o)^{-1}$ well known from the theory of glasses [77] might be not necessarily applicable to protein globules. This formula may be checked for globules, by studying the variations of the low- temperature specific heat versus the protein hydration degree.

As is known from XRDA measurements [13,15], approximately 2/3 of atoms of the Mb globule has large $\langle x^2 \rangle$ and, therefore, a large number of QEPs. The RSMR date show (see Fig. 3) that the critical temperature T_c of the macromolecule (an analogue of the vitrification temperature T_g in glasses) in the crystalline met-Mb samples is close to 230 K. When further increasing the temperature atoms form regions whose features are close to that of the supercooled liquids. The atoms in the remaining 1/3 amino acid residues (with a small $\langle x^2 \rangle$) are characterized by small QEP number making a "solid" frame of the protein at room temperature.

Therefore, the LH, RSMR and XRDA data encouraged us to consider the native protein macromolecule (or DNA molecule) as a peculiar- heterogeneous glass - at room temperature biopolymer macromolecule contains regions whose properties are close either to crystal, or to glass, or to supercooled or usual liquid. (It is pointed out that the physical parameter used to identify the above various phases as inherent to the different regions of macromolecule is the free volume rather than their being at equilibrium or not.) So the main conclusion produced by our RSMR data (Fig. 2-8) and other data (Fig. 10-14) as qualitatively shown by the three dimensional plot, $f(T, h)$, (Fig. 9) favors peculiar "transition" regions ΔT_c (near $T_c \approx 200\text{-}220$ K) and Δh_c (near $h_c \approx 0.1\text{-}0.3$). At $T < T_c$ and $h < h_c$ the dynamical features of the biopolymer are mainly a combination of crystal and glass where as at $T > T_c$ and $h > h_c$ they are mainly crystal, glass and supercooled liquid. The specific volume of the biopolymer (and, hence, its inner mobility characterized in RSMR by the $\langle x^2 \rangle$ values) essentially increases with temperature at $T > T_c$ and with the hydration at $h > h_c$ (which seems to be evidenced by the experimental data of the Ref. [78]. When the biopolymer hydration increases the number of QEPs grows for each molecular region (in particular, QEP may now arise in the regions which at lower h were quite solid).

This water-protein interaction determines not only the glassy properties of the protein but especially those of the water bound to the protein which even below $0°C$ is in a peculiar supercooled state (this reveals itself, in particular, by a change of the dielectric relaxation rate in the bound water [79]). So in this sense the observed phase transition of the hydrated protein at the critical (temperature) point is a transition of the whole system-protein + bound water, i.e. at $T < T_c$ not only separate protein regions but also the bound water become glass-like.

5. Interpretation of RSMR Data Within the Glass-Like Dynamical Model of the Biopolymers. Tunneling in the Transitions Between QEPs [25]

Now to the description of the atomic group transition between QEPs and their effect on the changes in RSMR elastic fraction versus temperature. As already mentioned above, various modifications of the overdamped Brownian oscillator model are used [45,47,19] to explain experimental data on the f and line shape changes versus temperature. However, when applying these models, one must take into account two important circumstances: (i) As is known, viscosity of supercooled liquids does not obey the Arrhenius law but has a sharper temperature dependence [80]. (ii) The use of the overdamped Brownian oscillator model implies the infinite number of "instantaneous" QEPs. In this case the barriers between QEPs become very narrow and therefore tunneling should be taking into account.

Now a model is considered in which these circumstances are taken into account. As a first, rough approximation the protein is treated as a uniform dynamical system, i.e. homogeneous, rather than heterogeneous glass. The concept of QEP for atomic groups in the protein is illustrated at the example of an amino acid residue, Arg 31 in Mb (Figure 16) [25]. It should be noted that the microscopic nature of QEPs cannot be unambiguously described even in the simple glasses of the SiO_2 type [66]. (Hence a simple model suggested here should be looked at as a simplified illustration only). We shall consider $C_\alpha H, C_\beta H_2, C_\gamma H_2, N_\epsilon H$, etc. along the chain of covalent bounds $C_\alpha-C_\beta-C_\gamma-C_\delta$ etc. as elementary groups in the protein which have different QEPs. Each of these groups is bound to atoms of another amino acid residue by non-valence interactions (dispersion or electrostatic or may be hydrogen bonds). The energy of such interactions depends on the intergroup distances, but on the whole is much less than the energy of the covalent bonds.

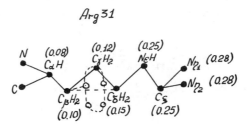

Figure 16. QEPs of atomic groups in a protein are illustrated at the example of Arg 31 residue in Mb. Bracketed are the mean-square deviations for C and N atoms obtained by XRDA 13.

Therefore, in the plane normal to the chain of the covalent bonds each such group, hinge-linked to the neighboring hydrogen or nitrogen atom might have energy-degenerate QEPs whose appearance is associated with non equilibrium, i.e. metastable state of the protein structure. The number of QEPs for each group depends on its position in the protein, temperature and hydration degree. Some atomic groups, $C_\gamma H_2$ for example, are assumed to perform transitions between QEPs. A specific feature of the model is that at $T > T_c$ (for homogeneous glass $T_c \equiv T_g$), i.e. in the viscoelastic region the QEP, neighboring the initial one, exists only for some time and the lower the protein temperature the longer it exists. Therefore, the transition to the neighboring QEP proceeds in two stages (i) fluctuative preparation (creation) of a neighboring QEP (cavity), which requires a work $W'(T, h)$ and (ii) the transition of the neighbor to the prepared cavity (QEP), which can occur either by activation or through tunneling. Figure 17 shows a possible instantaneous spatial realization of two neighboring QEPs, while Figure 18a represents the energetic scheme of the corresponding two-level system.

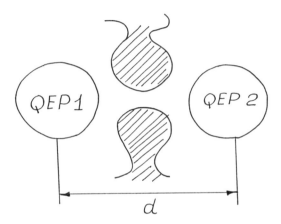

Figure 17. Possible spatial "instantaneous" realization of QEPs in a protein.

It is assumed now that the transition into the neighboring "instantaneous" QEP occurs by activation and consequently, requires surmounting of the activation barrier ε. The existence of this barrier follows at least from the fact, that even if a free volume actually exists (it means that the work $W'(T)$ is already performed) the group motion proceeds via a profile of a complicated and besides changing in a time conformational potential of the biopolymer.

Creation of QEPs and transitions between them are assumed to be independent events. Then the transition probability from QEP 1 to QEP 2 (see Figure 17) is the product of the probability of neighboring QEP 2 creation and the probability of surmounting the barrier ε. In this approximation the total transition energy $W(T)$ consists of two parts

$$W(T) = W'(T) + \varepsilon. \tag{13}$$

As $T < T_c$ the free volume strongly decreases within the framework of the glass-like dynamical model of the proteins and respectively the work necessary to create neighboring QEP, strongly increases (one can consider, approximately, that $W'(T) \to \infty$). In those regions, where some free volume remains it is frozen in which means that the stationary QEPs are formed for which $W'(T) = 0$. According to the model [74-76] in the regions, where the free volume remains at $T < T_c$, those QEPs are situated which contribute to C_1 when measuring LH of the protein. As told above, the number of such QEPs is very small – not more than two or three two-level systems per globule. This is the very reason why only vibrational degrees of freedom are seen in the RSMR experiments at $T < T_c$.

The situation drastically changes at $T > T_C$. Since the free volume swiftly increases and continues growing with the further $(T > T_C)$ increase of the temperature, this leads to a sharp decrease of the mean $W'(T, h)$ value, i.e. to the creation of more and more "instantaneous" QEPs, between which the transitions of atomic groups are possible. This circumstance leads in turn to gradual decrease of f (increase of $\langle x^2 \rangle$), registered by RSMR.

However, viscosity measurements in organic supercooled liquids show, that the creation of a neighboring cavity (free volume) always plays the role of the process limiting the mobility in the substance [81]. Therefore, usually $W'(T) \gg \varepsilon$, and only at $T \approx T_{room}$ both addends in (13) can become close.

Inequality $W'(T) \gg \varepsilon$ reflects the decisive role of the free volume for the protein dynamics - in the spirit of the "mobile defects" model due to Lumry and Rosenberg [6]. The concept of conformational substates [13] as well as the treatment of Mössbauer spectroscopy data within the framework of the overdamped Brownian oscillator model [19,46,48] do not consider the role of the free volume at all, and consequently, do not imply an existence of the phase transition in the biopolymer macromolecule at $T \approx T_c$. To be more correct, it is noted that the mobile defect model [6] does not also imply a phase transition.

Since one of the principal properties of the protein is its ordering, defects (in a simplest case - cavities) leading to the creation of QEPs are formed in the different globular regions with essentially different probability. Even if only one elementary group of $C_\gamma H_2$ type (see Figure 16) performs transitions between QEPs situated around $C_\beta - C_\delta$ axis, the probability of QEP creation on the circle along which the group C_7H_2 rotates has maxima and minima. Now the maximum distance between neighboring QEPs is estimated within the framework of the homogeneous protein model. For this purpose the solution [82] of the problem of the jump diffusion of atomic groups with the frequency ν over N equidistant sites is used. Fitting the formula in [82] for the incoherent RSMR, i.e. substituting $Q = \frac{4\pi \sin \theta}{\lambda}$ instead of $1/\lambda$, one can receive (we take N QEPs) the following expression for the spectral function.

$$g(w) = P\frac{\Gamma/2\Pi}{(\omega - \omega_o)^2 + \Gamma^2/4} + \sum_{\kappa=0}^{N-2} p_\kappa \frac{\Gamma_\kappa/2\Pi}{(\omega - \omega_o)^2 + \Gamma_\kappa^2/4} \tag{14}$$

where

$$\Gamma_\kappa = \Gamma + 4\nu \cos^2(\Pi\kappa/2(N-1))$$

$$P = \frac{\cos(d/2Q) \cdot \sin\left(\frac{Nd}{2Q}\right) \cdot \sin\left\{\frac{(N-1)d}{2Q}\right\}}{N(N-1)\sin^2(d/2Q)}$$

and

$$P_\kappa = \frac{1}{N(N-1)} \cdot \sum_{\kappa=0}^{N-2} \frac{\sin^2\left(\frac{d}{Q}\right) + (-1)^{N-\kappa} \cdot \sin\left(\frac{d}{Q}\right)\left[\sin\frac{Nd}{Q} + \sin\frac{(N-1)d}{Q} \cdot \cos\frac{\Pi\kappa}{(N-1)}\right]}{2\delta_{\kappa,0}\left[\cos\frac{d}{Q} + \cos\frac{\Pi\kappa}{(N-1)}\right]}$$

where d is the distance between QEPs. The elastic fraction is determined by means of the integration of the spectral function over frequency

$$F = \gamma \int\limits_{|\omega-\omega_o|\sim\Gamma} g(\omega)d\omega. \tag{15}$$

It follows from the Eqs. (14) and (15) that with the rising temperature (i.e. with the increase of ν) the elastic fraction can fall at most N times. Since the RSMR experimental data give maximum values of $\langle x^2 \rangle \sim 1\text{Å}^2$ and f falls by a factor of 4-5, hence the QEP number $N \geq 4-5$, and the distance between them $d \cong \frac{\sqrt{\langle x^2 \rangle}}{N-1} = 0.25 - 0.3\text{Å}$.

At that point our consideration of glass-like dynamical model of proteins which implies the phonon-assisted tunneling (PAT) of atomic group at low temperatures meet the investigation of tunneling chemical reactivity performed by one of us (V.I.G.) since 1959.

These investigations started [83,84] by rather simple calculations which led to the conclusion concerning the existence of low temperature plateau of the rate of thermally-activated exothermic chemical processes due to the contribution of tunneling from the zero-point level of initial state and to the introduction of a criterium of so-called "tunneling temperature" T_t. Tunneling temperature separates the regions of exponential predominance of Arrhenius-type overbarrier (at $T > T_t$) and quantum tunneling (at $T < T_t$) transitions.

Subsequent experimental studies highlighted in the discovery of quantum low-temperature limit of the rate of chemical reaction [85,86] at the example of spontaneous growth of polymer chains in radiation-induced polymerization of formaldehyde.

This observation demonstrated the first case of temperature independent kinetics of chemical reaction in the full sense of these words (i.e. the process involving rearrangement of atoms, change of lengths and angles of valence bonds) and led to the introduction of the concept of molecular tunneling and to the description of chemical reactions as radiationless transitions [87-89].

At present two types of tunneling in chemical conversions are discerned in the literature – electron-nuclear tunneling discovered by Chance and De Vault in the oxidation of cytochrome- C by chlorophyll [90] and above mentioned molecular tunneling. Just this latter type of tunneling observed for ca. ten low- temperature solid-phase chemical reactions of various types see [87-89, 91,92] and references therein) is the closest relative of the low temperature phonon-assisted tunneling mobility of biopolymers and ordinary glasses. One of important steps in the development of the molecular tunneling concept was the account of two kinds of motions double-mode Franck-Condon factors) – besides the high frequency atomic vibrations within the potential wells representing initial and final states the low frequency motion of these two wells relative to each other is also accounted for [92-94] – see Figure 18.

It is obvious that such a motion leads to the existence of various fluctuative transient states – from the most approached wells i.e. lowest and narrowest potential barrier between them - up to its complete vanishing) to most remote wells (highest and broadest potential barrier). It is also clear that the contribution

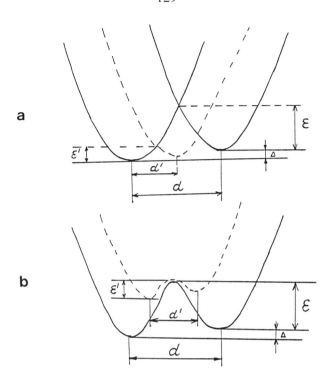

Figure 18. Simplest energy scheme of two QEPs. Possible fluctuative approach of wells (a) and tendencies of tunneling parameter changes (b) are shown.

of the transitions between the approached wells in the total reaction rate would strongly exceed that of the remote wells and hence the quantum transitions should gain additional advantage over the ordinary overbarrier (Arrhenius-type) transitions, and the temperature range of prevalence of quantum transitions should spread well above the simplest estimate of tunneling temperature [83,84]:

$$T_t = \sqrt{\frac{\varepsilon}{m}} \cdot \frac{\hbar\sqrt{2}}{\Pi \cdot \kappa_B \cdot d} \tag{16}$$

where ε is the barrier height, d is its width, and m is mass of tunneling particle. A comprehensive survey and treatment of tunneling phenomena in chemical physics are presented recently in refs. [95,96] - so now let us proceed directly to phonon-assisted tunneling of atomic groups in biopolymers.

At $d \approx 0.25 - 0.3\text{Å}, \varepsilon \approx 0.2eV$ (estimate based on RSMR data treatment according to refs. [46,48] and $m = 14$ daltons (CH_2), $T_t \approx 160 - 120K$. As mentioned above when the width d and the height ε of the barrier fluctuatively decreases (sometimes down to zero) the tunneling becomes several orders more probable. Detailed calculations of this mechanism done for usual glasses in [97,98] show that the frequency of atomic tunneling between QEP even at high temperature will be much larger than the frequency of activation transitions. Fluctuative approach of the wells can yield a strong increase of T_t.

Therefore, fluctuatively prepared tunneling can cause activated transitions between QEPs up to the room temperature. It is noted that for the above mechanism the limiting stage of the transition is not the actual tunneling, which occurs almost instantaneously $(10^{-10} - 10^{-11}s)$ between the approached wells, but the creation of the cavity (neighboring QEP) and its approach to the given QEP.

Then the temperature dependent rate of such fluctuatively prepared tunneling is activation-like

$$\nu = \nu_o \exp\left\{-\frac{W(t,h)}{\kappa T}\right\}. \tag{17}$$

When using the empirical expression for the temperature dependence of the supercooled liquid viscosity [80] we can write $W(T,h)$ in the form

$$W(T,h) = \frac{A(h)}{T^{n-1}}. \tag{18}$$

Internal microviscosity of the biopolymer is determined by the equation similar to (18)

$$\eta^{\text{int}} = \eta_o \exp\left\{-\frac{A'(h)}{T^n}\right\}. \tag{19}$$

Estimations yield the following values of parameters from $n = 2$ for trypsin to $n \gtrsim 1$ for DNA; $A(h) = 10^6$ (for trypsin $A(h) = 1.0x10^6$ at $h = 0.8$ and $A(h) = 1.4x10^6$ at $h = 0.14$). Internal microviscosity of the globular protein η^{int} strongly decreases with rising T and h, but quantitative estimations in this case are not so definite as for ν.

Considering more realistic heterogeneous glass-like model of the protein the existence of crystal-like and glass-like regions in the macromolecule is taken into account at $T > T_c$. At the same time, as is seen from Figures 3,4,6,7, the elastic fraction for strongly hydrated biopolymers at room temperature is close to $f \approx 0.1 - 0.2$, i.e. the "solid" regions (where $\langle x^2 \rangle \leq 0.04 \mathring{A}^2$) constitute not more than 10-20% of the protein. We believe that this fact can be explained in the following way. At temperature close to the room temperature the probability of transitions between QEPs in the "liquid-like" regions (where $\langle x^2 \rangle \gg 0.04 \mathring{A}^2$) is high, the average relaxation time $\tau_c = \frac{1}{\nu} \sim 10^{-10}$s, i.e. individual links of side chains often travel over the distances $\sim 1 \mathring{A}$, which provides the penetration of such small molecules as H_20, O_2, CO etc. These facts follow from the experiments on fluorescence quenching [7,8] and the H-D exchange [6]. Inner regions of α-helices – the most rigid regions of the proteins can be hydrated and, consequently either QEP is formed or atoms become more mobile while travelling over QEPs of α-helices. This picture seems to be similar to decreasing DNA mobility during its hydration (Figure 7). It is known that DNA is an analogue of the α-helix as to its torsional and bending strength [99]. High mobility of individual atoms of α-helix is also evidenced by XRDA of myoglobin [13].

Thus, the hierarchy in the structure organization of the protein is reflected in the hierarchy of its intramolecular mobilities. With the temperature rise in RSMR experiments all types of motions in proteins can be observed – small-scale vibrations of atoms and groups, $\langle x^2 \rangle_v$, then moderate-scale fluctuation transitions $\langle x^2 \rangle_{QEP}$ of atoms or atomic groups between QEPs, which in terms of the globule states correspond to fluctuation transitions between QDCs, then large- scale motions of the individual side chains, and, finally, a motion of domains $\langle x^2 \rangle_c$, described by the Langevin equation (of course, the globule as a whole becomes mobile as well)

$$\langle x^2 \rangle_\varepsilon = \langle x^2 \rangle_v + \langle x^2 \rangle_{QEP} + \langle x^2 \rangle_c \tag{20}$$

In accordance with the Eq. (20) the elastic fraction is determined by the expression

$$f = \gamma \cdot \exp\left\{-Q^2 \langle x^2 \rangle\right\} \cdot \exp\left\{-Q^2 \langle x^2 \rangle_{QEP}\right\} \cdot \exp\left\{-Q^2 \langle x^2 \rangle_c\right\} \tag{21}$$

where γ is defined in (6) and

$$\exp\left\{-Q^2 \langle x^2 \rangle_{QEP}\right\} = \int\limits_{|\omega - \omega_o| \sim \Gamma} g(\omega) d\omega \tag{22}$$

$g(\omega)$ is given by the Eq. (14) and the Debye-Waller factor for the conformational mobility is determined as in [46]

$$\exp\{-W_c\} = \left\{1 - a^2 \exp(-a^2) \int\limits_0^1 y^n e^{a^l y} dy\right\}. \tag{23}$$

It should be emphasized that the f fall with the temperature and hydration degree rise is mainly due to the second summand in (20), whereas the "wide" component and a weak broadening of the "narrow" component in Figure 8 is due to the third summand. This very incoincidence of the temperature f fall (the line intensity decrease) and the temperature dependent total line broadening of the RSMR spectrum is the most bright manifestation of the amplitude and correlation times hierarchy of the protein intramolecular mobility.

Conclusions

Thus § 1 of this review describes specific features of the RSMR technique as applied to the study of the biological systems [25]. Experimental data on the temperature dependencies of the RSMR elastic fraction f and RSMR spectrum for a number of biopolymers are described in §2. These experimental data don't fit either Debye or Einstein models of solids, or the picture of free, unbounded diffusion in liquids and demand for their explanation a multimode approximation. In other words the internal mobility of biopolymers at high temperature is characterized by a wide spectrum of correlation times (at $T = 293K$ from 10^{-6} s to $10^{-12} - 10^{-13}s$).

Experimental data on f and $g(\omega)$ dependencies versus hydration degree for a number of biopolymers are described in §3. Additional water turns out to increase the conformational mobility in all biopolymers studied (except for lysozyme) within the whole hydration range $0.05 \leq h \leq 0.75$. For lysozyme this process terminates at $h \cong 0.35$. Temperature and hydration are the main parameters which influence the internal biopolymer mobility (see Figure 9).

Basing on the RSMR, LH and XRDA data a glass-like dynamical model for biopolymers if formulated in §4 [23,24]. The analysis of these data and general considerations that the information macromolecule is necessarily non-equilibrium and, hence, occurs in a degenerate system of local energy minima, rather than in the absolute energy minimum, have led us to the assertion of a similarity between the biopolymers and glasses. This analysis yields an internal mobility of the biopolymer macromolecule even at $T \to 0$ and it notation in terms of the Table 1 as $(+ - -+)$.

Finally, a plausible interpretation of the RSMR data within the glass-like dynamical model of the biopolymer is carried out in §5. The possibility that the fluctuatively prepared tunneling between QEPs prevail thermally activated transitions up to the room temperature is pointed out [25].

We thank E.E. Gaubman, I.V. Kurinov, I.V. Sharkevich, I.P. Suzdalev and T.V. Zhuravleva for the assistance in RSMR experiments and A.Yu. Grosberg for the illuminating discussion of theoretical aspects of the problem.

References

1. Schroedinger, E., What is Life? Cambridge University Press, 1944.

2. Linderstrom-Lang K. Chem. Soc. (London) Spec. Pupl. 2, 1 (1955).

3. Koshland, D., Meet, K.H., Ann. Rev. Biochem. 37, 359 (1969).

4. Ravnoveshaya dinimika nativnoi struktury belka (Equilibrium Dynamics of the Native Structure of Proteins) ed. By E.A. Brushtein, Pushchino, 1977.

5. Perutz, M.F. and Mathews, F.A., J. Mole. Biol. 21, 199 (1966).

6. Lumry, R. and Rosenberg, A., Colloq. Intern. C.N.R.S. 246, 53 (1975).

7. Lakowicz, J. and Weber, G., Biochemistry 12, 4171 (1973).

8. Eftink, M.R. and Chiron, C.A., PNAS USA 72, 3290 (1974).

9. Cooper, A., PNAS USA 73, 2740 (1976).

10. McCammon, A., Gelin, B.R. and Karplus, M., Nature 267, 585 (1977).

11. Austin, R.H., Beeson, K., Eisenstein, L., Frauernfelder, H., Gunsalus, I.C. and Marshall, V.P., Biochemistry 181, 541 (1973).

12. Austin, R.H., Beeson, K., Eisenstein, L., Frauenfelder, H., Gunsalus, I.C. and Marshall, V.P., Phys. Rev. Lett. 32, 403 (1974).

13. Frauenfelder, H., Petsko, G.A. and Tsernoglou, D., Nature 280, 556 (1979).

14. Artymiuk, P.J., Blake, C.C.F., Grace, D.B.P., Callye, S.J., Phillips, D.C. and Sternberg, H.J.E., Nature 280, 563 (1979).

15. Hartman, H., Parak, F., Steigemann, W., Petsco, G.A., Ponsi, D.R. and Frauenfelder, H., PNAS USA 79, 4967 (1982).

16. Frolov, E.N., Mokrushin, A.D., Likhtenshtein, G.I., Trukhtanov, V.A., Goldanskii, V.I., Doklady Akademii Nauk SSSR 212, 165 (1973).

17. Keller, H. and Debrunner, P., Phys. Rev. Lett. 45, 68 (1980).

18. Parak, F., Frolov, E.N., Mossbauer, R.L., Goldanskii, V.I., J. Mol. Biol. 145, 825 (1981).

19. Parak, F., Knapp, E.W. and Kucheida, D., J. Mol. Biol. 161, 177 (1982).

20. Bauminger, E.R., Cohen, S.G., Nowik, I., Ofer, S. and Yariv, J., PNAS USA 80, 736 (1983).

21. Van Hove, L., Phys. Rev. 95, 249 (1954).

22. Singwi, K.S. and Sjolander, A., Phys. Rev. 120, 1023 (1960).

23. Goldanskii, V.I., Krupyanskii, Yu. F. and Fleurov V.N., Doklady-Biophysics 272, 209 (1983).

24. Goldanskii. V.I., Krupyanskii, Yu. F., Sov. Phys.- Uspekhi 27, 462 (1984).

25. Goldanskii, V.I. and Krupyanskii, Yu. F., Proc. Intern. Conf. Mossb. Strectr. (Alma-Ata, 1983), Gordon and Breach, NY (1985) v. 1, p. 83.

26. O'Connor, D.A., Proc. Intern. Conf. on Mossb. Spectra., Cracow 2, 396 (1975).

27. Champeney, D.C., Rep. Prog. Phys. 42, 1017 (1979).

28. Albanese, G. and Deriu, A., Riv. D. Nuovo Cim. 2, 9 (1979).

29. Albanese, G., Deriu, A. Chezzi and Pegoraro M., Nouvo Cim. 1D, 313 (1982).

30. Gaubman, E.E., Krupyanskii, Yu. F., Goldanskii, V.I., Zhuravleva, T.V. and Suzdalev, I.P., JETP 45, 1141 (1977).

31. Krupyanskii, Yu, F., Parak, F., Hannon, J., Gaubmann, E.E., Goldanskii, V.I., Suzdalev, I.P. and Hermes, K., JETP 52, 31 (1980).

32. Parak, F., Mossbauer, R.L. Hoppe, W., Thomanek, U.F., and Bade, D., J. de Phys. 37, C6-703 (1976).

33. Gaubman, E.E., Krupyanskii, Yu. F. and Suzdalev, I.P., Pribori Tekhnika Exsperimenta 3, 22 (1981).

34. Champeney, D.C. and Sedwick, D.F., J. Phys. c5, 1903 (1972).

35. Champeney, D.C. and Bean, W.W., J. Phyus. C8, 1276 (1975).

36. Interm. Tables for X-ray Cryst., Kynoch, (1962).

37. Krupyanskii, Yu. F., Bade, D., Sharkevish, I.V., Uspenskaya, N. Ya., Kononenko, A.A., Suzdalev, I.P., Parak, F., Goldanskii, V.I., Mossbauer, R.L. and Rubin, A.B., Eur. Bioph. J. 12, 107 (1985).

38. Krupyanskii, Yu. F., Gaubman, E.E., Shaitan, K.V., Goldanskii, V.I., Rubin, A.B., Suzdalev, I.P., Frolov, E.N., Shvedchikov, A.P. and Shchukin, N.F., Molec. Biol. (Sov.) 15, 866 (1981).

39. Krupyanskii, Yu. F., Parak, F., Goldanskii, V.I., Mossbauer, R.L., Gaubmann, E.E., Engelmann, H., and Susdalev, I.P., Z. Naturforch 37C, 57 (1982).

40. Krupyanskii, Yu. F., Parak, F., Gaubmann, E.E., Wagner, F.M., Goldanskii, V.I., Mossbauer, R.L., Suzdalev, I.P., Litterst. F.J. and Vogel, H., J. de Phys. 41, C1-489 (1980).

41. Krupyanskii, Yu. F., Sharkevish, I.V., Khurgin, Yu. I., Szudalev, I.P. and Goldanskii, V.I., Molec. Biol. (Sov.) in press.

42. Krupyanskii, Yu. F., Kurinov, I.V., Sharkevish, I.V., Suzdalev, I.P. and Goldanskii, V.I., Proc. Intern. Conf. Appl. Mossb. Spectr. Alma-Ata (1983) Gordon and Breach, NY (1985).

43. Kurinov, I.V., Krupyanskii, Yu. F., Genkin, M.V., Davydov, R.M., Suzdalev, I.P. and Goldanskii, V.I., Biofizika, to be published.

44. Krupyanskii, Yu. F., Shaitan, K.V., Gaubman, E.E., Goldanskii, V.I., Rubin, A.B. and Suzdalev, I.P., Biofizika XXYI, 1037 (1981).

45. Zhuravleva, T.V., Krupyanskii, Yu. F., Uspenskaya, N. Ya., Kononenko, A.A., Suzdalev, I.P. Rubin, A.B. and Goldanskii, V.I., Biofizika, to be published.

46. Shaitan, K.V. and Rubin, A.B., Biofizika XXY, 796 (1980).

47. Kilyachkov, A.A., Shaitan, K.V., Chernavskii, D.S. and Rubin, A.B., Biofizika XXY, 981 (1980).

48. Khapp, E.W., Fisher, S. and Parak, F.J. of Phys. Chem. 86, 5042 (1982).

49. Krupyanskii, Yu. F., Shaitan, K.V. Goldanskii, V.I., Kurinov, I.V., Rubin, A.B., Suzdalev, I.P., to be published.

50. Reinish, L., Heidemeier, J., and Parak, F., Eur. Biophys. J. 12, 167 (1985).

51. Wise, W.W., Debrunner, P.G. and Wager, G.G. ILL-(EX)- 84.

52. Kurinov, I.V., Krupyanskii, Yu. F., Suzdalev, I.P. and Goldanskii, V.I., to be published.

53. Singh, T.P., Bode, W. and Huber, R., Acta Cryst. B36, 621 (1980).

54. Walter, J., Steigemann, W., Singh, T.P., Bartunik, H., Bode, W. and Huber, R., Acta Cryst. B38, 1462 (1982).

55. Mrevlishvili, G.M., Sov. Phys. - Uspekhi 22, 433 (1979).

56. Godovskii, Yu. K., Teplofizika polimerov (Thermal Physics of Polymers) Khimiya Moscow (1982).

57. Yang, P.H. and Rupley, J.A., Biochemistry 18, 2654 (1979).

58. Eisenstein, L., Frauenfelder, H., in Frontiers in Biological Energetics, Vol. 1, A.P.H.Y. p. 680 (1978).

59. Palmer, R.G., Stein, D.L., Abrahams, E. and Anderson, P.W., Phys. Rev. Lett. 53, 958 (1984).

60. Likhtenstein, G.I. and Kotelnikov, A.I., Molek, Biologiya, 17, 505 (1983).

61. Karplus, M. and McCammon, Ann. Rev. Biochem. 52, 263 (1983).

62. Vol'kenshtein, M.V., Biofisika, (Biophysics), Moscow, 1981.

63. Blumenfeld, L.A., Problemy biofiziki (Problems of biophysics), Moscow, 1974.

64. Phillips, W.A., J. Low Temp. Phys. 7, 351 (1972).

65. Anderson, P.W., Halperin, B.I. and Varma, C.M., Phil. Mag. 25, 1 (1972).

66. Amorphous Solids. Low Temperature Properties. Topics in Current Physics, Berlin, Springer, V. 24 (1981).

67. Mizutani, U. and Massalskii, T.B., Nature 259, 505 (1976).

68. Finegold, L. and Cude, J.L., Nature 238, 33 (1972).

69. Fanconi, B. and Finegold, L., Science 190, 458 (1975).

70. Verkin, B.I., Suharevskii, V. Ya., Telezhenko, Yu. V., Alapina, A.V. and Vorob'eva N. Yu., Fizika Nizk. Temp. (Sov. Low Temp. Phys.) (Russ.) 3, 252 (1977).

71. Verkin, B.I., Blagoi, Yu. P., Curevich, A.M. and Eropkin, V.I., Fizika Nizk. Temp. (Sov. Low Temp. Phys.) (Russ.) 10, 1225 (1984).

72. Singh, G.P., Schink, H.J., Lonsysen, N., Parak, P. and Hunklinger, S.Z., fur Phys. B55, 23 (1984).

73. Fleurov, V.N., Mikheikin, D.I. and Trachtenberg, L.I., Solid State Comm. 43, 142 (1985).

74. Cohen, M.H. and Grest, C.S., Phys. Rev. Lett. 45, 1271 (1980).

75. Cohen, M.H. and Grest, G.S., Phys. Rev. B 20, 1977 (1979).

76. Grest, G.S. and Cohen, M.H., Phys. Rev. B 21, 4113 (1980).

77. Cohen, M.H. and Grest, G.S., Solid State Comm. 39, 143 (1981).

78. Low, B.W. and Richards, F.M., J. Am. Chem. Soc. 76, 2511 (1954).

79. Singh, G.P., Parak, F., Hunklinger, S., Dransfeld, K., Phys. Rev. Lett. 47, 685 (1981).

80. Barlow, A.J., Lamb, J. and Matheson, A.J., Proc. Roy. Soc. A 292, 322 (1966).

81. Doolittle, A.K., J. Appl. Phys. 22, 1471 (1951).

82. Shaitan, K.V., Vestnik Mosk. Gos. Univers. (Phys. Astron.) (Russ.) 23, 15 (1982).

83. Goldanskii, V.I., Doklady AN SSSR (Russ.) 124, 1261 (1959).

84. Goldanskii, V.I. Doklady AN SSSR (Russ.) 127, 1037 (1959).

85. Goldanskii, V.I., Frank-Kamenetskii, M.D. and Berkalov, I.M., Science 182, 1344 (1973).

86. Goldanskii, V.I., Franck-Kamenetskii, M.D. and Barkalov, I.M., Doklady AN SSSR (Russ.) 211, 133 (1973).

87. Tunneling in Biological Systems, ed. B. Chance et.al. Academic Press (New York, San Francisco, London), 1979.

88. Goldanskii, V.I., Chem. Scripta. 13, 1 (1978-1979).

89. Goldanskii, V.I., Nature 279, 109 (1979).

90. De Vault, D., Chance, B., Biophys. J. 6, 825 (1966).

91. Kiryukhin, D.P., Barkalov, I.M. and Goldanskii, V.I., Doklady AN SSSR (Russ.) 238, 388 (1978).

92. Benderskii, V.A., Goldanskii, V.I. and Ovchinnikov, A.A., Chem. Phys. Let. 73, 492 (1980).

93. Klochikhin, V.L., Pshezhetskii, S. Ya. and Trakhtenberg, L.I., Doklady AN SSSR (Russ.) 239, 879 (1978).

94. Ovchinnikova, M. Ya., Chem. Phys. 36, 85 (1979).

95. Goldanskii, V.I., Trakhtenberg, L.I., Fleurov, V.N., Tunneling phenomena in chemical physics (in Russian) Nauka Moscow, 1986.

96. Goldanskii, V.I., Trakhtenberg, L.I., Fleurov, V.N., Sov. Sci. Rev., 1986 (to be published).

97. Trakhtenberg, L.I. and Fleurov, V.N., JETP 56, 1103 (1982).

98. Trakhtenberg, L.I. and Fleurov, V.N., JETP 58, 146 (1983).

99. Zhurkin, V.B., Molek. Biolog. (Russ.) 17, 622 (1983).

Acknowledgement

Figure 12 (on page 115) is reprinted from Yang, P.H. and Rupley, J.A., Biochemistry 18, 2654 (1979), with the permission of the American Chemical Society, Washington, D.C.

Dynamics and Function in Cytochrome c by Two-dimensional NMR

S. Englander and H. Roder

Department of Biochemistry & Biophysics

University of Pennsylvania

Philadelphia, PA

A. Wand

Institute for Cancer Research

Fox Chase Cancer Center

Philadelphia, PA

1. Introduction

We are working on an approach that promises to provide a great deal of information on protein structure-function relationships. Our immediate target is cytochrome c. We want to use 2D NMR methods to resolve the protons in cytochrome c, and then measure the behavior of those that exchange with solvent protons. We expect this information to provide a great deal of insight into the secrets of protein function.

There are two problems to start with in this approach. First one has to face an NMR assignment problem of a size that has never been accomplished before. By using recently developed 2D NMR methods, other workers have now found it possible to assign many protons in several proteins, but these proteins range up to a size only a little more than half that of cytochrome c. Nevertheless, we are well along with this problem. The second problem is that there is as yet no general agreement concerning what it is that hydrogen exchange can tell us about proteins. We are well along on this issue too.

We are interested in getting at a number of protein functional issues. For example, how does the cytochrome c protein set the redox potential of its heme? This function, though one of the simplest in protein chemistry, is not understood as yet. A similar statement can be made for most protein structure-function issues. Many of these questions can be profitably phrased in a sense that hydrogen exchange (HX) methods can deal with directly. For example, for cytochrome c the question becomes how does the protein interact differentially with the reduced and oxidized forms of the heme. If the protein interacts more favorably with the reduced form, as it does, this will, in the linked function sense, shift the redox equilibrium toward the reduced form. Experimentally then one wants to learn what parts of the protein interact differentially with the heme, and by how much - i.e. by how much in terms of free energy. If what we think about hydrogen exchange is correct, then it may be possible to read out structural free energy changes directly from HX data.

2. Two Dimensional NMR of Cytochrome c

As a result of earlier work using 1D NMR methods, a number of proton resonances in cytochrome c have been assigned, especially by Moore and Williams and by Keller & Wüthrich. Figure 1 shows a stacked plot of a 2D COSY NMR spectrum of horse ferrocytochrome c.

The spectrum can be examined more effectively in the contour plot representation shown in Figure 2. In a 2D NMR experiment, one imposes on the sample a sequence of rf pulses designed to elicit a signal that provides information about magnetic interactions between nuclei, either through-space NOE interactions or through-bond J-couplings. For example, the NOESY experiment in Figure 2 depicts nuclear Overhauser interactions between protons that are close together in space, in this case within about 3.2 A. Both axes in Figure 2 are in terms of chemical shift. On the diagonal is the familiar 1D NMR spectrum. The major informational content in the 2D spectrum is in the off-diagonal cross peaks, each of which has two different chemical shift coordinates, indicating the presence of a dipolar interaction between two protons at these two chemical shift values. The 2D aspect of this kind of spectroscopy helps to resolve the protons that are, in the 1D representation, hopelessly overlapped. Still the spectrum is tremendously complicated. Over 2000 cross peaks appear in the complete NOE spectrum of cytochrome c.

One wants to assign the resonances resolved in this way to the individual protons that they represent. In smaller proteins, the assignment procedure can

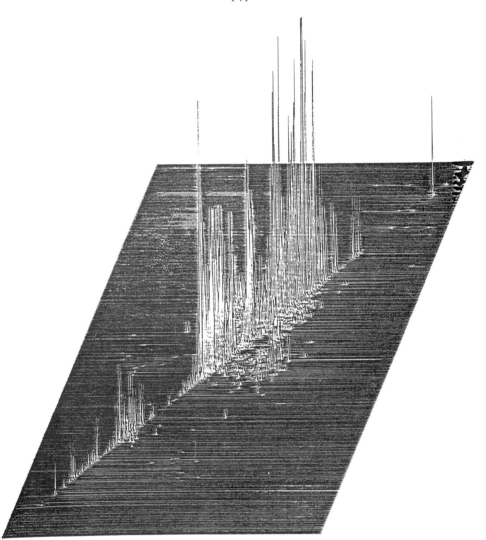

Figure 1. Stacked plot COSY spectrum of reduced horse heart cytochrome c in D₂0. This and other spectra shown below were taken at 500 MHz on a Bruker WM-500 spectrometer.

be carried forward in two steps, according to the sequential assignment method elaborated by Würthrich and colleagues. Starting with the 2D COSY spectrum, which displays cross peaks between protons that interact by through-bond J cou-

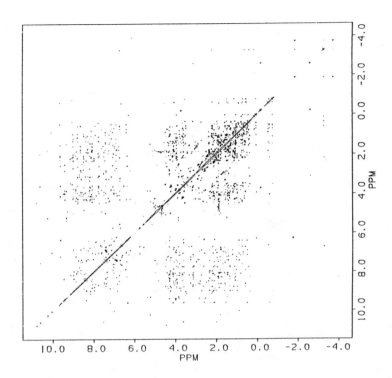

Figure 2. Contour plot of a ferrocytochrome c phase sensitive NOESY spectrum.

pling, one attempts to recognize patterns of cross peaks that represent particular spin systems, that is sets of J-coupled protons in particular amino acids. For example Figure 3 is a segment of a COSY spectrum, far upfield, containing a typical methionine pattern. This represents Met 80, which ligands the heme iron of cytochrome c. The lines drawn indicate how the cross peaks relate to the on-diagonal resonance positions of the J-coupled $C_\alpha H$, $C_\beta H$, and $C_\gamma H$ protons. Suppose you can now recognize in the COSY spectrum another amino acid spin system type, say an Ile, and then it turns out by examining the NOESY crosspeaks that the Met and Ile residues together at positions 80 and 81, these observations would tend to assign those residues. This general approach works well for small proteins.

Figure 4 shows what the COSY spectrum really looks like in the side chain region of cytochrome c, and begins to give some idea of the true complexity of the assignment problem. It is not possible to pick out amino acid spin systems

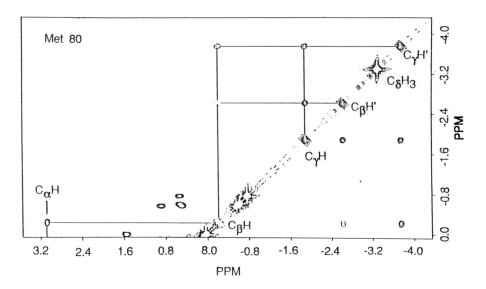

Figure 3. Section of an absolute value COSY spectrum showing the J-coupling network assigned to Met 80, one of the axial heme ligands.

with any degree of confidence. One of the steps in the attempt to recognize true spin system patterns in this very complicated side chain region was to compare a number of cytochrome c species. As an example, Figure 4 shows COSY spectra for horse and cow cytochrome c cut through the diagonal and placed together for comparison. The horse and cow proteins differ by three amino acids. Thr 47 in horse is replaced by Ser 47 in cow. Also Lys 60 and Thr 89 in horse do not exist in cow. One can look for just these cross peaks by focussing on differences in the otherwise nearly identical spectra. Arrows in the Figure indicate the gamma methyl to $C_\beta H$ cross peaks of Thr 47 and Thr 89 in horse and their absence in cow. The lines drawn identify, among other features, the spin system of Ser 47 in cow. Similar comparisons with cytochrome c from cow, dog, rabbit, chicken and tuna led to over a dozen spin systems in the horse protein.

Figure 5 shows a segment of the NOESY spectrum containing the main chain peptide NH protons. We want to look at the detailed H-exchange behavior of these. The lines drawn in Figure 5 identify NOE connectivities between peptide NH in the N-terminal helix. In a standard α helix, the peptide NH protons are within NOE distance of each other, and so produce a string of NOE-derived

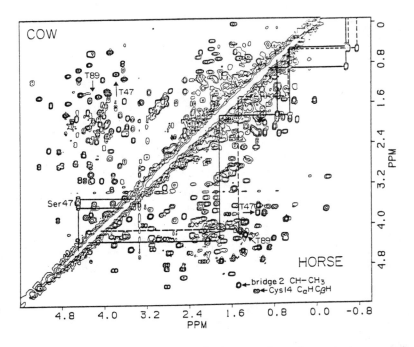

Figure 4. Absolute value COSY spectra for horse and cow ferrocytochrome c. Separate spectra for the different species were cut along the diagonal and placed together for comparison. Arrows indicate cross peaks of Thr 47 and Thr 89 present in horse but absent in cow. Spin systems of Leu 32 and Ile 57 in horse and Ser 47 in cow are also indicated.

cross peaks connecting each one to its neighbors. Note the Lys 5 NH resonance indicated by the filled circle on the diagonal at 8.09 ppm. Arrows lead to the cross peak connecting Lys 5 to Gly 6, Gly 6 to Lys 7, then to Lys 8, further to Ile 9, then Phe 10, and so on out to Ala 15. Of course in the real world it doesn't just come out that easily. But these assignments have in fact been very highly documented. In all these amino acids except for the lysines, the entire spin system has been worked out in the COSY spectrum. Also almost all the many NOE connectivities between the neighbor amino acids have been identified.

In assigning the cytochrome c protons, we have made use of COSY and NOESY spectra in H_2O and D_2O at different ambient temperatures, the relayed COSY spectrum which provides second neighbor J-couplings, other 1D NMR approaches, and multiple species comparisons. All this data constitutes a mas-

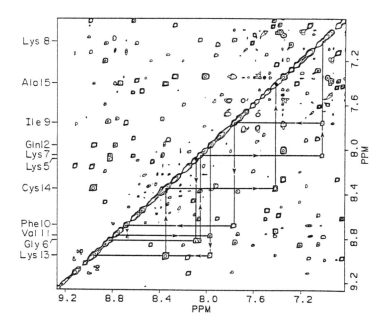

Figure 5. Segment of the phase sensitive NOESY spectrum in the amide region. NOE connectivities between sequential peptide NH in the N-terminal helix are shown.

sive puzzle with several interlocking networks (COSY, relay COSY, NOESY) of complicated connectivities. One starts with what is known and moves outward into the unknown, testing tentative possible assignments until a whole multiply connected cluster of assignments shows full consistency with all the data. In cytochrome c these clusters generally have turned out to be much larger than the single residue spin systems that serve as the basic unit for stepwise sequential assignment of smaller proteins. Joshua Wand in his dissertation research has managed to assign the proton resonances in 60 of the 104 residues in cytochrome c, and we look forward to soon having just about all of them.

3. Hydrogen Exchange

Now that many protons all through a protein can be individually monitored, these can serve as probe points to study structure, dynamics, structure change, and the interaction of these with function. Among other approaches, we plan to do this by using the special capabilities of hydrogen exchange (HX). Unfortunately, in the HX field there is a lot of disagreement about how HX data relates to these structural parameters.

The reader is aware that peptide NH and protons of the polar side chains exchange with water protons, and that their exchange rates are spread over an extraordinary range, about 10 orders of magnitude wide. What determines these rates? There are two general models for the mechanism of hydrogen exchange in proteins. Accessibility-penetration models take the common sense point of view that protons on the surface exchange fast, while those more and more buried are slower and slower. The effective HX catalysts, hydroxide and hydronium ions and water, find their way into the protein by way of the small, rapid fluctuations that one hears a lot about these days, reach the buried protons, and carry them back out to solvent.

A different model originated with Linderstrom-Lang back in the 1950s. Lang supposed that slow hydrogens are slow not because they are out of contact with water, but because they are H-bonded to other protein groups. Initially that preconception had something to do with the fact that Lang invented and used HX methods specifically to look for the H-bonds postulated in the Pauling-Corey α helix, but a lot of subsequent information supports his insight. Linderstrom-Lang further supposed that the exchange process involves a dynamic breaking and remaking of H-bonds, and that only during the small fraction of time when an H-bond is broken can the hydrogen be exchanged. From this point of view, a lot of protein HX data now available shows that in most cases H-exchange rates in proteins must be in proportion to the fraction of time H-bonds are reversibly broken, i.e. essentially to the equilibrium constant for the H-bond breakage, multiplied by the normal exchange rate of the fully exposed proton. We have accurately calibrated the latter. This means that if you can measure the exchange rate of a particular hydrogen, you can compute the equilibrium constant for the particular opening that frees it.

An added wrinkle to this model was proposed by us some time ago on the basis of quite a lot of data which suggests that H-bonds tend not to break individually. Rather we expect that H-bonds in regular structures, like an α helix, tend to break as a result of some cooperative structural transition, which we call local unfolding. For example to break any H-bonds in the middle of a helix, it may be necessary to pull out a whole turn of the helix so that three or four or more H-bondings break together.

If that is true, then an unprecedented capability falls into the protein chemist's hands. K_{eq} for an unfolding equilibrium must depend upon stuctural free energy in the usual way, i.e $\Delta G_o = $ -RT ln K_{eq}. Since we can read out the unfolding K_{eq} from HX rate data, this yields the free energy of structural stabilization against the operative local unfolding reaction. Suppose we now produce a local change in the protein, for example by changing cytochrome c from reduced to oxidized, and some segment then interacts differently with the heme. Then the stability of this segment must change, its equilibrium constant for local unfolding must change accordingly, and the HX rate of the protons on the segment must change. If we now compare in reduced and oxidized forms the HX rates of many individual hydrogens all over cytochrome c, then by noting the HX changes we should be able to tell what parts of the protein are involved in setting the redox potential, and how much each involved segment contributes to the redox equilibrium in terms of thermodynamic free energy.

Well, that is a lot of ifs. Does protein hydrogen exchange really work that way? We have done the HX experiment just alluded to. Reduced and oxidized samples of cytochrome c in H_2O were placed into D_2O so that H starts to be replaced by D. We took samples in time (20 min, 1 hr, 3 hr, 9 hr, and so on), put them aside under conditions that essentially stop the HX reaction, and later read out the degree of change of H for D at each point in the protein. Hydrogen exchange has this nice archeology-like characteristic. Any interaction can be carried out under any chosen conditions while H-D exchange is allowed to proceed. Changes in structure or stability that occur during the interaction become imprinted in the HX pattern, and this can be read out later. It is not at all necessary to face the difficult, sometimes impossible, problem of making NMR measurements during the reaction or under the reaction conditions.

Figure 6 shows a number of COSY spectra, in the peptide NH-$C_\alpha H$ cross

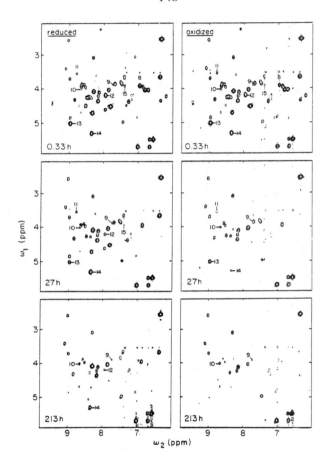

*Figure 6. Sections of COSY spectra in the NH-C$_\alpha$H region taken for samples of cytochrome c after exchange against D$_2$0 for various times pH*20°C in the reduced and oxidized forms. Cross peaks representing residues in the N- terminal helix are indicated. Before recording the spectra, all samples were changed to the reduced state.*

peak region, for samples taken from the exchange experiment at various times. At the first time point, after 20 minutes of exchange, about 52 amide NH, just under the total number of H-bonded amide NH in cytochrome c, are still unexchanged. Of these we know the assignment of 40. Cross peaks representing amide NH in the N-terminal helix are indicated. The time course of exchange of each, read from the decrease in cross peak volume, is shown in Figure 7 for both reduced and oxidized forms.

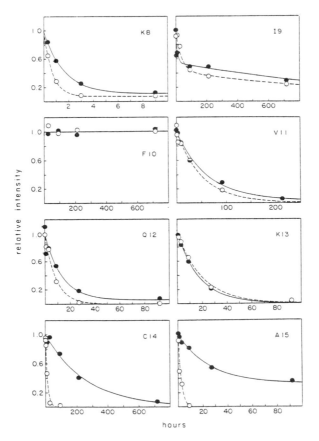

Figure 7. Exchange kinetics of peptide NH in the N- terminal helix in both the reduced (filled circles) and oxidized (open circles) protein.

These results are summarized in Figure 8 which plots the slowing factor relative to the characteristic free peptide rate for each peptide NH, represented by its residue number (5 through 15). For example, protein structure slows the exchange of the Lys 8 NH by a factor of about 10^5.

Consider first the exchange behavior in the oxidized protein, indicated by the open data points in Figure 8. Residues 5 and 6 have fully exchanged before the first 20 minute time point - not surprising since they are not H-bonded. Residue

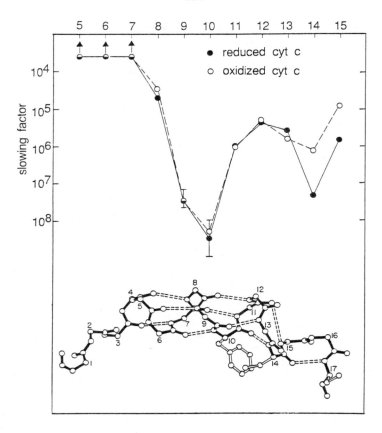

Figure 8. Exchange rate pattern through the N-terminal helix for reduced (filled circles) and oxidized (open circles) cytochrome c. The slowing factor relative to the free NH rate is plotted for each residue, identified by its residue number. The diagram of the N-terminal helix is viewed from within the protein, the mass of which lies in front of and below the helix. NH to CO H-bonds are indicated by dashed lines. The N-terminus is to the left, the Phe 10 ring shown is buried in the hydrophobic core, and the Cys 14 and Cys 17 side chains form covalent bridges to the heme (not shown).

7, the first H-bonded NH in the helix, is also relatively fast exchanging. The NH of residue 8 is slower, residue 9 is slower still by a large factor, and residue 10 is still slower. A similar though less steep gradient, moving in from the helix end, has been seen in other proteins. We believe this represents simple dynamic fraying, as dictated by the statistical mechanical theory of helix-coil transitions. In these terms the data suggest that Lys 7 is reversibly separated from the helix perhaps 10^{-3} of the time to produce an exchange rate reduced from the free peptide rate by about 1000-fold. During the small fraction of time when Lys 7 is swung out, Lys 8 breaks away about 10^{-2} of the time (overall unfolding $K_{eq}10^{-5}$. During the small fraction of time when both these are out, Ile 9 reversibly swings out with a K_{eq} of 10^{-3}. Phe 10 is in the H-bond-broken state about 1/10 of the time that the prior three residues are open. This is the essence of cooperativity.

Another aspect of cooperative behavior is exhibited by residues at the other end of the helix. The fraying behavior presumably continues inward, and would allow the next residue, Val 11, to exchange somewhat more slowly than Phe 10. However another faster pathway is available involving a different opening reaction. Four sequential peptide NH, from residue 11 to 14, all exchange about 1 million-fold slower than the free peptide rate, but they are all within a factor of 6 of each other (within a factor of 15 if Ala 15 is included). It is a little hard to think that these neighboring residues just happen to adopt these same rates by sheer coincidence. The fact that a similar flat exchange pattern has also been seen in other helices further speaks for some underlying mechanistic explanation. We strongly suspect that this exchange pattern reflects a single conformational transition, which breaks all these H bonds in some cooperative unfolding and exposes the amides to solvent more or less equally, so that they all exchange at about the same rate. It is possible that residue 15 should be included in this group. The rates observed for these NH are not precisely identical; there is a little bit of "fine structure" in the overall pattern. We suspect this means that in the open state not all the protons are equally exposed. It may be that here we begin to see something of the structure of the open state.

A diagram of this helix, focussing mostly on the main chain atoms and H-bonds, is included in Figure 8. The length of the helix lies along the protein surface. In this view the body of the protein lies in front of and below the helix. The N-terminus is to the left; the right end of the helix is covalently joined to the heme (not shown) through the other bridges at Cys 14 and Cys 17. Among the peptide NH just discussed, the NH of Gln 12 is is, no other atom is between it and solvent water. The NH of residues 11 and 13 are more occluded, and the Cys 14 NH is fully inside, between the helix and the protein body, remote from solvent water. Nevertheless all these protons exchange at essentially the same rate, 1 million-fold slower than free peptide NH. It is hard to reconcile this behavior with a mechanism that connects exchange rate with solvent accessibility. Rather the result suggests that all these protons exchange from some kind of transiently distorted state that renders all of them more or less equally accessible to exchange. This is just what the local unfolding model leads one to expect. The exchange pattern could be produced by an unfolding reaction that allows the segment of helix between residues 11 and 14 to collapse, breaking the connecting H-bonds and also the Ala 15 H-bond. The data interpreted in this way suggest that this unfolding occurs with an equilibrium constant about 10^{-6}, so that the protons are exposed to solvent about 10^{-6} of the time and experience exchange rates slower than fully exposed models by just that factor. (This would correspond to an opening free energy of 8 kcal.)

We submit that these observations, the apparent fraying at the N-terminus and the equivalent unmasking of structurally disparate residues at the other helix end, provide firm evidence for a cooperative local unfolding mechanism. It is especially significant that this same kind of behavior has been seen in other cases, the flat HX rates for sequential helical amides in other proteins and the fraying behavior at the ends of helices in both proteins and nucleic acids.

4. Function Related Changes

Figure 8 provides a comparison between HX rates in the reduced and oxidized forms. At the helix N-terminus, the protons in reduced cytochrome c display the same rates and the same fraying behavior as in the oxidized form. This part of the helix is far away from the heme. At the heme-associated residues, differences appear; on reduction residues 14 and 15 are slowed by about 15-fold. In the reduced form residues 11, 12, 13 and 15 now exchange at very similar rates (within a range of 4) while 14 is somewhat slower. It is interesting to notice that an unfolding of the segment of chain between residues 11 and 13 could break these H-bonds and also the H-bond from Ala 15 while leaving the Cys 14 NH still bonded.

Similar data for protons all through the cytochrome c molecule are being obtained. A sizeable number of protons display sensitivity to the redox state, while many others do not. In most cases the NH that change are slower in the reduced protein.

5. Looking Ahead

It seems to us that the HX results described here and other results in the literature rather convincingly document a role for local unfolding mechanisms in the determination of protein H- exchange rates. One wants to know next what opening reactions look like and what factors determine their parameters. Such information can provide important insights concerning protein structure, dynamics, and stabilization. There is also here the promise for discerning the bases of protein function. For cytochrome c one can expect the total picture derived from the beginnings just described to display the parts of the protein that help to set its redox potential and perhaps to reveal even the quantitative free energy that each involved segment contributes to this function. The very same approaches should be directly applicable to other functions and other proteins.

This work was supported by NIH grants AM 11295 and GM 31847.

Panel Discussion: How Rigid are Proteins and Do We Care?

Peter Debrunner Session Leader

Question: I guess the title of this is "Protein flexibility, so what?" Actually that may be a meaningful way of putting it, because I think you can certainly talk to some groups for whom the concept that proteins are flexible things is a revelation and you can also talk to other groups, for example, many physical chemists who say, well, we knew these things were at room temperature and therefore they had to move and so why are you telling us anything. One of the questions in this conference is the role of flexibility in chemical reactivity. Presumably some problems in chemical reactivity have to do with very fast processes, but the processes that are presumably of physiological significance are ones which are relatively slow on the time scale of molecular dynamics. In any event, things on the microsecond to millisecond time scale, and for these processes, it has certainly become clear that flexibility of the protein is important, if the protein was an absolutely rigid structure, for example, oxygen would never get in, that was shown by the case in Karplus' studies, however, it's not so clear yet whether the dynamics of this flexibility is important or not because most of the studies we've seen have been of processes much, much faster than these microsecond/millisecond events. Still, these rapid processes are important to study because they certainly modulate the time dependence on the longer time scales and they also tell us what the actually equilibrium state is. Many of the things that we're interested in are just what does the protein look like, how is information stored in it. So I think that in many cases, the studies that we're doing dynamically are just trying to act as a probe of equilibrium structure and we're doing dynamics because we don't know how to do anything better. We don't have x-ray microscopes.

There are certainly many questions then that arise as to what the correct simple model of a protein; I think we need a simple model because if we looked at a millisec of film, we're going to be bored during a certain fraction of the time. A millisecond of film, I guess, would have taken us about ten years to watch. So, I think we need some models. I think the emerging models have to do with the role of multistates in these systems. Can we just think of a protein as something that has one single minimum and we perturb it and we slightly perturb away from that minimum, that's sort of an elastic model which is a very old model of proteins, or do we really need to know every individual potential well. I think we're starting to see that there are lots of potential wells there. I think a few years ago, that was less clear. On the other hand, - well, we're going to have to find out a lot more about how the different motions between these different wells are related to each other. And, I think now that we've established that the motions are there, or all the experiments and simulations have established that they are there, that will be our next problem.

Anyone want to argue with that?

Warshel: Many people knew that proteins are flexible, at least chemical physicists, and I think the important question from the biology point of view is how much it costs to move along the reaction coordinate? Not what is the dynamics. The importance is terms of order of magnitude. How much it costs in terms of an allosteric energy to push the heme to the plane.

Answer.: I think that's true, I think that's the basic point that most of the things we're interested in are really structural questions and questions of free energy. I think the multiple minima do present a possible problem for using simulations to find these free energies since you have to sample all of those states. Actually, the results showing that the protein actually looks like a liquid and is above the glass transition temperature, I think is rather heartening in that respect. I think something we haven't talked about too much here is that there have been some very recent experiments on processes such as recognition and particularly immunological recognition of proteins by Klug and Phillips which indicate that things are flexible on the surface tend to have more immunological effect. That's somewhat controversial, but I think still it shows that whether the dynamics is important or not, I think the question of how easy it is to push these things around and find appropriate minima for some other group is an important one.

Watt Webb: I really do come from outside this field so I'm a bit of a ringer here in that my research has to do with supra- molecular structure, cellular biophysics and these issues such as the conditions under which antibodies can contort themselves to cross-link receptors under certain circumstances on cell surfaces are very important. In our business, the movement of proteins is a key process and we recognize that in the course of their motion, they must be moving within themselves. I've come here to learn about a particular kind of fluctuation in proteins which I see is not on the program, so I think I'll take the chance to advertise the problem.

Transmembrane channels, acetylcholine receptor, sodium channels, potassium channels, all of these channels change their conformation every time they do their thing and some of these channel molecules have many "equilibrium conformational" states and they shift from one to another. Many of those states are not yet known. We would like to be able to understand how the binding of the ligand, say acetylcholine, to the receptor, can under some circumstances open the channel and under other circumstances put it in a state where it's very reluctant to open. We think that we have one, and now this is the advertisement, we begin to think we have one direct experimental probe of fluctuations in some of these channels and that probe is simply measurement of the open channel, electrical conductivity fluctuations. These channels are ion channels and the conductances are the order of picosiemens, pico amperes per volt.

One can measure these small currents in single channels and we're beginning to be able to measure them with bandwidths that are wide enough to see the conductance fluctuations. So you can imagine this piece of sewer pipe that fluctuates and allows a larger or smaller number of ions to pass through the channel while its open. And, this kind of experiment is just beginning. Fred Siegworth at Yale is the leader in the field at the moment and many other people are beginning to look at it. But clearly, people from outside the field of protein fluctuations care what's learned about protein fluctuations, because they do matter to us when we deal with protein movements.

Burt Bronk: It would be interesting to look from a biological point of view at numerous homologous proteins and try and come up with an average potential which presumably would be smoother than the potential we see, and then we might have suggestions about groups that we can eliminate and still have the enzyme work.

Answer: I think that is a very good point. One is not really doing biology unless you have evolution in your problem, otherwise you are just doing physics or chemistry and I think that a lot has been learned about sequences and how they evolve, a lot has been learned about three dimensional structures and how they evolve, but I think what we need to know is whether dynamics and flexibility play any role in evolution. And, unless we look at some sort set of similar things, that's going to be difficult.

Fortunately, a big future multi-million dollar industry is site directed mutagenesis which would give a lot more information about these kinds of things.

Bill Eaton: Since there's been so much discussion today about time scales for conformational changes, I thought it would be useful just to show you in two or three slides what's actually known about the system that's been discussed today. There's an enormous effort in twelve or fifteen laboratories around the world doing pulse and probe experiments to see what the structural response is experimentally, to the photodissociation of hemoglobin. Martin Karplus mentioned earlier that there's really very little detail that one can get from the experiment and he's quite right. What people have been able to do is look at time resolved optical absorption and time resolved resonance Raman, both of them are subject to a large number of different types of interpretation. The experiment then which you've heard over and over again is to break the carbon monoxide bond with a short light pulse, and light pulse from 240 femtoseconds to 10 nanoseconds have been used, and then to look at the spectrum of the heme group, either the optical absorption spectrum or the resonance Raman spectrum. And, we've actually started to do a new type of experiment which is to look at the spatial localization in addition to the time scale of the conformational changes by using a hybrid molecule prepared by Ikada Iseto and Yonetani here at Penn which has cobalt on two of the four hemoglobin subunits. Cobalt doesn't bind carbon monoxide so that one can ask a very interesting question. We break the iron carbon monoxide bond and then look at the spectrum of the cobalt heme which is perturbed when the structure changes and ask, when does the cobalt subunit know that its neighbor has been photodissociated.

The next slide shows the type of experimental data that we get. This is work of Eric Henry and Jim Hoffrichter where we've developed a very sensitive nanosecond spectrometer. The time scale is running from a few nanoseconds out to 100 milliseconds. This is just part of a data set. So we collect this spectral surface and in this spectral surface in the amplitude of the large spectral change is the information on ligand rebinding and once when one normalizes out that effect, one can look at the conformational changes. The next slide just shows that if one uses this hybrid molecule then there are three optical absorption bands, the one on the right is the deoxy iron, the one in the middle is the CO peak, and the one on the left is the cobalt which is only slightly perturbed when you go from deoxy to CO in the iron, and in the lower panel you see there's this minimum which is the probe of the transmission of the conformational change from one subunit to another. The next slide just shows the analysis of one of these experiments. The top panel shows the ligand rebinding curve; in this case, we're photodissociating the iron subunit and looking at the subunit to look for the conformational change. The solid curves are if this hybrid molecule that has two cobalts in the beta subunit and irons in the alpha subunit and the dotted curve is to reference the all iron molecule. And what you can see is that there are five relaxations in the ligand rebinding curve, which I won't go into in detail, but the first one is the most interesting one and that's the geminate recombination. That's rebinding of ligands which have not yet gone out into the solvent and rebind to the hemes from which they are photodissociated. If you look to the right side of the spectral changes, these are the spectra changes of the iron and there are several interesting features to this. One is that in all the relaxations, that the 50 nanosecond relaxation, which is relaxation 1, and the 1 microsec relaxation, which is relaxation 2, that these spectral changes all have a very similar shape. It's as though the molecule is doing the same kind of thing in each one of these relaxations but that there's no response of the cobalt to the first relaxation. So that says that whatever the conformational change is representing in the alpha subunit, the beta subunit has not felt it yet. We have independent kinetic evidence that nothing is transmitted in that first relaxation as well.

The third relaxation is the famous quaternary change at 20 microsec which John Hopfield had an awful lot to do with in the early '70's, and if you like, that's the most functionally significant conformation change known in proteins because its THAT conformational change which changes the reactivity of the molecule which controls ligand affinity and is essentially responsible for the cooperative behavior of hemoglobin. And you can see that the cobalt responds very clearly to that conformational change as it is expected from the x-ray structure.

What was a little bit surprising is that there's a very similar spectral change in the cobalt that's in the second relaxation which occurs at about a microsecond. So that essentially answers the question that we were asking, that it takes about a microsecond for the conformational change that's initiated at the heme to be transmitted to the neighboring subunit. It's however a much smaller change if one assumes that there's some kind of linear relation between the amplitude of the spectral change and the magnitude of the conformational change. So what we've learned from this kind of an experiment is that the first kind of conformational change is localized, the conformational change of relaxation 1; the conformational change relaxation 2 is delocalized but the molecule hasn't yet changed its packing between the subunits and then we see the clear transmission of the effect of photodissociation in the quaternary conformational change which is relaxation 3. Now there's no comparable data on the subnanosecond time scale and that's in fact the next stage in this investigation and Robin Hochstrasser at Penn, in collaboration with Tom Spiro, is looking at resonance Raman spectra on the subnanosecond time scale. We hope to also convince Robin to do the transient absorption to see what the magnitude of the spectral changes that are occurring on the subnanosecond time scale.

So, we're sort of filling in the gaps; we have the experiments which tells us that the deoxy heme is forming in 350 femtoseconds, we look at the ligand rebinding all the way out to close to 100 millisec, so we're looking at nine order of decades in time and we're looking at both the ligand rebinding and the conformational relaxation. The fundamental problem which is not answered in these experiments is what is the relationship between these conformational relaxations that take place prior to the quaternary change and the ligand rebinding curve. And I think that that's still a major issue in the low temperature myoglobin work and is an issue in the hemoglobin work as well, and I think John Hopfield will have something to say about possible relationships between the conformational relaxations and the ligand rebinding in his talk tomorrow morning.

B. Honig: It's hard not to notice that there were two different types of talks given today. One breakdown might be physicists and chemists but I could perhaps think of others. Some of the talks have focussed on the exchange of a particular proton or ligand bind and others have focussed on glass transitions and fractals, and I'm wondering if anyone sees the two types of talks ever getting together and having one make a contribution to the other?

Watt Webb: I'm glad you asked that question. One of the objectives of the panel, I suppose, is to look at one of the reasons for having this kind of a meeting, namely, can the physicists and the biologists and the chemists get together, so after listening today there is some suggestions as to whether common areas exist and where these generalized models and specifics might come together, however, if you'll allow me, let me hear your difference in philosophy which becomes very clear when a physicist begins to get into this field. There was a story about the sinking ship and the captain was trying to separate those who could swim from those who couldn't swim and with respect to priorities on loading the lifeboat. When the physicist came to the table to check out, he was asked that question, yes or no, he said, "well I know the theory of it". That's one point and I think it's also true that one realizes that physicists in many cases are interested in general models and the generic aspects of things and biologists are really interested in differentiation and we see that comparison between the two fields. But, having thought about that, it is possible to make a list, at least from some of the things today that come from a background, for example, my background of condensed matter physics and the issues that had to do with the interpretation of structure and dynamics in biomolecular systems. First of all, as Peter said, I think that there's developing evidence that we have the possibility of not only simple fluctuating modes but conformations fluctuating back and forth and without getting into that in detail, there are some general things that the physicist can say about that. One of them is that the Debye-Waller factor, or the logarithm of the Debye-Waller factor, is no longer simply proportional to mean $\langle X^2 \rangle$. It's much more complicated than that. In the higher temperature limit, ultimately it is that, but in general it's not. And furthermore, another aspect of this, if one looks into the work that's been done in structural phase transitions. There are experimental examples and theoretical examples where the Debye-Waller factor does not even increase with increasing temperature, it goes down. So, the question is to find a specific example in the case of biomolecules to which that might apply or might not apply. I don't think there are any general rules. It may apply, it may not apply; it will depend very much on the system and the variations in biological systems are so vast compared to what we look at in conventional solid state. I wouldn't be surprising to find both.

Well, going on now, let me speculate about a couple of other things. One area of interest in condensed matter physics today is that of the dynamics of nonlinear systems and people have become very aware of the fact that the mechanics that you see in Goldstein's book just isn't that simple. All you have to do is to introduce the simplest amount of nonlinearity between two coupled oscillators and depending on the parameters, you can have both chaotic behavior and systemic behavior. There has been some simulation work done at Texas simply on a driven double well oscillator and depending on the driving conditions you can get oscillation more or less in one well of systems, of beating back and forth, but if the amplitude of the driver gets large enough, then you get a very, very chaotic behavior. What goes intermittently from one well to another, back and forth, and the interesting thing of it of course is that the oscillation frequency when it flops into one well, the oscillation frequency in the other well (if you were to put an oscillator in well) would be different and affected by what's already done, and I think we've heard some suggestion of that today in biomolecules that these parameters may actually depend on the process through which it's going through. So, this is reaching for that kind of a correlation between the two fields.

Then there's the issue of spatial correlation. These nonlinear systems, if they are nice and homogeneous and extended can propagate something called solitons. But, that doesn't mean that these don't exist, or that there can't be a cooperative local unfolding in the nature that we've heard about from Englander. What stabilizes this? Well, what stabilizes it is that there may be a nonlinearity that very much wants to have a few of the molecules do the same thing at the same time and Karplus' simulations also show that this is not something which happens on just one residue at a time, but that there is a spatial correlation. Here also is a question then, are the correlations of intermittency in these random systems; or are there correlations actually having to do with a cooperative link in unfolding. Now, these are questions where the physicist comes in and has some general idea, and perhaps has seen these phenomena in general models independent of the details and that comes back to universality in the some of the problems the condensed matter physicists look at. It doesn't make any difference exactly what the short range interactions are, this concept of universality and behavior is an important concept; it happens. And, so I think that what this is saying is that there's a tremendous opportunity now in the interface between these two fields because the amount of detail that's being developed dynamically and in proteins is not sufficient so that the general theories can be questioned.

Let me just conclude on just one piece of long ancient history at this point, the question of whether the density of modes in tunnelling gives a linear specific heat: Many years ago it was discovered that the specific heat of graphite goes as T^2. But after all, crystalline graphite is a three dimensional material. Why did it go as T^2? Well, everybody said, it's just two dimensional in the layer planes and that's all that counts. The fact of the matter is that if we take the argument though, sooner or later, the three dimensionality comes into play and to fix the parameters quantitatively. The interaction between layer planes is a van der Waals interaction. The interaction with the layer of plane is strong covalent interaction. I think that if you work out the theory, you'll find that there's a cross over between a two dimensional specific heat and a three dimensional specific heat and that cross-over occurs experimentally at about 5K. So that what this says is that simple van der Waals interactions can make something three dimensional all right, but it is two dimensional over a much more extended range of temperature. So, a question then: Is the linear heat capacity over an extended range due just to tunneling modes, or is it a mixture of both? I think this now gets back to the question of getting particular about particular molecules and there's no question in my mind that there's a lot of tunneling, on the other hand, in these linear molecules depending on the strength of the cross interactions. Getting back to fractals then, the strength of the cross interactions will essentially determine the cross over between the molecular vibrational linear specific heat and something that is really tunneling.

Well those are some of the ideas that come to mind after listening to both sides of the street here.

Tom Spiro: I wonder if I could just return to Barry Honig's question for a moment. It does seem to me, it's very important in this question of the joining of substates in spin glasses with photodissociation experiments to remember that proteins generally have a function, at least the interesting ones do. This is the point of course of Hans Frauenfelder's euphonious term "fins", I mean, that there are motions which are important for the function of proteins and there are other motions which are presumably not. Proteins are subject to both kinds of motions, but those of us studying hemoglobin photodissociation are primarily interested in the fins, the actual coordinate which begins to shop up in the kinds of curves that Bill Eaton was showing, resulting from his absorption transient work. I'd like to put in a plug here for resonance Raman spectroscopy in the absence of Joel Friedman whose plane was missing, that one can with the Raman spectrum, a technically more difficult experiment, really start to see specific structural changes rather than the more diffuse kinds of alterations observed in the optical spectrum and in fact, it is possible to distinguish two different structural processes in Bill Eaton's states I and II, or stages I and II in that process because in the early time Raman spectrum which we've associated with an expanded porphyrin core, which we have then interpreted it in terms of a restricted motion of the iron out of the plane. One can discuss whether that's an appropriate interpretation. But in any case, that's a different relaxation from the one seen later on , primarily in Joel Friedman's work, where the 1 microsec relaxation stage II where it involves motion, or change at least, in the frequency of the iron imidazole bond itself as well as different frequency of the porphyrin ring, so called oxidation state marker. So that one sees in fact structurally, two distinct processes going on in that case.

I do have a question of Bill Eaton actually, connected with the molecular dynamic simulation of this process that Eric Henry discussed and this has to do with the question of the iron out of plane displacement which we are asserting is the likely limit on the porphyrin frequencies at the early times. One starts out with photodissociated heme having some energy stored in the nonbonded interactions and we guesstimated that was worth 3-20 kilocalories depending upon the potential function chosen for the heme and then this is somehow sent into the protein. I mean, this must be the primary force heading the protein on its functionally important motions. Now, it would appear from the calculations that, although there's a bit of uncertainty as to how far the iron goes out, all of that's been dissipated within the first few picoseconds, and although there are some indications of it showing up elsewhere in the protein, how does one think about the necessary coupling of the structural change of the heme with the rest of the protein?

Bill Eaton: There's two parts to the answer: One is that the only energetic restraint in the calculation is to ensure that the energy that's put into the molecule and the potential function is switched, is not in excess of the photon energy minus the bond energy. So that's something that one has to worry about; that we're not just artificially putting a huge amount of energy into the system. So, that was controlled by putting in this very weak interaction potential with the carbon monoxide. If you put a steep potential surface in or one that's too steep and change the potential function of the carbons in one position you can easily put a couple of hundred kilocalories into the molecule.

As far as the coupling, what we can say is that we don't see any change in the half time of the relaxation. The tail end of the heme conformational transition, I don't think, is really well enough documented to know how much the tail is affected and also, as Eric pointed out, it's not really certain that the iron has gone completely out of the plane. That adds up to possibly one to two kilocalories which of the order of the sort of potential coupling energy between the heme and the protein. We don't see any of it though in this particular simulation, which is a little bit curious, which could mean that what we're basically looking at is an entropic effect, that the iron is rigidly connected to the interface between the subunits and when the iron moves, as Perutz envisioned, then the subunits would have to change but the iron moves, it finds itself in a metastable state, a lot of atoms have to rearrange, so there's basically an entropic barrier then to the structural change. But.

Question: Its the directionality that really matter, isn't it? If it's really entropics then it's all over the place, how you ever get it back to some particular reaction channel which leads to the final change in the quaternary structure. That seems to me to be the problem.

Question: I wanted to say something about Barry Honig's question also. We're being, or I'm being, a little disingenuous to simply say that the idea in the glass model is that you have many, many states and that seems to be rather clearly established, but the glass model tries to say something I think more than that; namely, that there are motions on a wide variety of time scales and I think Hans Frauenfelder will probably tell you about some experiments about that. Now the question is, are those motions fins, functionally important motions, or are they what Hans calls, bums (biologically unimportant motions). I think if you're looking at covalent bonding, almost by definition, changing the covalent bonding will be things that you can pin point in time more precisely than these conformational changes, but I think the lesson that there are processes that are uniformly interesting on lots of time scales, is something that may apply in other biological context. As for example, the folding problem where we saw in Englander's data that there were many things happening on very many time scales in the system and they were happening in a correlated way. And, certainly the folding problem is one that we would like to have answer to, although it's not one that we're talking much about at this conference.

Webb: Honig's question has really stimulated us all I think. He reminds me of many years past. I have the feeling from what I've heard today that this field is in a situation that's parallel with the physics of metals in the 1950's. Quantum mechanics could be solved and one could work out band structures but they weren't really good enough to get you there. And what happened in the next ten or twenty years after that was that some new ideas, some cooperative processes, got built into the theory and then, by golly, you could do things. You can understand superconductivity, you can understand cooperative phase transitions, and that sort of thing. I think I hear the new ideas here, but I have the feeling that you all who are experts in this field have yet to find out which of those new ideas are the ones that are going to win and it sounds like a magnificent opportunity and one hell of a lot of fun. But when something like that's going on, the experimentalists have an extra obligation too. I'm particularly concerned about things I've heard today about atomic displacements deduced from x-ray crystallography data, having done in the 1950's that sort of thing myself. It seems to me that there is no such form of theory. When you are in a field like this where there is (gee! I'm giving a sermon) – where you're not quite sure how the theory is going to go, it seems to me that it's extremely important that the analysis of data like the mean square displacement data from x-ray diffraction experiments be presented in detail with all of the detail of what assumptions are going into the analysis. And all of the detail of this statistical analysis that gives you the uncertainties have to be out there, so that when the new theory that's really going to work comes out two years later, you can see what's going on. End of Sermon.

Low Temperature Biophysics and Room Temperature Biology

J.J. Hopfield

California Institute of Technology, Pasadena, CA 91125

and

AT & T Bell Telephone Laboratories, Murray Hill, NJ 07194

Preamble

I was challenged by the organizers to open this session with a provocative address. This paper is a reduction to written English of that talk, rather than a conventional review of work in progress. I may step on a few toes – my own, unfortunately, included – but this is the cost of being frank.

Introduction

What is the purpose of the enterprise which we at the conference are pursuing? I can't answer for others, but my motivation is chiefly to learn how biological molecules *function*. Thus I want to be able to explain how the biomolecular structure (including dynamic structure) gives rise to the important properties used by biology. This explanation should be at a level of sophistication and accuracy satisfactory to a physical scientist.

I want to emphasize *that which is important to biology*. There is an arbitrarily large amount of static and dynamic structure in biomolecules which is of no interest whatsoever to biology. Of the total possible physical studies, I want to select at the beginning aspects which have a reasonable chance of being of serious interest in biology. In order to do this, one must begin by at least *thinking* about what the biological question or problem is for a particular molecule or assembly. Better, one might actually *start* from a biological problem.

Each experimentalist and theorist alike has a limited repertoire of techniques and viewpoints at his or her personal command. When told about some new molecule or situation, each of us tends to go to his collection of techniques and pull out a – say trick# 17 - and apply it. The situation has more than passing resemblance to the old story of the drunk looking for the car keys under the streetlight. Asked where they were lost he replied, "it was probably somewhere over there in the parking lot, but there is no point in looking in there – there isn't enough light there to see by." Similarly, we all look where we find a little light, but we should sometimes ask whether the regions where we are looking actually have any possibility of yielding discoveries of real significance to biology.

For some scientists here – including me when biology is not going well – significance to chemistry is itself a major interest. Again, to achieve this in any effective way demands an analysis of what directions in pure chemistry would represent significant advances. My own record is rather less than ideal in biological directions, but for those of you to whom this approach seems radical, references [1-3] contain in their introductions at least a description of how some of the problems I have worked on in biophysics are connected to the problem of an organism.

Myoglobin

Myoglobin is a favorite molecule of many physicists interested in biology because all the major experimental physical techniques for studying biomolecules can be readily applied. The streetlight is very bright. It is not a particular favorite of mine, because myoglobin is a candidate for the world's dullest biological molecule. If you are going to study myoglobin protein for its biology you might as well study coal, which also has something to do with biology. What does myoglobin do for physiology? It reversibly binds O_2, and diffuses with the bound O_2. An almost naked heme – a heme with an appropriate ligand in the 5th Fe site synthesized and studied by Collman et al [4] – binds oxygen with essentially the same binding constant. The protein does little with respect to the modulation of the site affinity for O_2.

What about kinetics? Myoglobin is used in such a dull fashion that if oxygen were bound three times more rapidly or three times more slowly (while maintaining constant affinity) there would be little physiological effect. The binding rate must be adequate, but does not seem a critical parameter.

What *is* myoglobin about? Given the fact that the machinery for making proteins is in a cell, protein synthesis provides a "simple" way of putting the 5th ligand imidazole on the heme Fe. In addition, myoglobin *does* have a kinetics problem, which Bill Eaton recently reminded me of. Ferrous hemes in water do *not* react reversibly with O_2, but instead tend to oxidize, in such reactions as

$$2Fe^{++} + H_2O + 1/2O_2 > 2Fe^{++} + 2OH^-.$$

Myoglobin functions only in the ferrous state. One important *kinetic* function of the protein is to immensely slow this oxidation process, chiefly by keeping the iron atoms of two different hemes from approaching close enough to share a common ligand. This is achieved in a mundane fashion by isolation of the Fe^{++} sites. Biological kinetics concerns not only what happens rapidly and how it happens, but also concerns what does *not* happen at an appreciable rate [2,3]. But biochemists and molecular biologists tend not to regard this other kinetic half as a problem worthy of attentions. Why "wrong" reactions don't happen is generally neglected.

Low Temperature Kinetics

Low temperatures were originally used in studying biochemical kinetics as a means of slowing down chemical reactions so that more kinetic intermediates could be resolved [5-6]. As decades have gone by, more and more detailed properties have been resolved at low temperatures, both because of the trapping of intermediates at low temperatures and because spectroscopic probes have higher resolution at low temperature. If you are interested in biology, which unfortunately takes place at room temperature, the question arises as to how to relate the low temperature studies to room temperature phenomena.

To take an example from engineering, suppose you wish to build cross-country electricity transmission lines. What interests you then should be the electrical conductivity of aluminum near room temperature. This is the *kind* of mundane engineering problem which most biological molecules are about. But as a physicist, you love to do high resolution experiments. So you study the conductivity of aluminum at $1°K$, and find superconductivity and a wealth of exotic associated phenomena.

There are now two viewpoints which can be taken. One is to tacitly forsake interest in the engineering problem and study superconductivity for its own sake. Some low temperature biophysics is done with this view. The other viewpoint is to realize that electron-phonon interactions are central to superconductivity, are involved in electrical conductivity at room temperature, and that it may be possible to make quantitative predictions of room temperature properties from low temperature studies. The problem then becomes one of extrapolation. When the changes of structure with temperature are small and anharmonic motions not major, there is a reasonable chance of making such an extrapolation.

The problem of low temperature biokinetics is to learn to make reliable extrapolations from low temperature studies to room temperature properties. Real biochemistry goes on in a liquid environment with large non-harmonic motions of some parts of biomolecules. These anharmonic motions prevent an extrapolation from low temperature "normal modes" to room temperature "normal modes" [7]. Some proteins have abrupt changes on freezing, and some go into frozen non-equilibrium configurations below a glass temperature. We know that even gentler effects in solids between $1°K$ and $300°K$ could have prevented the use of low temperature studies to predict room temperature properties (e.g., in V_3Si). Entropy plays a somewhat unpredictable role in a system whose structure and "elementary excitations" are temperature dependent, as clearly will be the case in proteins over the range from $4\text{-}300°K$. There is no reason to believe that the free energy of a chemical state can be adequately represented by a form A + B →T over such a range as is often assumed.

Since the protein in myoglobin is responsible for so little with respect to O_2 binding kinetics, myoglobin may prove a particularly simple case to try to make such an extrapolation. But alas, the same simplicity which might make the problem relatively tractable relieves it of much significance to biology.

Thus endeth the sermon. I now join the congregation of sinners.

Ligand Binding and Electron Transfer

What is the effect of a protein on what would otherwise be a "simple" problem of the binding of two relatively small molecules? In small molecule chemistry, one would at a naive level attempt to describe such a reaction by a single reaction coordinate as sketched in Fig. 1a. There are two potential wells in such a scheme, one representing the trapping of the pair in a weak non-covalent binding, and the second (deeper) well describing the state in which a chemical bond is present between the two.

Ligation events in heme proteins can be more complex, even at the naive level. In the binding of oxygen or CO to hemoglobin, cytochrome oxidase, myoglobin, etc., there are spin considerations also involved [8-11]. In the case of CO binding, for example, the lowest energy states of the system without a covalent bond will have the Fe^{++} atom in a spin state $S = 2$. There will be a manifold of $S = 2$ states, which presumably do not yield an appreciable covalent bond. The $S = 0$ state characterizes the bonded state, but lies at higher energy then $S = 2$ in the non-bonded configuration. $S = 1$ states (presumably) lie even higher in energy. Energy versus reaction coordinate curves for $S = 0$ and $S = 2$ states are indicated by the solid and dashed lines in Fig. (1b) in the absence of spin-orbit coupling (SOC). In the presence of SOC, the curves mix, resulting in the two solid lines of Fig. (1b). The lower line looks generically like the curve of Fig. (1a), but its mate is inevitably present in ligand binding to heme proteins. For many aspects of kinetics, one needs to understand whether the gap in Fig. (1b) is sufficiently large that in the chemical reaction to the lower solid curve is followed adiabatically, or whether a non-adiabatic process involving the two dashed curves is relevant.

The problem of O_2 binding may be even more complicated. In the initial state, the oxygen also has a spin, and there may be exchange coupling between the Fe^{++} spin and the O_2 spin. Anything from $S = 1$ to $S = 3$ might result as the ground state in the unbonded configuration. Too much depends on the sign and magnitude of this exchange splitting to make the rate of O_2 binding an easy theoretical subject at low temperatures.

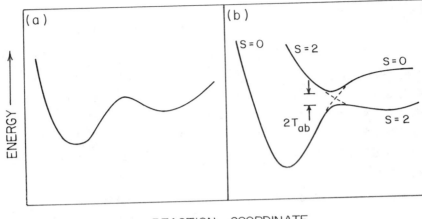

Figure 1 a) The elementary energy-reaction coordinate scheme for a simple monomolec-ular binding. b) Similar energy-reaction coordinate curves for CO binding to a ferrous heme.

The effect of the thousands of other coordinates of the protein on this simple picture is what we are fundamentally after – how the protein modulates the reaction rate and binding by both changing the reaction coordinate from the small molecule description and by controlling the rate of motion along the resultant reaction coordinate. And if the system is to be studied at low temperature, an adequate conceptual generalization to room temperature must be found.

The ligation process in heme proteins is related to a simpler process. From the physicists's viewpoint, the simplest ligand to deal with is no ligand at all – simple electron transfer. It involves exactly the same description as that of Fig. (1b). One of the dashed curves - say the right hand one - could be labeled "electron on donor" and the other dashed curve "electron on acceptor". The donor and acceptor are to be thought of as held at a fixed distance from each other in this description. There is a small electronic matrix element between these two states because of the overlap of the wave function tails of the initial and final electron wave functions on donor and acceptor. The electron transfer reaction for chromophores embedded in proteins is in concept isomorphic to the

CO-heme binding problem because it generates the same kind of reaction surface, though it differs considerably in detail.

A theorist, examining the electron transfer problem, would like to be able to modify in experiments the parameters fundamental to the theory and test experiment against theory [2]. In the case of electron transfer, this would mean running the positions of the two "parabolic" wells of Fig. (1b) up and down (change redox levels), changing the size of the electronic matrix element T_{ab}, and altering the vibrational coupling due to the material surrounding the electron transfer groups (the effect of the "protein" or solvent). The problem is conceptually similar in electron transfer and ligand-to-heme binding. But the electron transfer system comes close to a case in which the desired alterations of parameters can be carried out.

We have been working on such a program for electron transfer at Caltech. G. Geller, A. Joran, and B. Leland have, in Professor P. Dervan's laboratory, synthesized a set of idealized electron transfer molecules. The electron transfer experiments on these molecules have been carried out in Professor A. Zewail's laboratory in collaboration with P. Felker, and D. Beratan has made the relevant theoretical studies. I will describe the nature of these studies, and a few of the results so far obtained.

A typical such molecule [12,13] is sketched in Fig. 2. It has four regions, the chromophore (porphyrin), a phenyl ring attached to the meso position and lying perpendicular to the porphyrin, the linker (units of bicyclo [2.2.2] octane), and the quinone. The linker is rigid, and the only appreciable molecular flexibility of the entire system is the rotation around the colinear single bonds between phenyl group, linker, and quinone. The relevant electronic levels are also sketched in Fig. 2. Electron transfer is studied in this system by photo-exciting the porphyrin and observing the electron transfer from the excited porphyrin to the quinone. The back transfer from quinone to the porphyrin ground state is also potentially observable, but has not yet been experimentally studied.

All the interesting theoretical parameters can be experimentally manipulated. The energy levels of the donor can be modified by altering the choice of the metal in the porphyrin.

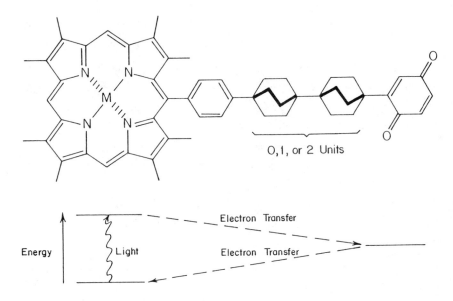

Figure 2. The generic porphyrin-phenyl-linker-quinone molecule which was synthesized for electron transfer studies. The phenyl ring and the quinone are joined by bicyclo [2.2.2] octyl spacers. Either 0, or 1 or 2 such units can be inserted.

The energy level of the acceptor can be modified by peripheral substituents on the quinone. The electronic matrix element can be changed by using 0, 1, or 2 units of the [2.2.2] linker. The effect of the "protein" can be charged by altering the solvent environment. (The solvent and solvent motions do for this system what the protein interior does for heme proteins.) In the case of the Zn porphyrin molecule, it may even be possible to insert the synthetic molecule into the heme site of Mb or Hb.

These lovely materials have been difficult to learn to synthesize, but the tools for a fairly flexible synthesis of related compounds are now in hand. The cumulative price of 2,000,000 dollars a gram is rapidly dropping.

Experimental results are accumulating. I will here summarize the motivation and results of some of them. Preliminary results have been published [12] and more complete and accurate findings will be published within the next year [13].

In the absence of an electron acceptor group, the fluorescence decay of the excited state of photo-excited porphyrin is accurately monoexponential, with a decay rate of 0.68×10^9 sec $^{-1}$ in benzene at room temperature. Experiments by the group show that when the quinone is present in a molecule with a single [2.2.2] linker, the fluorescence decay remains monoexponential but the decay rate is increased to 1.56×10^{10} sec $^{-1}$ under the same solution conditions. [The decay is not absolutely simple - a couple percent of material with the lifetime of parent porphyrin is present (presumably unreduced hydroquinone, which will not be an electron acceptor). In addition, a convolution must be done with the known shape of the laser pulse and photodetector jitter. This latter is standard technique in single-photon counting fluorescence decay studies, and this system has been previously shown to perform with high accuracy in this time domain.] At room temperature and in simple liquid organic solvents, the electron transfer process is well described by a constant rate of transfer and an exponential decay law. We have as yet no direct experimental test that electron transfer is the source of the fluorescence quenching. But the quinone is so far away that the only plausible mechanisms by which it can quench the fluorescence are electron transfer and excitation transfer. Excitation transfer is not possible in this system because the quinone absorption lies at much higher energies than the porphyrin fluorescence.

How does the rate of transfer change with the separation of the donor and acceptor? The same fluorescence decay experiment has been done in the same solvent for molecules with two [2.2.2] linker groups. The observed decay is again exponential. The decay rate (below) is so slightly larger than the decay rate of the parent porphyrin that it is difficult to guarantee that the difference is not due to small systematic errors. It is thus found that the electron transfer rates (in benzene, at room temperature) are

	1 linker	2 linker	ratio
experimental	$1.5 x 10^{10}$	$\leq 9 x 10^6$	$\leq 1/1600$
theory	–	–	$1/1500$

The theoretical number of Beratan [14] comes from a calculation of the propagation of wave functions in the bonding and anti-bonding orbitals of the [2.2.2] linker, and the supposition that the electron transfer is non-adiabatic. It will be soon possible to obtain a real experimental number, rather than a bound, in such systems. The theoretical ratio for the *back-transfer* in the same system [15] is only 1/60. The difference comes from the theoretical conclusion that this transfer process is dominated by hole transfer through the filled linker orbitals rather than by electron transfer through empty orbitals.

Is this electron transfer non-adiabatic when one [2.2.2] linker is present? This question of non-adiabaticity occurs in electron transfer and in heme ligand binding problems. In this particular electron transfer system, there is a direct experimental test which indicates that the transfers are non-adiabatic. The molecules do not have a single conformation. The rotation around the single bonds at the ends of the [2.2.2] linker are only slightly hindered. The electron transfer matrix element (between the π excited state orbital on the porphyrin and the lowest π empty orbital on the quinone) depends on the angle ϕ between the porphyrin and the quinone plane. If the matrix element is T_0 when porphyrin and quinone are parallel, its general dependence is $T_0 \cos \phi$. (There is a fine-structure dependence on the orientation of the [2.2.2] unit also, but it does not alter the general line of argument.) At $77°K$, the solvent has frozen into a glass, and the system is a frozen distribution of rotational conformers. At room temperature, the molecular rotation is rapid, and each molecule averages the rotation angle.

In an adiabatic situation, the electronic matrix element is adequately large so that making it 1/10 the size will not change the electron transfer rate appreciably. If the transfer is a diabatic, the modulation of the matrix element with angle will have little effect. At $77°$ *or* $300°$, whether the configurations are frozen or averaged, the decay should be *exponential*.

In a non-adiabatic situation, the square of the electronic matrix element is a prefactor in the rate. At room temperature, the rapid rotation of the quinone means that all molecules will have an averaged squared matrix element $T_o^2/2$, all molecules should have the same decay rate, and the decay should be exponential. At $77°K$, rotational configurations are frozen, different molecules will have different electron transfer rates, and the fluorescence decay should be non-exponential. The decay should be approximately

$$\frac{I(t)}{I(0)} = \frac{1}{2\pi} \int_0^{2\pi} \exp -[\,(k_0 + k_1 \cos^2 \phi)t\,] \, d\phi$$

where k_0 is the fluorescence decay constant in the absence of electron transfer and k_1 is the electron transfer rate when the quinone and porphyrin are parallel. This expression gives a decay which is non-exponential in form.

A fit of this expression to the data [13] at $77°K$ is excellent, showing no trends in the residuals. A fit to a single exponential is poor. Since k_0 is known, the fit is really a fit into a *single parameter*, just as in the case of single exponential decay. We conclude that this electron transfer process is non-adiabatic. (The fit to the room temperature data is very good, but not *quite* as good as the fit to the $77°$ data. The rotation rate even at room temperature may not be sufficiently rapid compared to the electron transfer rate to completely average the result to a single exponential.)

In ligand binding, the protein has a modulating effect. There is an analogous aspect to this electron transfer system which can be controlled and studied. The protein contributes polar and non-polar interactions to a chromophore embedded in it, but it is not easy to change or measure those contributions, even though the structure, the solvent contributes such interactions. The structure is not known in detail, but in return, the polar and non-polar aspects of the solvent are freely adjustable.

In vacuum, the nuclear motion dominantly associated with electron transfer is an anti-symmetric linear combination of the breathing coordinates of the quinone and of the porphyrin. In a non-polar solvent, there will be relatively little coupling between the motion of solvent molecules and the electron transfer process. The dominant effect of the non-polar solvent is a renormalization of the donor and acceptor energies due to the interaction of the electron to be transferred with the high frequency dielectric constant $\varepsilon_\infty = n^2$ of the solvent. In the particular system at hand, the major effect of the non-polar solvent motion is to control the dynamics of the rotation angle ϕ, as described in the previous paragraphs. For simple liquid solvents at room temperature, this results only in an uninteresting rotational averaging of the transfer rate. In a polar solvent having the same high frequency dielectric constant, all these effects exist. In addition, the electron transfer process is coupled to the orientation of the solvent molecules [15]. A collective coordinate can be used to describe these orientation effects. The energy surface for electron transfer is then described in terms of two reaction coordinates, the breathing motion of the chromophores and the polarization of the polar solvent. This second reaction coordinate is intrinsically *diffusive* (i.e., overdamped) and relaxes with the characteristic dielectric relaxation times.

In simply theory, the electron transfer rate $P^*Q \rightarrow P^+Q^-$ in non-polar solvent would be controlled by the dynamics of the breathing modes. The rate of transfer W_{da} is a simple Gaussian function of the driving energy, [2,16,17] (the difference $E_a - E_d$ between the electron initial and final states) as sketched in Fig. (3a). The shift parameter Δ is determined by the vibronic coupling to the breathing modes.

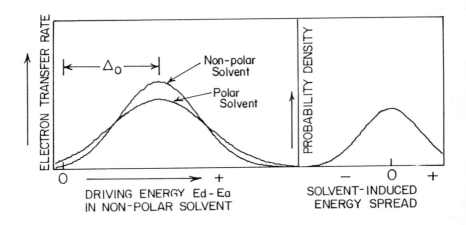

Figure 3 a) The rate of electron transfer versus the value of $(E_a - E)d)$ measured in non-polar or polar solvents. b) The shape of the distribution of effective transfer energies generated by polar fluctuations.

What happens if the solvent is polar, with the same ε_∞? The particular molecules and transfer process we have been studying are neutral before charge transfer. Thus in the initial state, there is *no* coupling between the polar solvent motions and the electron transfer molecule. The fluctuations in solvent polarization are the same as they would be in the absence of the electron transfer molecule. Due to these polar fluctuations, there will be a thermal probability distribution of electrical potential difference between two sites in the solvent.

This distribution is also Gaussian, centered at zero potential difference, and has a mean square width proportional to $(1/\varepsilon_\infty - 1/\varepsilon_0)$ where ε_0 is the low frequency dielectric constant.

In non-adiabatic electron transfer from an initial uncharged state, electron transfer takes place and *then* dielectric relaxation occurs. Thus the amount of dielectric relaxation which will take place in the final state – and even whether this relaxation can be described as a linear effect – *does not influence the forward transfer rate*. The polar solvent effect acts through the distribution (3b), which makes the effective $E_a - E_d$ not exactly that of non-polar solvent, but gives it a broadening line (3b). The net forward electron transfer rate including polar solvent motion is thus given (as a function of driving energy $E_a - E_d$) by the *convolution* of (3a) and(3b), which is the broader Gaussian shown by the dashed curve in (3a). The more polar the solvent, the larger is the convolutional broadening. $(E_a - E_d)$ is this plot refers to the redox difference measured *in non-polar solvent*.

This simple viewpoint using the reaction coordinates for the solvent and a second one for the small molecule nuclear coordinate agrees with the usual manipulation of elementary electron transfer theory when the solvent is involved [15]. But in addition, it makes clear why the solvent polarity has a minor effect on the forward rate from a very physical viewpoint. (The polar solvent will have large effects on the *reverse* rate, from an initial charged state to a final uncharged state, because there will be a large initial state polarization.)

We can see from a glance at Fig. (3) that the solvent polarity effect on the forward transfer rate will be

a) Small unless $(E_a - E_d - \Delta)$ is very large

b) Systematic, either increasing or decreasing steadily with solvent polarity. *Which* systematic trend is displayed will depend on $(E_a - E_d - \Delta)$. Experimentally, a small and systematic solvent polarity dependence has been found on the forward transfer rate, as expected.

Because the polar response of the solvent and the vibrational coupling are assumed to be linear, it is possible to think of this problem as convolutions. In a general problem, it would be necessary to construct a two-dimensional energy surface.

Let us consider, finally, the viewpoint toward the rate of ligation of hemoglobin (and incidentally, other heme proteins) which Agmon and I developed [18]. There were two reaction coordinates, an "Fe-CO" small molecule coordinate and a "protein coordinate" expressing the effects of the protein on the heme geometry. This protein coordinate is the analog of the solvent polarization coordinate in our model electron transfer system. Just as in the electron transfer case, one must consider dynamics on at least a 2-dimensional reaction surface to understand what is happening. Of course, one could go to 10^3 coordinates and do the modeling "exactly" (within certain quantum problems of frozen-out vibrational modes). But the object is to find the right kind of simplification for understanding experiments and biology, rather than to simulate by brute force a very complex situation and see that it gives some particular answer. (Such simulations do have a major role to play, however, as a substitute for doing experiments.) Theory to me has to do with conceptual and mathematical frameworks for *interpreting* experiments or simulations in some *unifying, predictive,* and *simplified* fashion.

In CO-Fe binding as in electron transfer, the binding reaction is *possible* for any fixed value of the protein coordinate. The kinetics which will be seen experimentally will be determined by the motion of the system in both coordinates.

A contour map of the kind of energy surface constructed for such systems is illustrated in Fig. 4. The reaction was taken as a diabatic. This point might be challenged, but it does not really enter the major conclusions of the study. Such energy surfaces were used to describe quantitatively various aspects of binding kinetics. Here I will only comment on some qualitative aspects of such an energy surface.

First, below about $200°K$ the *protein* becomes frozen, and different protein molecules are frozen in different configurations. Within a description on this kind of energy surface, each molecule is frozen with some particular value of the protein coordinate. Flash photolysis results in displacement horizontally along the Fe-CO reaction coordinate, and recombination moves back along that horizontal line. Each different protein coordinate results in a different trajectory of rebinding. The overall system then appears to have a *distribution* of activation barriers. Reasonable parameter choices for such energy surfaces can be made to

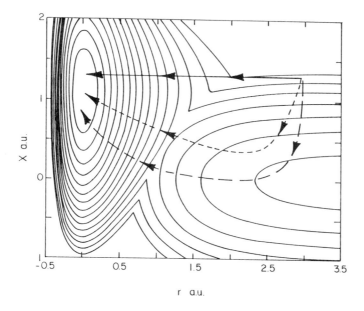

Figure 4. Energy contours for CO binding to a heme in Hb or Mb. The deep well in the upper left represents the bound state, and the shallow well to the right is the CO free in the heme pocket. Flash photolysis removes the CO without changing the protein coordinate X. At low temperatures, the rebinding process follows the solid line. At higher temperatures, the protein coordinate will partially or totally relax before the CO rebind (dashed lines).

fit in detail the low temperature rebinding kinetics of such systems.

At higher temperatures, the protein coordinate is no longer frozen. More accurately, as temperature is raised, a point is reached for which the "mobility" of the previously frozen coordinate is large enough that protein relaxation is on the same scale or faster than the kinetic rebinding time. One major qualitative effect expected from this model is that such motions allow the protein to readjust during the time the ligand is off to a state which is lower in energy in the deoxy configuration. (Remember that in the usual experiments, the protein is frozen in a ligated configuration). The apparent activation enthalpy (or if applicable, free energy) is therefore expected to *increase* when the system is then through this temperature range, since the rebinding will take place from geometries which have been allowed to relax toward the unligated state. A sketch of the reaction pathway in this higher temperature circumstance is also shown in Fig. 4.

The nature of the motion of the protein coordinate in this intermediate temperature range is not clear from first principles. Agmon and I described it as a diffusive motion, but this need not be the case; there may be a hierarchy of structures and relaxations [19], and far more complicated kinetics of this relaxation process are possible.

Whatever the nature of the motion of the protein coordinate(s), and regardless of the detailed mechanism of the coupling between CO binding and protein coordinate, one view stands out. It is usual in chemistry for an activation barrier height to move in away which follows to some reasonable extent both the reactant energies and the product energies. When an unligated heme is prepared by photodissociation, the initial product state will be unrelaxed. When the relaxation isn't permitted (as at low temperature) the free CO – deoxy heme state is a higher energy initial state for the recombination reaction than when relaxation is permitted (at high temperature). While exceptions may occur, *in general one therefore expects the apparent activation free energy for recombination from the heme pocket to increase with temperature in the glass transition region due to the effects of the protein coordinates.* I expect a problem in extrapolating from low temperature "frozen glass" kinetics to room temperature kinetics by any means other than constructing energy surfaces whenever the protein has a substantial modulatory effect on the small molecule reaction – i.e., whenever the protein is doing an interesting and subtle task. The Illinois group for a long time attempted to describe the recombination events in heme proteins with a conceptually inadequate single coordinate [20] model but has recently also moved in the direction of an explicit representation of a protein coordinate [19,21].

Part of the difficulty of evaluating the experimental situation in ligand rebinding is that there are no well-developed probes to give physical meaning to postulate kinetic intermediates. In the case of electron transfer, theory relates *spectroscopy* to rates of transfer [22] in an unambiguous and checkable [23] fashion. In electron transfer, theory predicts functional relations of electron transfer rates with modified molecular structure and energy levels. No such quantitative checks of theory and experiment have yet been produced in the case of ligand binding so the structural identity of postulated intermediate states is yet rather ambiguous.

Summary

If you are interested in the fundamental chemistry of protein molecules, the question of how a protein modulates the "small-molecule" properties of a group embedded in it has fundamental interest. If you are interested in the biology of protein molecules, as I happen to be, you must also select cases in which the modulations solve a problem posed by a biological need. I have described systems in which such problems may be tractable because the problem approximately factors. There actions can be described on a reaction surface of two coordinates, one referring to the intrinsic small molecule problem, and the other referring to the effect of the protein. Electron transfer, the best understood chemical reaction in biology and perhaps in chemistry as a whole, can explicitly be written in such a fashion. And it is plausible that simple ligation processes can also be so written.

In the ligation case, care must be taken to distinguish between myoglobin and interesting molecules. In particular, the description given by Agmon and Hopfield was motivated as an effort to link low temperature kinetics to room temperature modulation effects in hemoglobin, though alas, most experiments and analysis of those papers have emphasized myoglobin. There are large quantitative, and perhaps qualitative, differences in the binding kinetics of Hb and Mb at intermediate temperatures. Similarly, Friedman has found that the protein has much larger modulatory effects in transient resonant Raman scattering [24] in Hb than in Mb. The simple theory as described in explicitly constructed only for a description of the binding from the hemepocket, and does not deal with additional sequential processes involving other locations of the ligand.

I believe it is constructs such as these energy surfaces which are likely to be able to be carried from low temperatures to room temperature. At low and intermediate temperature, reaction kinetics will be complex due to the "glassy" nature of proteins. Such complexities tend to disappear by being averaged over as the temperature is raised. There is at present no strong indication that the glassy effects in proteins are important in biology. The most useful aspect of the glassy behavior will probably be that it permits the experimental study of details of energy surfaces, structures, and modes which would otherwise be averaged over and not resolved. Such knowledge, carried to room temperature, may contribute greatly to our understanding of interesting biological processes.

Acknowledgements

The author thanks B. Leland, A. Joran, P. Felker, A. Zewail and P. Dervan for permission to cite some of their illustrative results prior to publication, and R. Austin for the challenge to be provocative. This paper was written during a stay at the Aspen Center for Physics. The research was supported in part by NSF #PCM-8406049.

References

1. Hopfield, J.J., *Mol. Biol.* **77**, 207 (1973); also, in "Modeling and Analysis in Biomedicine", 101 ff, ed. C.Nicolini, World Scientific Pub. Co., Singapore, (1984).

2. Hopfield, J.J., *Proc. Natl. Acad. Sci.* **71**, 3690, (1974); also in "Electrical Phenomena at the Biological Membrane Level," 471 ff, Elsevier, Amsterdam (1975); also in "Oxidase and Related Redox Systems," p. 35ff, King, T.E., ed., Pergamon Press, New York (1982).

3. Hopfield, J.J., *Proc. Natl. Acad. Sci.* **71**, 4135 (1974).

4. Collman, J.P., Braumann, J.I., and Suslick, K.S., *J. Am. Chem. Soc.* **97**, 7185 (1975).

5. Broda, E.E. and Goodeve, C.F., *Proc. Roy. Soc. (London)* **B130**, 217 (1941-42).

6. See for example, Devault, D. and Chance, B. *Biophys. J* 6, 825 (1966) and Chance, B., Sardonio, C. and Leigh, J.S., *J. Biol. Chem.* **250**, 9226 (1975) and earlier works of the Johnson Foundation group.

7. There is a recent viewpoint on this problem which *might* help. Bialek, W. and Goldstein, R.F., *Biophys. J* (November 1985).

8. Hopfield, J.J., in Tunnelling in Biological Systems, p. 646, Chance, *et.al.*, eds., Academic Press, New York (1979).

9. Redi, M.H., Gerstman, B.S. and Hopfield, J.J., *Biophys. J.* **35**, 471 (1981).

10. Jortner, J. and Ulstrup, J., *J. Am. Chem. Soc.* **101**, 3744 (1979).

11. Frauenfelder, H. and Wolynes, P.G., *Science* **229**, 337 (1985).

12. Joran, A.D., Leland, B.A., Geller, G.G., Hopfield, J.J. and Dervan, P.D., *J. Am. Chem. Soc.* **106**, 6090 (1984).

13. Leland, B.A., Joran, A.D., Felker, P.M., Hopfield, J.J., Zewail, A.H. and Dervan, P.B. (to be published).

14. Beratan, D.N., Ph.D. Thesis, California Institute of Technology, p. 58 ff [submitted June 11, 1985].

15. Marcus, R.A., *J. Phys. Chem.* **67**, 853,2889 (1963).

16. Marcus, R.A., *Discussions Faraday Soc.* **29**, 21 (1960).

17. Ulstrup, J. and Jortner, J., *J. Chem. Phys.* **63**, 4358 (1975).

18. Agmon, N. and Hopfield, J.J., *J. Chem. Phys.* **78**, 6947 (1983); **79**, 2042 (1983).

19. Ansari, A., Berendzen, J., Browne, S.F., Frauenfelder, H., Iben, I.E.T., Sauke, T.B., Shyamsunder, E. and Young, R.D., *Proc. Natl. Acad. Sci.* **82**, 5000 (1985).

20. See for example, Beece, D., Einstein, L., Frauenfelder, H., Good, G., Marden, M.C., Reinisch, L., Reynolds, A.H., Sorensen, L.B. and Yue, K.T., *Biochemistry* **19**, 5147 (1980).

21. Young, R.D. and Browne, S.F., *J. Chem. Phys.* **81**, 3730 (1984).

22. Hopfield, J.J., *Biophys. J.* **18**, 311 (1977).

23. Goldstein, R.F. and Bearden, A., *Proc. Natl. Acad. Sci.* **81**, 135 (1984).

24. Scott, T.W. and Friedman, J.M., *J. Am. Chem. Soc.* **106**, 5677 (1984).

Protein Vibrations Can Markedly Affect Reaction Kinetics: Interpretation of Myoglobin-CO Recombination

W. Bialek[*] and R. Goldstein[**]

Department of Biophysics and Medical Physics, and

Biology and Medicine Division, Lawrence Berkeley Laboratory

University of California, Berkeley

Berkeley, CA 94720 USA

Chemical kinetics are governed by the flow of energy from electronic states through molecular vibrations to the solvent. Small changes in vibrational frequencies may have large effects on reaction rates. These effects are illustrated in an analysis of the non-exponential low-temperature kinetics of ligand binding in myoglobin.

The past decade has witnessed an explosion of information regarding protein structure, with the goal of explaining a protein's function in terms of its architecture. At the same time, many experiments have provided evidence for dynamic fluctuations in protein structure, but the functional importance of these motions is not yet clear. What we wish to consider here is not that proteins have a dynamical nature, but rather the nature of their dynamics, with a view toward relating motion to function.

A particularly revealing experiment was the demonstration by Austin et al. [1] that the rebinding of CO to myoglobin (Mb) following flash photolysis is significantly non-exponential below 160K. This was interpreted as due to a "freezing in" of an inhomogeneous distribution of reaction rates. That is, molecules that are ostensibly identical at room temperature become non-identical at low temperature at least to the extent of having rate constants that differ by many orders of magnitude.

* Present address: Institute for Theoretical Physics, University of California, Santa Barbara, CA 93106 USA

** Present address: Department of Cell Biology, Sherman Fairchild Center, Stanford University School of Medicine, Stanford, CA 94305 USA

These differences in rate constants were attributed [1] to differences between protein conformations. At low temperature, a molecule is trapped in a particular conformation; therefore an inhomogeneous distribution of conformations (and reaction rates) is obtained. At high temperature, a molecule can quickly convert among various "conformational substates" and therefore sample a wide range of rate constants, producing an apparent homogeneous reaction rate.

This link between reaction rates and protein dynamics is important historically conceptually in motivating studies of molecular dynamics in relation to protein function. In this regard, we shall present an alternative explanation of the non-exponential kinetics, one that can be distinguished from the conformational substates model by experiment. As we discuss below, the non-exponential decay argues strongly for a distribution of reaction rates, but it does not indicate the physical origin of this distribution and does not, by itself, force us to adopt any particular model of protein dynamics.

In addition to the anomalous kinetics at low temperature, a wide variety of approaches have provided complimentary insight into the unusual dynamics of Mb, and include techniques such as Mossbauer spectroscopy [2-4], infrared absorption [5-7], x-ray crystallography [8], and isotopically substituted kinetics [9]. Because of this general wealth of data, we shall base our discussion primarily upon Mb-CO rebinding, although our results should be of much wider applicability.

The conformational substates model postulates that a protein's dynamics is dominated by diffusive "hopping" between localized conformations; each conformational substate has its own set of high frequency, localized vibrations. We shall contrast the conformational substates model with a *quasi-harmonic* model of protein dynamics, in which atomic motions are underdamped, not diffusive, and in which an atom may participate in low-frequency, breathing motions as well as high-frequency localized vibrations. We show how the quasi-harmonic model can also explain the non-exponential kinetic data.

The two models present radically different views of protein dynamics and very different views of how dynamics can affect reaction rates. In the quasi-harmonic model a reaction is governed by the rate at which energy can be transferred from the electronic states *through* the vibrations to the solvent, while in the substates model a reaction rate depends on the conformation involved, but no specific physical link has been specified. Most importantly, however, we wish to stress that the quasi-harmonic model provides a physically reasonable picture of a specific relation between function and dynamics, one that can be tested experimentally.

Harmonic Vibrations vs. Conformational Substates

The potential energy of a protein is a function of approximately 10^4 atomic coordinates, and defines a multi-dimensional "potential surface". If we take an arbitrary slice through this surface, we obtain the energy as a function of one "generalized coordinate", as shown in Figure 1.

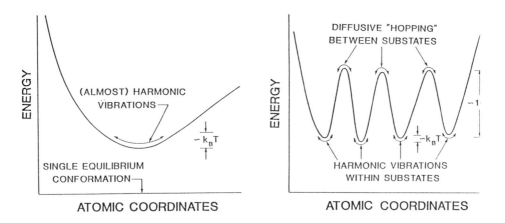

Figure 1. (a) Cross section of a protein potential surface in the nearly harmonic model. There is only one minimum, and motion is smooth and underdamped. (b) Cross section of a potential surface in the conformational substates model. There exist many minima (conformations) with large energy barriers between. Thus motion is diffusive.

In the conformational substates model (Figure 1a), the potential surface contains many local minima which arise because each atom can occupy several stable positions. Adjacent substates differ by very small atomic motions [8], $d \approx 0.01nm$, but are separated by large barriers [1], $\Delta \approx 100kJ/mole$ or $1eV/molecule$. At room temperature, hopping among substates is assumed to dominate the protein dynamics, while at low temperatures each molecule is trapped in a different substate and executes only small amplitude, high frequency vibrations.

In the harmonic model (Figure 1b), in contrast, the energy surface is parabolic and therefore has only a single minimum. The most appropriate coordinates are normal modes [10], denotes Q_a, which are superpositions of atomic coordinates; each normal coordinate executes sinusoidal oscillations at frequency ω_a. The normal modes range from localized high frequency vibrations of a small number of atoms to low frequency, large scale "breathing" motions. Different normal modes are independent, although motions of different atoms are *not* independent.

Neither of these models can be exact. For example, the harmonic model predicts that the energy of each normal mode is constant in time; in fact vibrational energy is ultimately dissipated to the solvent. In addition, the potential surface cannot be precisely parabolic, particularly for large scale motions such helix-coil melting transitions, ring flipping, and binding or unbinding of solvent molecules. Finally, the oscillations of each normal mode should be treated quantum mechanically.

The harmonic model can be extended to a "quasi-harmonic approximation" in which some of these problems are solved without sacrificing a simply physical picture. Rigorous quantum-mechanical solutions are available for damped harmonic oscillators [11-13] characterized by vibrational frequencies ω_a and damping constant γ_a. These solutions can be extended to weakly anharmonic potential surfaces, including those which describe melting transitions [14], by assuming the vibrational frequencies to be slowly varying functions of excitation energy. In this case the average frequency decreases with increasing temperature, while the atomic mean-square displacements increase more rapidly than linear with temperature.

A *small* number of coordinates, such as the motion of solvent molecules on the protein surface, can occupy several stable configurations separated by *large* ($\approx 0.1nm$) displacements. In each solvent shell configuration the protein exhibits the full range of normal modes, although possibly with different parameters in each configuration. The internal vibrational frequencies of the protein thus exhibit small fluctuations on a time scale given by the hopping rate from one configuration to another, but the basic feature of a quasi-harmonic model are preserved.

The problems of the conformational substates model are not as easily solved. First, even at absolute zero, quantum-mechanical tunneling causes hopping among substates at a rate [15] $k \approx \nu e^{-d\sqrt{2m\Delta/\hbar}}$, where $\nu > 10^{13} s^{-1}$ is the vibrational frequency within a substate m is the mass of the atoms which are displaced when the substate changes, and d and Δ are the distance and barrier between substates given above. For typical values of these parameters [1,8] the tunneling time must be $k^{-1} < 10^{-8}s$, but the kinetic data require [1] that the distribution over substates be frozen on a time scale *greater* than $10^3 s$, a discrepancy of six orders of magnitude. Second, if the only true vibrational modes are high frequency, localized motions, then for all accessible temperatures, $\hbar\omega \gg k_B T$ the rates of transition among substates cannot be strongly temperature dependent, which is again inconsistent with the substates interpretation [1] of the kinetic data.

These discrepancies can be resolved either by increasing the mass of protein involved in hopping (e.g. $0.01nm$ motions of entire alpha helices rather than single side chains) or by increasing the distance between substates (e.g., $> 0.1nm$ rather than $< 0.01nm$ motions of amino acid side chains). The former solution requires that individual inter-atomic potentials within the helix are significantly anharmonic for displacements smaller and $0.001nm$, which is implausible [16]. The large displacements associated with the latter solution are inconsistent with crystallographic data indicating that atomic positions do not fluctuate by more that [8] 0.01 to $0.05nm$, except for special cases such as residues at the protein surface. While some of these problems might be resolved in more complex versions of the substates model [17] these approaches are not sufficiently developed to give quantitative predictions.

These problems do not preclude the existence of some conformational substates; solvent shell configuration, for example, may play a significant role. The crucial question is whether the predictions of either the substates or harmonic models are in accord with experiment.

We next show how protein reaction rates can be related to harmonic vibrations, and then discuss our calculations for the case of Mb-CO. We will conclude that the quasi-harmonic model *does* provide quantitative relations among different experimental results, and can provide a physical connection between the vibrational dynamics of a protein and its reaction rate.

Reaction Rates in a Harmonic Model

Ligand binding in heme proteins consists of several reactions, of which only one is observed below 160K [1]. This reaction is accompanied by a change in spin state of the heme iron [18,19] and thus involves at least two distinct electronic state; for myoglobin and carbon monoxide we denote the liganded (low-spin) and photolyzed (high-spin) states $(Mb \cdot CO)_{S=0}$ and $(Mb \cdot \cdot CO)^*_{s=2}$, respectively. We think of this transition as occurring in two stages (Figure 2). First a molecule in the nth vibrational state of the photodissasociated species, state $|n; (Mb \cdot \cdot CO)^*_{S=2}\rangle$, makes a transition to the n' vibrational state of the ligand bound species, $|n'; (Mb \cdot CO)_{S=0}\rangle$. Second, the molecule vibrationally relaxes into states $|n''; (Mb \cdot CO)_{S=0}\rangle$ with $n'' < n'$.

Two general effects influence the overall reaction rate. First the rate depends on the overlap of the n and n' vibrational levels, and on the thermal population of the reactant levels. The rate constant then becomes an approximately Gaussian function of the energy difference between reactants and produces an energy gap. This Gaussian form is called the energy gap law [20], and this reduces to the standard Arrhenius expression at high temperatures $k_B T \gg \hbar\omega$.

Second, conservation of energy in the initial step requires that the reaction rate be larger when $\varepsilon \approx (n'-n)\hbar\omega$, where ε is the energy gap and $\hbar\omega$ is the energy of a vibrational quantum. That is, for the ligand binding to take place, ε joules of electronic energy must be transferred into vibrational energy and ultimately dissipated to the solvent. This may happen more quickly if ε can be divided

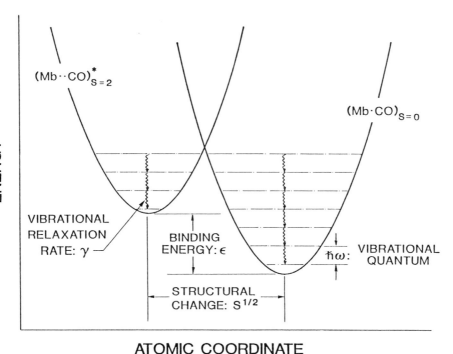

Figure 2. Energy level diagram for Mb · CO rebinding reaction.

almost exactly into an integral $(n' - n)$ number of vibrational quanta, but less quickly if half a quantum is left over. Thus the rate constant k_i may exhibit a sequence of "resonances" as a function of either ε or ω. Taken together, the energy gap law and the resonances lead, in some parameter regimes, to a very strong dependence of reaction rates on vibrational frequencies.

At absolute zero, the reactants are in the vibrational ground state, $|0; (Mb \cdot CO)^*_{S=2}\rangle$. If the temperature remains below $\hbar\omega/k_B$, there is not appreciable population of excited vibrational states, and the rate constant is temperature independent. As the temperature increases above $\hbar\omega/k_B$, higher vibrational levels are populated and because the reaction depends on the initial vibrational level, the rate constant becomes temperature dependent.

If the reaction rate depends critically on the vibrational frequency, then the vibrational frequency fluctuations which result from solvent interactions will give rise to large fluctuations of reaction rates. In frozen solvent, these dynamics fluctuations become a static distribution. Each molecule therefore has a different but fixed rate constant and the overall reaction time course becomes

$$F(t) = \int d\omega P(\omega) e^{-k(\omega)t}, \tag{1}$$

where $k(\omega)$ is the rate constant corresponding to frequency ω and $P(\omega)$ is the probability of a molecule having vibrational frequency ω. The "inhomogeneous lineshape" $P(\omega)$ is generally Gaussian, with standard deviation $\Delta\omega$, and can in principle be measured by infrared absorption or Raman Spectroscopy.

Equation (1) predicts that for broad inhomogeneous lineshapes the time course of a low temperature reaction will be strongly non-exponential. Are the parameters required to account for the observed non-exponential behavior in myoglobin consistent with the known structure and dynamics of this molecule?

There are rigorous methods for calculating the rate constant $k(\omega)$ within the quasi-harmonic approximation [21-25], based on quantum mechanical perturbation theory. This allows the rate constants to be expressed in terms of electronic energy levels, vibrational frequencies, and "coupling constants" which are proportional to the square of the difference in atomic positions between reactants and products. In order to estimate these parameters for the $(Mb \cdot \cdot CO^*_{S=2}) \rightarrow (Mb \cdot CO)_{S=0}$ reaction, we identify several important vibrational modes:

[1] The Fe-c stretching modes [26] ($\omega \approx 500 cm^{-1}$) changes frequency upon photodissociation (the bond is broken) and also changes its equilibrium length by [27] $\delta x \leq 0.005 nm$.

[2] The C-O stretching mode ($\omega \approx 2000 cm^{-1}$) changes frequency [5] by $\delta\omega \approx 220 cm^{-1}$ and the bond length may also change.

[3] The Fe-His stretching mode [28] ($\omega \approx 200 cm^{-1}$) changes length by [27] $\delta x \approx 0.004 \pm 0.002 nm$. These modes have $\hbar\omega \gg k_B T$ and therefore cannot contribute to the temperature dependence of the reaction rate.

[4] The "F" alpha helix in Mb appears to move in a collective mode, being rotated and possibly stretched when oxygen binds [29]. An average main chain atom in the F helix is displaced by $\delta x \approx 0.033 nm$.

We expect these collective motions to correspond to normal modes with frequencies [30] $\omega \approx 20cm^{-1}$, or $\hbar\omega/k_B \approx 30K$, the temperature at which the rebinding reaction changes [1] from temperature independence $(T < 30K)$ to the more conventional Arrhenius behavior $(T > 30K)$. Also, light-scattering measurements on alpha helical polymers [31] and on amino acid crystals [32] demonstrate that breathing motions can have small damping constants, $\gamma \approx 10^9 s^{-1}$, suggesting that the resonance effect in reactions coupled to these mode may be large. As a first approximation we therefore analyze a single mode model with $\omega \approx 20cm^{-1}$ and $\gamma \approx 0.01cm^{-1}$.

We have calculated $k(\omega)$ for this single mode model by the method of Ref. 24; the reaction time course from Eq. (1) is shown in Figure 3. Theory is in good agreement with the data of Ref. 1 for the following choice of parameters: The transition between electronic states has an electronic matrix element $V = 1.1cm^{-1}$, consistent with the estimate of Redi et al [33]. The transition is strongly coupled to a vibrational mode with frequency $20cm^{-1}$ and linewidth $\gamma \approx 0.01cm^{-1}$, although the fit is not very sensitive to the exact value of γ. The binding energy is $\varepsilon \approx 0.9eV$, which is consistent with the kinetic data [1]. Given these parameters, the extreme non-exponential behavior of Figure 3 can be generated with only a small amount of inhomogeneous broadening, $\Delta\omega \approx 1.6cm^{-1}$. The coupling constant, $S = 170$, is equivalent to a $\approx .013nm$ displacement of an alpha helix in a collective stretching or bending mode, some three to four times the observed value for the F helix upon oxygen binding [29]. That S is somewhat too large probably reflects limitations of the single mode model. If some high frequency modes effectively reduce the energy gap as seen by the low frequency mode, that S will be smaller. Alternatively, if there are several low-frequency modes with similar frequencies, the collective S may be large, but no single atom need move a large distance. Nevertheless, all other features of our model would remain.

This agreement between theory and experiment allows several conclusions to be drawn. First, non-exponential kinetics *can* be interpreted in terms of quasi-harmonic protein vibrations, where the only discrete hopping motions are those of the solvent molecules. Therefore, the data do not provide evidence, *a priori*, for

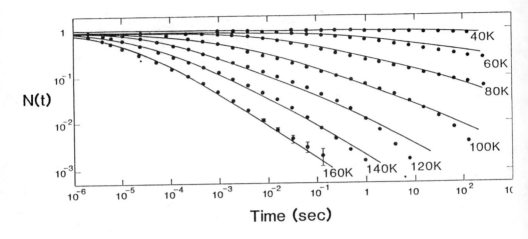

Figure 3. Amount of unbound CO after a flash. Data (points) taken from refer-ence 1. The line is a theoretical fit, based on the single vibrational mode shown in Figure 3. We assume vibrational relaxation to be exponential except a very short times where the correlation time of the heat bath must play a role.

conformational substates of the protein. Second, temperature-dependent kinetics can be directly related to vibrational spectra which characterize protein dynamics in the quasi-harmonic model. Finally, a reaction rate can depend very strongly on the protein vibrational spectrum; a small change in vibrational frequency may change the reaction rate by several orders of magnitude.

In addition to this discussion, we have found [34] that the Mossbauer [3,4] infrared [5-7], x-ray diffraction [8], and EXAFS [27] results revealed to myoglobin dynamics can all be interpreted in terms of the quasi-harmonic theory outlined here. Most importantly, this analysis leads to the same dynamical parameters as used here in the interpretation of the kinetic data. Further, some of these spectroscopies done at higher resolution, for example in absorption, might be used to experimentally distinguish between the substates model and the quasi-harmonic model of protein dynamics.

Conclusions

The application of the quasi-harmonic model to protein dynamics provides a clear physical picture that accounts for the observed non-exponential kinetics and other spectroscopic data with a consistent set of parameters. From this picture it can be seen that the details of a protein's vibrational spectrum may produce large functional effects. These vibrational effects are independent of other structural features which affect reaction rates and may thus explain how enzyme substrate complexes with very similar structures can have very different catalytic activities [35,36]. In the extreme case of myoglobin at low temperatures, a distribution of vibrational frequencies leads to a broad distribution of reaction rates and hence non-exponential kinetics. Similar kinetics have now been reported in other systems [37].

We thank Professor John Hopfield for telling us about his related work prior to publication, and Professor Lubert Stryer for a thorough critique of this manuscript. Ken Krieg kindly aided us with our computer calculations. We thank Professor Alan Bearden both for helpful discussions and a productive environment in which to carry out this research. This work was supported by the office of Basic Energy and Science, Office of Energy, Research, U.S. Department of Energy under contract DE-AC03-76F00098, by the National Science Foundation Biophysics (PCM 78-22245) and predoctoral fellowship program (to W.B.) and by the National Institutes of Health (National General Medical Institute GM 24032) and Bank of America – Gianni Foundation postdoctoral fellowship program (to R.G.). W.B. also acknowledges the hospitality of the Physics Faculty of the Rijksuniversiteit te Groningen during the final stages of this work. R.G. thanks the Department of Cell Biology of Stanford University for support during the latter part of this research.

References

1. Austin, R.H., Beeson, K.W., Eisenstein, L., Frauenfelder, H and Gunsalus, I.C., Biochemistry *14*, 5355-5375 (1975).

198

2. Parak, F., Frolov, E.N., Mössbauer, R.L. and Goldanskii, V.I., J. Mol. Biol. *145*, 825-833 (1981).

3. Parak, F., Knapp, E.W. and Kucheida, F., J. Mol. Biol. *161*, 177-194 (1982).

4. Keller, K. and Debrunner, P., Phys. Rev. Lett. *45*, 68-71 (1980).

5. Alben, J.O., Beece, D., Bowne, S.F., Doster, W., Eisenstein, L., Frauen-felder, H., Good, D., McDonald, J.D., Marden, M.C., Moh, P.P., Reinisch, L., Reynolds, A.H., Shyamsunder, E. and Yue, K.T., Proc. Nat. Acad. Sci. (USA) *79*, 3744-3748 (1982).

6. Makinen, M.W., Houtchens, R.A. and Caughey, W.S., Proc. Nat. Acad. Sci USA *76*, 6042-6046 (1979).

7. Caughey, W.S., Shimada, H., Choc, M.G. and Tucher, M.P., Proc. Nat. Acad. Sci. (USA) *78*, 2903-2907 (1981).

8. Frauenfelder, H., Petsko, G.A. and Tsernoglou, D., Nature (Lond.) *280*, 558-563 (1979).

9. Alben, J.O., Beece, D., Bowne, S.F., Eisenstein, L., Frauenfelder, H., Good, D., Marden, M.C., Moh, P.P., Reinisch, L., Reynolds, A.H. and Yue, K.T., Phys. Rev. Lett. *44*, 1157-1160 (1980).

10. Goldstein, H., *Classical Mechanics* (Addison-Wesley, Reading, MA 1959).

11. Senitzky, J.R., Phys. Rev. *119*, 670-679 (1960).

12. Haken, H., in *Encyclopedia of Physics* Vol. 25/2c (ed. Flugge, S.) (Springer-Verlag, 1970).

13. Caldeira, A.O. and Leggett, A.J., Physica *121A*, 587-616 (1983).

14. Lovesey, S.W., *Condensed Matter Physics: Dynamic Correlations* (Benjamin/Cummings, Reading, MA 1980).

15. Schiff, L.I., *Quantum Mechanics* (McGraw-Hill, New York 1968).

16. Morse, P.M., Phys. Rev. *34*, 57-64 (1929).

17. Frauenfelder, H., in *Structure and Dynamics for Nucleic Acids, Proteins and Membranes* (ed. Sarma, E. Clementi, R.H.) (Adenine Press, New York, 1984).

18. Nakano, N., Otsuka, J. and Tasaki, A., Biochim. Biophys. Acta *236*, 223-233 (1971).

19. Theorell, H. and Ehrenberg, A., Acta Chem. Scand. *5*, 823 (1951).

20. Englman, R. and Jortner, J., Mol. Phys. *18*, 145-164 (1970).

21. Huang, K. and Rhys, A., Proc. Roy. Soc, A*204*, 406-423 (1950).

22. Kubo, R. and Toyozawa, Y., Prog. Theor. Phys. *13*, 160-182 (1955).

23. Soules, T.F. and Duke, C.B., Phys. Rev. B *3*, 262-274 (1971).

24. Silbey, R., Ann. Rev. Phys. Chem. *27*, 203-223 (1976).

25. Goldstein, R.F. and Bialek, W., Phys. Rev. B *27*, 7431-7439 (1983).

26. Tsubaki, M., Srivastava, R.B. and Yu, N.-T., Biochemistry *21*, 1132 (1982).

27. Powers, L., Biochim. Biophys. Acta *683*, 1-38 (1982).

28. Kitagawa, T., Nagai, K. and Tsubaki, M., FEBS Letters *104*, 376-378 (1979).

29. Phillips, S.E.V., J. Mol. Biol. *142*, 531-554 (1980).

30. Levy, R.M. and Karplus, M., Biopolymers *18*, 2465-2495 (1979).

31. Randall, J. and Vaughn, J.M., Philos. Trans. Roy. Soc. Lond. A*293*, 341-347 (1979).

32. Kosic, T.J., Cline, R.E. and Dlott, D.D., Chem. Phys. Lett. *103*, 109-114 (1983).

33. Redi, M.H., Gerstman, B.S. and Hopfield, J.J., Biophys. J. *35*, 471-484 (1981).

34. Bialek, W., Thesis, UC Berkeley (1983).

35. Sielecki, A.R., Hendrickson, W.A., Broughton, C.G., Delabaere, L.T.J., Brayer, G.D. and James, M.N.G., J. Mol. Biol. *134*, 781-804 (1979).

36. James, M.N.G., Sielecki, A.R., Brayer, G.D., Delabaere, L.T.J. and Bauer, C.A., J. Mol. Biol. *144*, 43-88 (1980).

37. Kleinfeld, D., Okamura, M.Y. and Feher, G., Biophys. J. *41*, 121a (1983).

Adiabaticity Criteria in Biomolecular Reactions

P. Wolynes

School of Chem. Sci.

University of Illinois

Urbana, IL 61801

1. Introduction

There is a frustrating psychological complementarity presented to the physical scientist trying to understand biomolecular reactions. Ultimately one desires a complete mental picture of the atomic scale events involved in a reaction. On the other hand one seeks an understanding sufficiently general that one can see that only a few details matter. If this is possible, sweeping predictions can be made and simple experimental correlations can be adduced. Until a few years ago the first goal seemed the more quixotic but with the advent of computer simulations of biomolecules [1] and a variety of laboratory dynamical probes [2] such a picture seems almost in reach. It is now an appropriate time to ask which details of molecular motion will surface in determining overall reaction kinetics. This question has occupied many condensed matter theorists in recent years [3], most of this attention focussing on simple isomerization reactions in liquids. In many biological reactions another feature enters, namely the comparative role of electronic dynamics and molecular motion. The erudite way of saying this is, "which is more appropriate: an adiabatic or nonadiabatic description of reactions?" This issue was hotly debated eight years ago at another meeting here in Philadelphia [4]. In this lecture I would like to indicate how this issue of adiabaticity criteria is related to the issue of molecular motion in biomolecules. Indeed a simple "Gedankentheorie" shows how the electronic motion can act as a clock for the molecular motion. I will first review these ideas for thermally activated reactions. These have been more leisurely presented in a paper by Hans Frauenfelder and myself [5]. These notions are buttressed by some simulations carried out by Ray Cline in my research group [6]. Finally, I will make some speculations on adiabaticity criteria for reactions in which nuclear tunneling is involved.

2. Transition State Theory vs. The Golden Rule

As befits its longevity, the transition state theory of reaction kinetics suppresses the maximum number of details in calculating a rate constant.

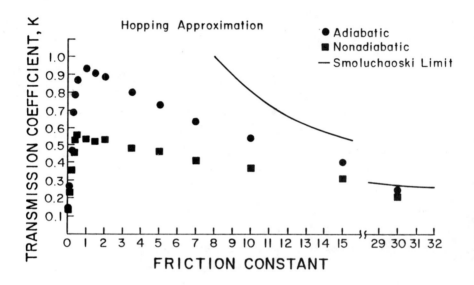

Figure 1

The forward rate constant k_{AB} counts the number of paths per unit time that carry the system from well A to well B. Every reactive trajectory must cross through the high energy bottleneck or transition state bottleneck at least once. If this passage is unhindered the crossing will occur only once on each path. Thus a count of paths per unit time would be equivalent to a count of the flux across the transition state. The number of reactions per unit time is therefore

$$\frac{dN_A}{dt} = \frac{1}{2}v_{th}N^{\neq} = \frac{1}{2}v_{th}\frac{N^{\neq}}{N_B}N_B$$

where v_{th} is the thermal velocity and $N^{\neq}dx$ is the number of systems within a range dx of the transition state configuration. The ratio $N^{\neq}/N_B$ is essentially an equilibrium constant and thus depends only on the free energy difference of the transition state configuration and the reactant. Thus the rate constant depends only on equilibrium quantities

$$k_{AB} = \frac{\omega_B}{2\pi} e^{-(G^{\neq} - G_B)/k_b T}$$

Because of the quasi-thermodynamic nature of this result it can explain a huge number of experimental correlations, most notably the usual Arrhenius temperature dependence and the linear free energy relationships of organic chemistry [7]. All motional details have been subsumed under the ansatz that reactive trajectories cross the transition state only once.

The transition state theory also assumed that the reaction occurs on a single energy surface. Often, however, the barrier to reaction arises from the crossing of two electronic energy surfaces. The exact crossing will be avoided because there will usually be a resonance energy between the two electronic configurations, Δ. If Δ is large the motion will remain on the lower surface. If Δ is small, however, the electronic motion may not be able to keep up with the nuclear motion and the system cannot transit from one diabatic surface to the other. In biological electron transfer reactions this resonance energy arises from electron tunneling and if the centers are far separated the energy splitting can be quite small. Δ can also be small if the interacting potential surface differ in spin quantum numbers thus requiring the intervention of weak spin-orbit forces. In these cases nonadiabatic effects become important.

How small must Δ be? The transition between surfaces can only occur when there energies do not differ by too much. Essentially this is in a region of size $\ell_{LZ} = 1/|F_2 - F_1|$ of the crossing. $F_2 - F_1$ is the difference in slope of the two adiabatic surfaces. The probability of making a transition is (for small Δ) proportional to the time spent in this region $\tau_{LZ} = \ell_{LZ}/v_{th}$ multiplied by the rate of electronic motion $\Delta/\hbar$. The degree of adiabaticity [8] is determined by the parameter $\gamma_{LZ} = \Delta/\hbar\tau_{LZ} = \Delta^2/\hbar F v$. If $\gamma_{LZ} \gtrsim 1$ the reaction should be adiabatic. On the other hand if $\gamma_{LZ} < 1$ only a fraction of the attempts at crossing the barrier are successful. The transition state result must be modified by a transmission coefficient:

$$k = \gamma_{LZ} k_{TST} = \frac{\Delta^2}{\hbar F v} \omega_B e^{-(G^{\neq} - G_B)/k_B T}.$$

This is essentially the same result that would be obtained through application of quantum mechanical perturbation theory i.e. the Golden Rule [9].

Although the Golden Rule contains an additional parameter Δ that characterizes electronic motion, still the details of the nuclear motion are suppressed. Most of the correlative successes of transition state theory are preserved but there is a new possibility of control through Δ. As we hear at this meeting, it is believed that many undesirable electron transfer reactions that are thermodynamically possible in the respiratory chain and photosynthesis are inhibited by the smallness of Δ due to large distances between the charged centers.

3. Molecular Motion and Reaction Rates

Both transition state theory and the Golden rule (for activated processes) have finessed the question of molecular motion by focussing attention on the bottleneck configurations for the reaction. If the bottleneck has been properly identified the details of molecular motion can modify the reaction rate by changing the number of times a reactive trajectory crosses the transition state. If a typical reactive trajectory crosses $\underline{n}_c$ times in the forward direction each of those crossings was incorrectly counted as a successful trajectory by transition state theory. Thus for an adiabatic reaction there should be an additional transmission coefficient $f_{ad} = 1/\underline{n}_c$. Similarly for a nonadiabatic reaction, if there are $\underline{n}_c$ crossings rather than one, there are more possibilities to hop from one electronic surface to the other thus modifying the adiabaticity criterion. Essentially $\gamma_{LZ} \underline{n} = \gamma^*$ is the appropriate parameter. γ^* can be much larger than γ. If $\gamma^* \ll 1$, it is still unlikely that a transition has occurred. The total transmission coefficient is $\gamma^* \cdot f_{ad} = \gamma_{LZ}$, the same as would be obtained assuming only a single attempted crossing. If $\gamma^* > 1$ there is no problem making the electronic transition and the transmission coefficient should be the adiabatic one $1/\underline{n}_c$.

What dynamical features can make $\underline{n}_c$ much larger than 1? To understand this we need to examine the nature of motions in a biomolecule. Let us first suppose the reaction coordinate is a simple motion such as a side group rotation. Computer simulation studies suggest that at short times such motions are very much like atomic motions in a simple fluid, but with some constraint [10]. Thus to get a qualitative picture we can think of the motion as a Brownian motion in a potential field. The simulations suggest that on these short times the amount of friction on say, a ring rotation, is roughly what it would be in an ordinary fluid. The friction and random forces on the reaction coordinate can be crucial. This was first noted by Kramers in the 1940's [11].

If friction is too small the number of recrossings will be large because the system will ricochet off the opposite wall of the potential and recross. The time for the system to recross may be rather short or long depending on the topography of the potential surface. Let's call the time between ballistic recrossings is τ_R. If τ_E is the time required for friction to trap the particle, then the number of recrossings $n_c = \tau_E/\tau_R$, in this limit. The time τ_E is inversely proportional to the friction constant ς. Thus in this regime the adiabatic rate coefficient is proportional to ς.

If the friction is very large, the number of recrossings is again large. The motion is like that of a drunk on the top of a hill and the number of recrossings is proportional to the friction constant which measures the rate at which the trajectory is interrupted. Thus in this regime the adiabatic rate will be inversely proportional to the friction constant. This is sometimes called the "diffusion limit."

If there is a happy medium (amount of friction) in between where neither ricocheting or drunk walking play a role there will be a small plateau in which transition state theory will be valid. Kramers in his early work thought this would be the case. Thus his paper deemphasized the role of friction. This is doubtless responsible for the small immediate effect of his work. It is through the dusting off of this theory by recent workers that it has been seen that the plateau may be more restricted than Kramers believed [3]. Friction in a solvent depends on collision rate which is a strong function of pressure. Systematic measurements of pressure effects on a isomerization indicate that the major part of the pressure dependence of the transmission coefficient [12] may be friction. Also magnetic resonance measurements of ring rotations in proteins have been interpreted in the same way [13].

A plot of transmission coefficient versus friction is shown in Figure 1, for an adiabatic and nonadiabatic reaction. In the very low and high friction regimes they agree. But when the adiabatic rate comes up to the Golden rule value (for the nonadiabatic process on the same surface) there *is* now a plateau, as indicated by our arguments. That this Gedankentheorie is at least correct within the model is shown by the data points which are simulations performed by Ray Cline based on surface hopping model with Brownian motion effects [6]. Similar results based on different arguments have been obtained by Friedman and coworkers [14].

The large plateau for reactions with small Δ would indicate that Golden rule calculations should be safe for large distance electron transfer if only small readjustments are needed in the protein. This last caveat is not at all a priori satisfied. A considerable part of the activation energy can come from relatively large scale protein movements ($> .1\text{\AA}$). These can be quite slow as evidenced by Mössbauer measurements [15]. In fact for these distance and time scales the protein may behave more like low grade engine grease or even a glass than an ordinary liquid! In that case all bets based on the Golden rule are off.

Of course with such a high effective friction we may wish to change our perspective. The molecular dynamics studies show that friction of this magnitude *must* come from some kind of activated events. That is, there are additional barriers to progress along the reaction coordinate. Thus, in some sense, we may have misidentified the bottleneck. However finding a proper bottleneck (and there may not be a unique one) is far from trivial.

4. Molecular Motion and Reactions in the Tunneling Regime

Although nuclear tunneling usually plays only a secondary role at physiological temperatures, much of the information about the potential energy surface in biomolecular reactions is gleaned from low temperature measurements where tunneling is important. It is interesting to understand how electronic and configurational time scales compete in this regime. At first this seems mysterious because motion in tunneling is classically forbidden and we must think how to characterize its time scale. While a perennial (and slightly controversial) question [16], for the present purposes this query can be reasonably answered. The WKB theory of tunneling indicates that there is a classical trajectory in imaginary time that determines the escape rate [17]. This tunneling path is known for simple barrier transmission problems both without friction [16] and with friction [18]. The concept of friction in the tunneling regime requires some subtlety but this is an issue that has received considerable attention recently [18,19]. At $T = 0$ this path has a typical velocity $q_o/\tau_o = v_{\text{tun}}$ where $\tau_o = \omega_o^{-1}$ when there is no damping and $\tau_o = \varsigma/m\omega_o^2$ when the damping is large. In this expression q_o is of the order of the separation between the wells in configuration space.

Once the tunneling velocity is known the adiabaticity criterion can be found using this velocity $\gamma = \Delta^2/\hbar F v_{\text{tun}}$. Amusingly for large damping this is the same result as is found for over the barrier motion.

If large scale motions with very large friction because of extra barriers are involved again one must look carefully at these results. Tunneling with intermediate states has been studied in other contexts such as localization in one dimensional wires [20]. Perhaps some of these results may be of value in the biophysical context.

5. Summary

Our increasing sophistication about motions in biomolecules allows us to see the limits of existing theories of reactivity. Often these limits are broad and simple theories such as Golden Rule calculations will be valid. Sometimes these limits are very constraining and the simplest theories will fail. In these cases, where longer time scale and distance scale motions in the protein couple strongly to the reactive event, a much more complete understanding of biomolecular dynamics will be needed to understand the chemistry.

Acknowledgements

I would like to thank Hans Frauenfelder for a most delightful collaboration on some of the ideas presented here. Thanks are also due to Ray Cline for his simulation studies. My research in this area has been supported by the National Science Foundation.

References

[1] J.A. McCammon and M. Karplus, Ann. Rev. Phys. Chem. *31*, 29 (1980).

[2] P. Debrunner and H. Frauenfelder, Ann. Rev. Phys. Chem. *33*, 293 (1982).

[3] J.L. Skinner and P.G. Wolynes, J. Chem. Phys. *69*, 2143 (1979); J.R. Montgomery, B.J. Berne and D. Chandler, J. Chem. Phys. *70*, 4065 (1979); R.F. Grote and J.T. Hynes, J. Chem. Phys. *64*, 4465 (1981) and many other references too numerous to mention.

[4] *Tunneling in Biological Systems*, B. Chance et.al, eds. (Academic Press, New York, 1979).

[5] H. Frauenfelder and P.G. Wolynes, Science, to appear 1985.

[6] R. Cline and P.G. Wolynes, in preparation.

[7] W.J. Albery, Ann. Rev. Phys. Chem. *31*, 227 (1980).

[8] W. Kauzmann, *Quantum Chemistry*, (Academic Press, New York, 1957); L.D. Landau and E.M. Lifshitz, *Quantum Mechanics* (Pergamon, London, 1958).

[9] See J. Hopfield in Ref. 4.

[10] J.A. McCammon, P.G. Wolynes and M. Karplus, Biochemistry *18*, 927 (1979). ·

[11] H.A. Kramers, Physica (Utrecht) *7*, 284 (1940).

[12] D.L. Hasha, T. Eguchii and J. Jonas, J. Chem. Phys. *75*, 1750 (1981); J. Am. Chem. Soc. *104*, 2290 (1982).

[13] M. Karplus and J.A. McCammon, FEBS Lett. *131*, 34 (1981).

[14] B.L. Tembe, H.L. Friedman and M. Newton, J. Chem. Phys. *76*, 1490 (1982); H.L. Friedman and M. Newton, Faraday Disc. Chem. Soc. *74*, 73 (1982).

[15] F. Parak, E.W. Knapp and D. Kucheida, J. Mol. Biol. *162*, 177 (1982).

[16] M. Buttiker and R. Landauer, Phys. Rev. Lett. *49*, 1739 (1982).

[17] C.G. Callan and S. Coleman, Phys. Rev. *D16*, 1763 (1977).

[18] A. Caldeira and J. Leggett, Ann. Phys. *149*, 374 (1983) and references therein.

[19] P.G. Wolynes, Phys. Rev. Lett. *47*, 968 (1981).

[20] B. Ricco and M. Ya Azbl, Phys. Rev. B *29*, 1970 (1984).

Structural Dynamics and Reactivity in Hemoglobin

J. Friedman and   B. Campbell

AT&T Bell Laboratories

Murray Hill, NJ 07974

1. Introduction

Proteins are macromolecular machines. In cases where the reactive centers are relatively small chromophoric units embedded in a protein matrix, they bear a resemblance to more familiar condensed phase materials such are organic mixed crystals or impurity centers in a lattice. Unlike these systems, the protein matrix has a structure that has been fine tuned over the course of evolution for a specific biological function. A major goal of biological research has been to understand how energy is both stored and utilized within these structures for the purpose of biological function. Despite the remarkable success of structural chemists and biochemists in determining the equilibrium structures and the functional properties of many protein, this problem has not been answered on a microscopic level. A major deficiency of this approach is the absence of information on how structural dynamics relates to reactivity. As a case in point, the status structure and functional properties of hemoglobin (Hb) are exceedingly well characterized [1-2], yet even for this well-studied system a detailed microscopic account of the relation between structure and function is lacking. Part of the problem is that, although high-resolution x-ray crystallographic techniques provide detailed information about equilibrium structures, they cannot be readily used to probe the transient structures that link the initial and final functional states of a protein. In recent years significant progress has been made in addressing such problems through the applications of a variety of laser based time resolved spectroscopies. At least in part these advances are the result of a chemical physics perspective based on the analogy with the impurity centers or the organic mixed crystal.

The equilibrium structures of ligand-free and ligand-bound Hb are well characterized [1,3]. In going from the ligand-free to ligand-bound protein, the local tertiary structure about the binding site responds to the change in both the quarternary structure (low-affinity T state → high-affinity R state) and the state of ligation at the heme (deoxy-liganded). To fully explore how protein structure controls the ligand- binding properties of the heme, it is necessary to determine how the local tertiary structure responds to functionally important perturbations (change in quarternary structure, ligand dissociation) and to establish correlations between specific structural features associated with the tertiary structures and parameters of ligand binding.

In this work, we first review the results of room temperature time resolved Raman studies that have provided dynamical information about specific structural degrees of freedom in Hb. These results are compared with those obtained using transient absorption and cryogenic trapping techniques. Based on these results a model is presented to account for the various relaxation processes. Finally an attempt is made to show how these different processes contribute to the reactivity of Hb thus providing a working microscopic model relating structural dynamics to ligand binding dynamics. The focus will be largely on the ligand off rate. The contributing elementary steps comprising the macroscopic ligand off rate are analyzed in terms of reaction coordinate diagrams that can be assigned to transient structures occurring over the time course of the reaction.

2. Structural Dynamics: Transient Raman Studies

The initial goal of most equilibrium and transient Raman studies on Hb is to determine the influence of the protein structure upon the ligand binding site, i.e. the heme. The protein structure responds to a variety of perturbations including ligand binding at the heme and solution conditions such as pH and phosphate concentration. If one views the heme as a solute molecule or an impurity center, the protein can be viewed as either a complex solvation sphere or a matrix respectively. The studies to be discussed are geared towards exposing the influence of the protein solvent or matrix upon the heme. To compare the influence of different protein structures it is necessary that the impurity center (the solute) be the same electronic species. Since the binding of a ligand to the heme alters the electronic makeup of the heme, comparisons between Hb's having ligand bound hemes and deoxy hemes are at best difficult. What is required is a fixed heme species with the surrounding protein structure as the variable. In these studies we use deoxy (i.e. high spin five coordinate Fe^{+2}) heme as our chromophore. This choice enables us to compare protein structures that have equilibrated with either deoxy or liganded hemes by virtue of the fact that rapid photodissociations of liganded Hb's results in the appearance of a deoxy heme on a time scale that is fast compared to protein relaxation [4]; consequently, one is left with a deoxy heme being influenced by the protein structure that was initially in equilibrium with the starting liganded heme. For example, the resonance Raman spectrum of the photoproduct of R state HbCO at "early" times will reflect the influence of the liganded R state protein structure upon photo-generated deoxy heme. Such non-equilibrium forms are often designated as Hb*(Mb*).

2.1. ν_{Fe-His}

The bond between the iron and the proximal histidine represents the only covalent link between the heme and the protein in both Hb and Mb. In high spin Fe^{+2} five coordinate forms of Hb and Mb, the frequency of the iron-proximal histidine stretching has been unambiguously assigned to a specific Raman band that is enhanced with blue excitation frequencies that are resonant with the Soret absorption band [5]. The frequency of this Raman band, ν_{Fe-His}, exhibits a systematic dependence upon protein tertiary structure [6,7]. A comparison of equilibrium forms [6] of deoxy Hb derived from different vertebrates, mutant human forms, chemically modified HbA and kinetically isolated [7h] doexy forms of HbA reveals that T state deoxy Hb's typically have lower ν_{Fe-His} values than do R state deoxy Hb's. In addition, ns time resolved Raman [7a-g] studies show that the deoxy photoproduct, Hb*, occurring within 10 ns of photodissociating the parent liganded species exhibits a value of ν_{Fe-His} that is increased relative to the corresponding deoxy form having the same quaternary structure as the starting liganded species. Thus, T and R state Hb* at 10 ns typically have a higher frequency for ν_{Fe-His} than corresponding T and R state Hb's. This sytematic behavior and the associated species specific spread in values for ν_{Fe-His} are shown in Figure 1.

The responsiveness to tertiary structure that is observed for equilibrium and transient forms of deoxy Hb is not evident in the Fe^{+2} liganded species [8]. Al-

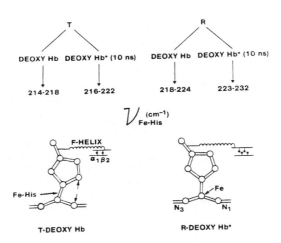

Figure 1. Systematic variation of the iron-proximal histidine stretching frequency with protein tertiary structure.

though the Fe-His stretching mode has not been clearly identified in the liganded forms, it appears that in most instances the environment about the iron is not very different between T and R state forms of liganded Hb's. Steady-state Raman, I.R. and EXAFS studies all indicate that differences in strain energy between R and T are not evident at the iron in CO derivatives. Picosecond time resolved Raman studies using the anti-Stokes H_2 shifted output of an active passive mode locked Nd:YAG laser (435 nm, 25 ps) indicates that within 25 ps of photodissociating carboxy or oxy Hb's, the R-T differences in ν_{Fe-His} are fully developed [9]. More recently, Campbell and Friedman using ps pulse probe techniques have shown that for photodissociated oxyHbA at pH 8, the line shape and frequency of ν_{Fe-His} does not change between 25 ps and 2.0 ns subsequent to dissociation [10].

Relaxation of ν_{Fe-His} to the frequencies characteristic of either deoxy R or T does not begin until several ns subsequent to photodissociation [11]. The relaxation of ν_{Fe-His} for R state HbA appears biphasic within the limits of the measurement. At pH 7.0 and lower, there appears to be a fast phase that ends at about 100 ns and a slower phase that persists out to μ seconds. At pH 9.0 no fast phase is observed. With decreasing pH both the rate and extent of the relaxation are increased as seen in Figure 2. Deoxy R state HbA values for ν_{Fe-His} (222-224 μm^{-1}) are reached at between 0.5 and 10 μ sec subsequent to dissociation depending upon the pH. The frequency and line shape for ν_{Fe-His} associated with the equilibrium T state deoxy species begin to be observable 10 μ sec subsequent to photodissociation which is consistent with the R to T switching times observed in transient absorption studies [13]. The effect upon ν_{Fe-His} of the pure T relaxation associated with the liganded to deoxy conformation has also been observed for photodissociated carboxy Hb from blue fin tuna at pH 5.8 [11]. In that instance the T state tertiary change is nearly complete by 10 μ sec.

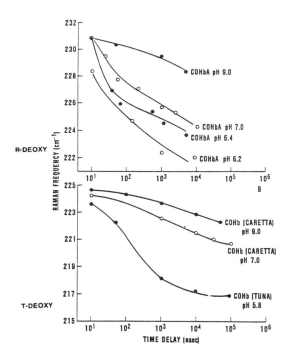

Figure 2. The relaxation subsequent to photodissociation of the tertiary protein structure as reflected in the frequency of ν_{Fe-His} for humans, see turtle (Caretta) and tune hemoglobins.

2.2. ν_4

The oxidation state marker band, ν_4, is strongly enhanced by blue excitation. The frequency of ν_4 is typically around 1353 cm^{-1} for deoxyhemes and $\sim$ 1375 cm^{-1} for liganded low spin Fe^{+2} hemes (e.g. CO, O$_2$, NO). As with ν_{Fe-His}, ν_4 for the deoxy Hb's exhibits a systematic dependence upon tertiary structure [12]. T state deoxy Hb's have higher frequencies than the corresponding R state species and the R and T state deoxy photoproducts at 10 ns are at lower frequency than the corresponding relaxed deoxy species. Indeed it has been observed that for both equilibrium R and T state deoxy Hb's [6b] and the 10 ns deoxy photoproducts [14], ν_4 and ν_{Fe-His} exhibit an inverse linear correlation in frequency. Variations in ν_4 for deoxy Hb's have been attributable to variation in the π electron density of the porphyrin macrocycle, primarily through π back donation from the iron [15]. Consistent with above is the hypothesis that the inverse correlation arises out of a modulation of π density, through the orientation of the proximal histidine. If this were so, one would anticipate an obligatory response in ν_4 with changes in ν_{Fe-His}. However, it has been noted [16] that the time evolution of ν_4 subsequent to photodissociating COHbA [17] does not follow that observed for ν_{Fe-His} [11]. The fast relaxation ($\sim$ 10 ns) observed for ν_{Fe-His} is not evident in the decay curves for ν_4. Furthermore in cryogenically trapped samples of photodissociated MbCO (vide infra), ν_4 is invarient both with temperature and deoxy species (i.e. Mb vs. Mb*), whereas ν_{Fe-His} is different for Mb vs. Mb* below 180K [18,19]. These observations indicate that there may be some common structural element linking ν_4 and ν_{Fe-His} but that this linkage is not obligatory and is not likely to be a result of direct electronic coupling between the porphyrin π system and the proximal histidine.

2.3. The 345 cm^{-1} Band

In all R and T state equilibrium deoxy Hb's, there is a well defined Raman band at $\sim$ 345cm^{-1}. The R and T state deoxy photoproducts both at 30 ps [9] and 10 ns [7b] do not typically exhibit this well defined 345 cm^{-1} peak. Instead, there is seen a weak shoulder at $\sim$ 355cm^{-1}. The observation that this Raman band is sensitive not to quaternary structure per se but to whether the heme had been liganded within 10 ns of the measurement suggest that this Raman feature is useful as an indicator of the ligation history of the deoxy Hb [7b]. It reflects the extent to which the environment about the heme retains memory of the ligation state of the heme. Subsequent to photodissociation of COHbA the 345 cm^{-1} band reappears on approximately the same time scale as the R state tertiary relaxation [11]. At pH 6.4 the 345 cm^{-1} is fully developed within 9.0 μ sec of photolysis whereas the ν_{Fe-His} still has the nonequilibrium frequency associated with R state deoxy Hb indicating that although the quaternary structure has not at this time switched, the heme environment has already lost memory of the initial ligand state of the heme. The structural parameter or feature responsible for this Raman band has not been clearly identified.

2.4. The Core Size Marker Bands

There are several bands in the Raman spectra of deoxy Hb's whose frequencies are known to correlate with the core size of the heme [15]. Transient Raman studies indicate that the deoxy photoproduct of COHbA at 10 ns has core size marker bands whose frequencies are shifted several ($< 4cm^{-1}$) relative to the stable deoxy form [20]. These shifts are consistent with an expanded core in the transient species. Kinetic isolation techniques show that the core is relaxed within 300 ns subsequent to photolysis (for COHbA) [7h].

2.5. Summary of Hb Dynamics

The picosecond time resolved Raman spectra of photodissociated COHb and O_2Hb indicate that within 25 ps of dissociation the heme is a high spin Fe^{+2} [9,21]. R - T differences in the deoxyphotoproduct are also evident within 25 ps of photodissociation [9]. Differences between the deoxyphotoproduct and deoxy Hb are attributable both to R - T differences as well as additional protein tertiary structure changes due to the protein response to ligation. The ligation induced tertiary structure begins to relax on a ns time scale and persists to a pH dependent degree out to several μ seconds. There is both a fast phase (< 100 ns) that effects ν_{Fe-His} and the core size markers and a slower phase (≤ 10 μ sec) that includes ν_4 and ν_{Fe-His}. The fast relaxation of the core may correlate with the fast phase of the ν_{Fe-His} relaxation. The slow component of the ν_{Fe-His} may follow the same time course as ν_4. Several μ seconds subsequent to dissociation the spectrum of the deoxyphotoproduct at pH 6.4 resembles R state deoxy Hb. This tertiary relaxation is then followed by a slower one (> 10 μ sec) that gives rise to the T state deoxy spectrum.

2.6. Comparison Between Mb and Hb

In contrast to Hb, Mb has a deoxy photoproduct at 10 ns that is indistinguishable from deoxy Mb both with respect to all the Raman bands discussed above for Hb and virtually tne entire rest of the spectrum as well [14,18b]. As 25 ps, the low frequency spectrum (including ν_{Fe-His} and the 345 cm^{-1} band) of the deoxy photoproduct are also the same as in deoxy Mb [22]. The high frequency regime of the 30 ps spectrum does however reveal a shifted core size marker band which relaxes on a severan hundred ps time scale [23]. Preliminary results in our laboratory indicate that the line shape and frequency of ν_4 and possibly the core size marker bands (ν_2) but not ν_{Fe-His} can be influenced by the energy density of the ps pulses [24].

2.7. Comparison with Cryogenic Raman Studies

A comparison between room temperature ns transient Raman spectra and cw cryogenic spectra of deoxy and photodissociated Hb indicates that the unrelaxed deoxy photoproduct observed at 10 ns can be trapped at cryogenic transient Raman are frozen in at cryogenic temperatures. In particular, the "ligation memory" band at 345 - 355 cm^{-1} shows the same behavior in the two temperature regimes. Lowering the temperature results in an increasing frequency for ν_{Fe-His}; however, the R-T differential for ν_{Fe-His} in the R state deoxy photoproduct at 300K over that of T state deoxy Hb (230 vs 215 cm^{-1}) is still maintained at 77K (240 vs 232 cm^{-1}).

As mentioned earlier Mb differs from Hb in that within 25 ps of photolysis the heme and heme environment with respect to ν_{Fe-His} and the 345 cm^{-1} band show no indication of the heme having been ligated. The heme core however, appears to relax to an equilibrium position within a fraction of a ns, although there is the possibility that the unrelaxed core may be an artificat photoproduct of Hb appears to correlate with that of the 25 ps of 10 ns 300K photoproduct, in Mb there appear to be differences.

At virtually all temperatures (1.2 to 300K) and at all probe times (25 ps and longer at 300K) the photoproduct of MbCO exhibits a "ligation memory" band in the 345 cm^{-1} - region [18,19]. This finding indicates that either ligand binding in Mb does not induce the changes in the heme pocket that are responsible for the

345 to 355 cm^{-1} shift as seen in Hb or the barrier for the relaxation of this ligand binding induced structural change is miniscule or non-existent in Mb. Either possibility indicates that on any meaningful time scale for ligand rebinding, the heme pocket in Mb will be relaxed with respect to this particular structural feature. There is evidence however for a qualitatively different type of ligand induced structural change in Mb. Below 100K the Mb* and the Mb do differ in that the 345 cm^{-1} band appears as a well defined doublet (338, 345 cm^{-1}) for the former but not the latter [18b,19].

The report [22] of no difference in ν_{Fe-His} between the 25 ps deoxy photoproduct of MbCO and Mb at 300K suggests that no difference is expected in ν_{Fe-His} at cryogenic temperatures. Surprisingly, it was observed [18b] and confirmed [19] that at temperatures between 4.2 and 77K, ν_{Fe-His} of the deoxy photoproduct is both shifted to higher frequency (232 cm^{-1}) and decreased in intensity relative to the frozen Mb sample ($\nu_{Fe-His} = 227$ cm^{-1} $T < 100$K). Temperature dependent changes in the frequency of ν_{Fe-His} for both Mb and the MbCO photoproduct are not observed until $T > 100$K [19]. Above 100K, ν_{Fe-His} in the cw spectra of both Mb and the MbCO photoproduct being to display a marked shift with temperature increases, both species exhibit a shift of ν_{Fe-His} to the ~ 22 cm^{-1} 300K value. From 2 to 100K ν_{Fe-His} does not shift in frequency for both Mb and Mb*; however the relative intensity of ν_{Fe-His} for Mb* exhibits a clear cut and pronounced temperature dependence. The intensity of this Raman peak decreases in what appears to be a linear fashion with temperature in going from 100 to 2K [19]. Rousseau and coworkers [18b] have shown that at 4.2K the incident laser beam can alter the intensity of ν_{Fe-His} presumably through local heating. This temperature effect is reversible in that the 2K spectrum is regained from a continuously illuminated sample that starts at either 80 or 40K and recooled at 2K [19].

Rousseau and coworkers [18a] have shown that at ≤ 4.2K the MbCO photoproduct has core size marker bands that are shifted to lower frequency by a couple of cm^{-1}s. This shift suggests an expanded core for the photoproduct that is consistent with lengthened Fe=pyrrol nitrogen distances and partially relaxed ($< .5$Å) out of plane displacement for the iron that were reported from EXAFS studies at 4K [26]. More recently [19] it was observed that the asymmetrcally shifted ν_2 core marker band in Mb* consists of two components from two distinct populations. One component has a deoxy Mb value for ν_2 and the other has a decreased frequency. The polarization properties of this shifted band differ from that of the unshifted peak indicating that the hemes in the two populations have different symmetries or that when the Raman spectrum reflects the fully photodissociated population, there are no obvious changes in this asymmetrically shifted ν_2 peak over a temperature regime from 2 to 100K. Above 100K the shifted component decreases with temperature. There are also preliminary indications that for partially photolyzed samples it is the shifted ν_2 population that tends to remain unphotolyzed. As the temperature increases, the rate of rebinding increases so that for a given cw laser intensity there is a progressive loss of the shifted component with increasing amounts of rebound MbCO in the laser induced steady state population. This effect appears to be due to a photoselectivity based on polarization. None-the-less, below 100K when fully photolyzed population is probed there does not appear to be any evolution of the ν_2 line shape.

2.8. Relationship Between Raman and Optical Absorption Studies

Pioneering transient absorption studies by Gibson and coworkers [27] on the photoproduct of liganded Hb occurring within μ seconds of photolysis revealed a deoxy heme species that has an absorption spectrum that is slightly different from that of the equilibrium T state deoxy species. This non-equilibrium doexy photoproduct is termed Hb*. In addition to the perturbed absorption spectra, Hb* also exhibits enhanced ligand rebinding rates relative to the on rates of deoxy Hb. Subsequent absorption studies on the ns [12], ps [28] and femto-second [4] time scale have shed considerable light on the dynamics of Hb*.

Instantaneous ($\leq$ 25 ps) bleaching of the absorption spectra of the parent liganded species upon illumination is followed by the appearance within 350 fsec of a spectrum indicative of a five coordinate high spin Fe^{+2} heme [4]. This result is consistent with the ps Raman studies that reveal the appearance well within 25 ps of a five coordinate high spin Fe^{+2} heme [9,21]. The absorption spectrum of this deoxy-like heme is both shifted in frequency and altered in intensity with respect to Hb [28b]. The spectrum appears to be a more exaggerated form of the μ sec

Hb* spectrum described above. Both the ps transient absorption and the Raman studies indicate that there is little or nor spectral evolution of this Hb* spectrum on the ps to several ns time scale. Eaton, Hofrichter and coworkers [12] have shown using ns transient absorption that this perturbed deoxy heme spectrum (at pH 7.0) relaxes in three stages. There is a $\sim$ 50 ns, a 0.8 μ sec and a 20 μsec relaxation at pH 7.0. Transient Raman studies show that at pH 7 and lower, the Hb* occurring at several μsec ($< 10\mu$ sec) subsequent to photodissociation has a Raman spectrum characteristic of an deoxy R state protein tertiary structure (e.g. deoxy Hb (Kempsey) at pH 9.0) [11]. Similarly the absorption spectrum of Hb* at 1 μ sec resembles that of R state deoxy Hb's. These findings support the claim that the ns relaxations and 20 μ sec relaxation are respectively the relaxation of the protein tertiary structure from the liganded R structure to the deoxy R structure and the relaxation of R state deoxy structure to the equilibrium T state deoxy structure.

The 50 ns relaxation observed at pH 7.0 using the transient absorption spectrum may be reflected in the transient Raman spectra. At pH 7.0 and lower, there is a fast component to the relaxation curve for ν_{Fe-His} that levels off at 100 ns [11]. Spiro and coworkers [7h] using kinetically isolated Hb* showed that the core is essentially relaxed by 300 ns. It is conceivable that the fast relaxing component of the ν_{Fe-His} decay and the relaxation of the heme core marker bands are correlated with each other and with the 50 ns absorption transient.

3. The Structural Picture

The structural data discussed above indicates that there are several categories of structural change that occur upon ligand dissociation in Hb and to a lesser degree in Mb. On the fastest time scale are the subpicosecond electronic reorganizations of heme that are involved in the transition from a six coordinate low spin Fe^{+2} system to a five coordinate high spin Fe^{+2} system. These changes are either accompanied by or followed by rapid elastic like responses on the part of those segments of the globin that are tightly coupled both to those heme coordinates that change substantially upon photodissociation and to the major structural determinants of the overall protein conformation (e.g. quaternary structure for Hb). The ps transient Raman data suggests [9,29,30] that the segment of protein composed of the iron-proximal histidine - F helix - α_1, β_2 interface is one such tightly coupled (vide infra) unit. Subsequent to these "elastic like" rapid responses within a given tertiary structure, are the slower "viscoelastic" changes in the tertiary structure which lead to those changes associated the R to T transition.

3.1. The First 25 ps

The R - T differences and IHP induced R state differences (COHbA $\pm$ IHP) at the heme in the 10 ns photoproduct of liganded hemoglobins are fully developed within 25 ps of photodissociation. This result is striking in that prior to dissociation, these liganded Hb's exhibit only minimal differences at the iron as probed by many different techniques [8].

The appearance within 25 ps of photolysis of differences in ν_{Fe-His} due to perturbations of those parts of the protein that are involved in the stability and definition of the quarternary state indicates a tight coupling between the structural elements associated both with the iron-histidine linkage and the determinants of quaternary state. The detectability of R - T differences at the heme appears to be dependent upon some structural or electronic parameter that changes within 25 ps of photolysis. The most obvious such change is the proposed rapid movement of the iron out of the heme plane (not necessarily to the equilibrium displacement of $\sim .5$Å). A working model to account for the above described rapid changes can be synthesized from the ligation dependent movement of the iron in and out the heme plane, the relationship between ν_{Fe-His} and the tilt of histidine [31] and the allosteric core model of Karplus and coworkers [32].

For a given hemoglobin the systematic variation in ν_{Fe-His} with solution conditions that affect the stability of the quaternary structure has been attributed to changes in the tilt of the histidine (in itw own plane) with respect to the heme plane [31, 7e,f]. For a given state of ligation the more T like the structure the lower frequency of ν_{Fe-His} and the greater is the tilt. Within a given quarternary structure, the tertiary structure changes induced by ligand binding result in an increase in ν_{Fe-His} (a decrease in the tilt) in the photoproduct relative to the corresponding deoxy species. For example deoxy R state hemoglobin has a ν_{Fe-His} that is lower than that of the R state deoxy photoproduct generated within ns of photodissociating the R state liganded species. For each category (e.g. deoxy T state) there is variability among hemoglobins from different organisms; nevertheless the basic pattern of shifts for ν_{Fe-His} is maintained as summarized in Figure 1. The correlation with tilt is probably overly simplistic and may only hold for a given species of Hb. The changes in the value of ν_{Fe-His} reflect changes in bond strength which can arise from several different contributions. In the tilt model the increased tilt reduces the distance between one of the histidine carbons and the same side pyrrol nitrogen thereby increasing the non-bonded interactions and hence the repulsive force between the two. Within the small tilt approximations such a tilt weakens the iron-histidine bond leading to a decrease in ν_{Fe-His} by changing the degree to which the histidine carbons eclipse the pyrrol nitrogens [33]. Nonetheless, the bottom line is that changes in $\nu + _{Fe-His}$ can be viewed as a change in bond strength due to protein induced tertiary structure change that alter the effective strain between the heme and the histidine. For simplicity we will use the tilt to represent the composite of potential contributions to the net strain.

The allosteric core defined and discussed by Karplus and corworkers [32] provides the link between the heme-histidine geometry and the quaternary structure. The allosteric core consists of the heme and its immediate environment (heme pocket), the proximal histidine, the $\alpha_1 - \beta_2$ interface and the segment of the F helix that links the proximal histidine to the $\alpha_1 - \beta_2$ interface. It is the $\alpha_1 - \beta_2$ interface that most likely defines and determines the quaternary state.

The net strain and hence the overall geometry between the heme and the histidine are determined by two elements of the allosteric core: the displacement of the iron from the heme plane and the structure of the $\alpha_1 - \beta_2$ interface. The interface via the helix pulls on the histidine favoring a tilted orientation. The more T like the structure the greater the pull. The tilt is countered by the repulsive force between the carbon of the histidine and the elements of the heme plane. The more in plane the iron, the greater is the repulsive force countering the tilt induced by the interface. Calculations indicate a dramatic increase in the repulsive force when the iron moves from the $\sim .5\mathring{A}$ to near $0\mathring{A}$ displacement upon ligand binding [32,34].

The observation by several techniques of little or no difference at the iron between R and T state carboxy hemoglobin strongly suggests that the increase in repulsive force generated by moving the iron in plane upon ligand binding is sufficient to dominate those other factors that contribute to the structure at the iron. In other words the repulsive force now overwhelms the influence of the interface induced pull on the histidine. The ps Raman studies show that the influence of interface on the histidine geometry is detectable within 25 ps of photodissociation. Both theory [35] and experiment [4] suggest that the iron moves out of the heme plane upon ligand dissociation within a few vibrations periods. Movement of the iron out of the heme plane substantially reduces the repulsive force between the heme and histidine. Within the allosteric core description the appearance of R - T differences within 25 ps implies tight coupling between the histidine and the $\alpha_1 - \beta_2$ interface. As the repulsive force diminishes the influence of the pull from the interface starts to be evident. The observed influence of the interface at the iron is modulated by the displacement of the iron. Since the state of the interface is communicated to the iron-histidine bond within 25 ps of dissociation, is also follows that the changes at iron should be communicated to the interface within 25 ps. Recent U.V. resonance Raman studies by Spiro and coworkers [36] seem to dramatically support this claim. They report a substantial change in the Raman spectra of tryptophan $\beta 37$ within 7 ns of photodissociating COHb.

The fast elastic like changes described above do not fall into the standard category of tertiary structure changes. They are more appropriately described as the extreme conformational endpoints for a given protein tertiary structure. These endpoints represent the elastic like response of the protein to the change in the displacement of the iron (or other rapid electronic or structural changes occurring with change in ligation state). Thus for every protein tertiary structure there are at least two conformational endpoints - one associated with an in the heme plane iron and the other with the fully or substantially displaced iron. Of course the tertiary structure may be unstable with respect to either endpoint and subsequently relax to a new tertiary structure which will have its own new set of endpoints. This scheme of conformational endpoints within a given tertiary structure is valid only because the conventional relaxation of tertiary structure occurs much slower than the movement of the iron. Within the allosteric core based model described above, each value of ν_{Fe-His} implies a tertiary structure which has its own unique set of endpoints. The ν_{Fe-His} values represent one such set of endpoints - those for the out of heme plane iron but they also imply the corresponding endpoints for the in plane geometry.

It is interesting to consider whether the sub 25 ps transition from the six coordinate iron in plane limit to the five coordinate iron at least partly out of the heme plane limit is truly an elastic like barrierless process without intermediates. To isolate possible intermediates it is necessary to either increase the time resolution or decrease the temperature. Two recent cryogenic cw Raman studies on the photoproduct of MbCO bear on this question [18,19]. In both studies it was observed that between 2°K and 80°K, ν_{Fe-His} for the MbCO photoproduct relative to deoxy Mb is both at higher freqneucy (233 cm^{-1} vs $\sim$ 227 cm^{-1} and has reduced intensity. One implication of this result is that if for this molecule time equals temperature then it is anticipated that this same transient form be observable at higher temperatures using pulsed Raman techniques. However, as cited earlier, the ps Raman results show the same ν_{Fe-His} for the 25 ps photoproduct and the stable deoxy form. A possible resolution of the paradox comes from the observation that ν_{Fe-His} in both the photoproduct and the chemical deoxy forms of Mb first start to shift at $T > 100°K$ [10]. X-ray studies on Mb crystals show an anisotropic change in the volume of the protein occurring somewhere between 80 and 300K [37]. It is conceivable that below 100K ν_{Fe-His} is constrained to values that are dictated at least in part by a contracted protein volume rather than transient intermediate structures (in the case of the photoproduct). This idea is supported by the observation that at temperatures ($T > 100K$) where the cw studies show a shifted ν_{Fe-His} for the photoproduct, the ns Raman spectra [38] show similar shifts without regaining the low temperature limit ($T < 100K$). If the below 100K spectra reflects a trapped intermediate then at early times above 100K one would anticipate being able to observe such intermediates. This expectation is not supported by the ns study where at 160K the cw and the ns spectra are nearly identical [38]. Thus there is still no indication of an out of plane (iron) intermediate having a shifted ν_{Fe-His} that might be trapped on times less than 25 ps at room temperature.

3.2. Tertiary Relaxation

Subsequent to the elastic like changes occurring within 25 ps of photodissociation are a series of visco-elastic relaxations involving several different degrees of freedom. The relaxation of ν_{Fe-His} in the Hb photoproduct indicates a progressive increase in strain in the iron-histidine bond that is interpretable within the simple limits of the allosteric core model as an increase in the tilt of the histidine with respect to the heme plane. This change can orginate from motion of either the F-helix or the heme with respect to each other. The U.V. resonance Raman work of Spiro and coworkers [36] suggests that the F helix may respond rapidly to photodissociation. Subsequent relaxations may involve the tilting of the heme with respect to the heme pocket. In any event at pH 7 the relaxation of the ν_{Fe-His} occurs in three segments: a $\sim$ 100 ns component, an $\sim$ 1μ sec component and a slower component of the order of 10's of μ seconds. The latter corresponds to the R - T switching time observed in transient absorption studies [12]. The 1 μ sec relaxation is also observed in both absorption [12] and for the $\nu_4\pi$ electron density marker band [17]. It appears that since the correlation between ν_4 and ν_{Fe-His} is not likely to be electronic in origin as discussed earlier, the 1 μ sec relaxation is probably due to a reorientation of the heme with respect to the heme pocket. Such a process is likely to effect the π electron system as well as the iron-histidine linkage. The fastest relaxation component ($\sim$ 100 ns) may also involve the core size marker bands. To understand the possible origins of this relaxation, it is important to consider the implication of a change in the core size marker bands.

3.3. The Heme Core

In the photoproducts of both MbCO and HbCO the core size marker bands are unrelaxed at early times. In Mb the relaxation to the equilibrium value occurs within a few hundred ps [23] whereas in Hb it is within a maximum of 300 ns [7h]. The shifted frequencies are consistent with an enlarged core for the photoproduct. The origin of the enlarged core is not as yet clear. Several possibilities exist.

Rousseau and Argade have shown that at 2K and 4.2K the heme core markers are shited to lower frequency in the photoproduct of MbCO compared to deoxy Mb [18a]. These shifts are similar to those observed in the 300K transient spectra [23]. Since at cryogenic temperatures and at early times the dissociated ligand might be trapped near the heme, it is plausible that transient trapping of the ligand near the heme may be responsible for the perturbed core. For the ν_2 core marker, the shifted spectrum of the fully photolyzed sample persists without change from 2 to 100K. Over this same temperature regime (4-400K) the photodissociated ligand is seen using EXAFS [26] to shift away from the iron by approximately 7 Å in a near steady state population generated by repetitive flashing. Concomitantly there is an increase in the population of a kinetically much slower (×100) - rebinding species. The changes in the relative amounts of fast and slow rebinding populations have been ascribed to the changes in relative distance between the iron and the dissociated ligand. Over this same temperature interval in which ν_2 does not change, IR studies show a loss with increasing temperature of a certain population of dissociated CO [39] which is likely to be those CO's that were initially in close proximity to the iron at the lowest temperatures. The shifted core marker seen in the transient Raman at higher temperatures (300K) is also not likely due to ligand induced perturbation because of the time scales for both core relaxation and the geminate rebinding. In Hb the core relaxation of the several 10's or 100's of ns [23] seems excessive for the movement of the ligand away from iron. In addition, the fast component of geminate rebinding which has been associated with rebinding from within the heme pocket is over by a few hundred ps [43]. The ns geminate rebinding is thought to originate from rebinding of the ligand from the bulk protein i.e. the matrix process which only starts at temperatures above 200K [40]. All of these considerations make it very unlikely that the CO remains sufficiently close to the heme for a period long enough to be responsible for the unrelaxed core at elevated temperature.

Structural relaxation involving the heme core is a more plausible explanation for the change in frequency of the core size marker bands at 300K. A reasonable scenario is that upon dissociation the iron moves rapidly out of the heme plane in response to the strong repulsive forces between the histidine and the heme. As these forces decrease, the restraining force of the F helix may inhibit the full excursion of the iron [21] thus creating a metastable state in which the iron has rapidly and elastically moved partly out of the heme plane to a displacement where there is a transient balance between the remaining repulsive force and the "viscous drag" of the F helix. Differences in the subsequent relaxation times of the core among different proteins could then be due to differences in either the rigidity of the F helix or the residual repulsive force. It seems unlikely that there are gross differences in the viscoelastic properties of the different F-helices. Instead there are clear indications of differences in the residual repulsive force between the heme and the histidine as reflected in ν_{Fe-His} [7,9,25]. The lower ν_{Fe-His} for the photoproduct the greater is the repulsive interactions which destabilizes the metastable intermediate. It follows that the relaxation time for the cores should shorten with decreasing values of ν_{Fe-His} in the photoproduct. This pattern is observed for Mb and Hb where Mb has the lower ν_{Fe-His} and the faster relaxation time. One would then preduct that T state photoproducts e.g. Hb(Kansas) should exhibit much shorter relaxation times compared to the corresponding R state forms. If, on the other hand, the location of the ligand is the determining factor than distal perturbations e.g. Hb (Zürich), Mb (elephant) should result in substantial changes in the core size relaxation rates.

If it is assumed that the core relaxation at 300°K is due to structural changes at the heme then we have four categories of structural change in Hb that involve coupling between the heme and the globin. The first is the rapid elastic like response to dissociation that involves the coupling of the movement of the iron to the instantaneous tertiary structure of the starting liganded protein. The influence of this tertiary structure is reflected directly in the value of ν_{Fe-His}. This process is followed by a small readjustment of the heme core due to the final movement of the iron to its full out heme plane distance. Variations in this relaxation are likely to be due to proximal strain induced differences in the stability of the expanded core intermediate. The slower pH dependent tertiary change occurring up to a few microseconds is probably a reorientation of the heme plane within the heme pocket in response to both the fast changes that have occurred at the interface and a release to the heme as reflected in a decreasing ν_{Fe-His}. The final steps are the last changes associated with going from a deoxy R to deoxy T state species. Again this change is reflected in a continued decrease in ν_{Fe-His} as well as the appearance of subunit asymmetry in ν_{Fe-His} [11]. We are now in position to assess the impact of these relaxation process upon functional events.

4. Function and Structural Dynamics

Ligand binding properties of Hb respond significantly to changes in the quaternary structure of the protein. The iron-proximal histidine linkage is the structural parameter at the ligand binding site whose static and dynamic properties most clearly responds to homotropically and heterotropically induced changes in the protein structure. The frequency ν_{Fe-His} has been shown to correlate with several parameters of ligand binding reactivity. Kitagawa and coworkers have shown a correlation between ν_{Fe-His} of several mutant, chemically modified and solution perturbed human deoxy Hb [41]. The higher ν_{Fe-His} the greater the affinity for the first bound ligand. The affinity however is not the ideal parameter for such studies since the off-rate step will depend upon the structure of the liganded or partially liganded species. The transient Raman studies clearly show that even within a given quaternary structure, ligand binding induces local tertiary changes that relax over many 100's of ns or longer upon dissociations [11]. Consequently, even for the first binding step, the on and off rate contributions to the binding equilibrium constant will depend upon two different tertiary structures. Ideally it is more appropriate to consider correlations between on or off rates associated with given structures. Even in these cases, structural relaxations induced by the change in ligation state complicates the analysis; nonetheless it is clear that changes in ν_{Fe-His} heterotropically induced by changes in solution conditions correlate with changes in the off rate [7a,e,g]. An IHP of pH induced decrease in ν_{Fe-His} for the 10 ns photoproduct of liganded Hb's is always associated with a corresponding increase in the off rate even when the IHP does not produce a change in quaternary state. Comparisons among Hb's from different species of vertebrates also indicate a correlation between ν_{Fe-His} and the off rate [7g]. Those Hb's with the highest ν_{Fe-His} values tend to have the lowest off rates. Deviations from this correlation are expected for those Hb's having distal perturbation. For example the ligand binding dynamics in Hb (Zürich) (His E7 $\rightarrow$ Arg) differ from HbA despite very similar proximal geometries [42].

In order to fully understand the nature and basis of the link between ν_{Fe-His} and ligand binding parameters, we reduce the off rates still further into more basic components that can be more closely associated with specific structures and structural relaxations. The off rate is proportional to the intrinsic rate of dissociation for the iron-ligand bond multiplied by the probability that the ligand escapes from the protein into the solvent. In the next sections we expose the relationship between structure and these two microscopic components of the off rate.

4.1. Protein Modulated Fluctuations and the Rate of Dissociation

There have been several studies that have shown that protein induced variation in the geminate yield cannot fully account for variations in the macroscopic off rate [12a,43]. This observation is most dramatic in the cases where there is no geminate rebinding (e.g. many fish Hb's). The obvious conclusion is that the intrinsic rate of dissociation must be a protein structure dependent variable. It follows that somehow the protein structure must couple into the steps leading to thermally induced ligand dissociation.

Thermally induced ligand dissociation presumably arises from thermal fluctuations in the iron-ligand distance. Let us focus upon fluctuations in which the iron moves out of heme plane on the proximal side. (The arguments put forward are readily extended to fluctuations involving distal movements.) If we assume that transition state theory is applicable for each protein structure (conformation, substate, etc.) then the rate of dissociation is determined by the dissociation energy barrier which is the energy difference between the initial most stable configuration of the iron (usually in the heme plane) and some transitions state which has some degree of proximal displacement for the iron coordinate. The protein structure will influence the dissociation energy barrier if the energy of the initial state and the transition state are affected differently by the protein. If the iron and ligand are totally decoupled from the protein or if the iron in plane and out of plane configurations are equally perturbed by changes in the protein then the dissociation energy will not be responsive to changes in the globin.

Several studies utilizing several different techniques including Raman, IR, LFIR and XANES all indicate that as a function of protein structure (e.g. R vs T) little if any variation is observed in the iron ligand distances (Fe-H_4, Fe-N_p, Fe-CO, Fe-O_2, Fe-NO) for the six coordinate Fe^{+2} heme [8]. This suggests either an absence of coupling between the iron and protein quaternary structure or a presence of coupling but with the differences in the energy of ligand binding within different protein structures not reflected in the local environment at the iron. The latter possibility does not mean that the changes at the iron associated with ligand binding are not the source of R-T strain differences rather it indicates that when the iron is ligand bound, protein induced strain energies are residing in some weaker links within the protein structure (vide infra).

The ps Raman results discussed earlier [9] suggest both the degree to which the iron coordinate is coupled to the protein structure and the nature of protein induced differences in the transition state leading to dissociation. The appearance within 25 ps of dissociation of relatively stable (> 10 ns) R-T differences at the iron indicates tight coupling between the iron coordinate and the protein determinants of quaternary structure. The earlier discussed model which is based to a large degree of these ps results accounts not only for the iron coordinate dependent localization of protein induced strain at the heme but also explains how protein structure induces variations in the rate of dissociation. In this model *variations* in the stabilization energy of the liganded and dissociated heme arise primarily from variations in the repulsive interaction between the proximal histidine and the heme. Two factors control the interaction: the iron displacement and the orientation of heme plane with respect to the F helix. The more T like proteins structures favor a more "tilted" (more strained) orientation of the proximal histidine with respect to the heme plane as reflected in the lower values of ν_{Fe-His}. For a given displacement of the iron the more tilted (strained) geometries are associated with a greater repulsive interaction and hence a greater destabilization. The repulsive interaction is however strongly modulated by the iron coordinate. For a significantly displaced ($> .3$Å) iron the repulsive interaction is reduced resulting in small energy differences at the heme between tilted and untilted geometries. As the iron moves into the heme plane for a protein structure favoring a tilted histidine. The movement of the iron is therefore modulating the heme related energy differences among the different structures. When the iron is out of plane, the energy differences are reduced whereas these differences increase for the more in plane configuration. The same arguments apply to the orientation of the histidine with respect to the heme as a function of iron displacement. Since the repulsive force counters the protein induced tilting of the histidine, the orientation of the histidine with respect to the heme will be the result of a balance between the protein induced pull on the histidine and the iron displacement dependent repulsive interactions (see Figure 3).

We now consider within the context of this model the influence of the globin upon the proximal fluctuations of the iron coordinates that lead to dissociation. For the six coordinate iron, an in plane configuration is favored. Under these circumstances, the repulsive interactions between the heme and the proximal histidine are dominant over the globin "pull" in terms of influencing the orientation of the histidine with respect to the heme plane. The respulsive interactions favors the upright histidine orientation, therefore for the in or near in plane configuration the upright geometry is favored regardless of the quaternary structure. Thus the lower portion of the potential will for the six coordinate heme will have a similar shape and curvature independent of the globin. However, the different potential wells will be displaced by the differing amounts of energy that are required to move the iron in plane for the different heme-globin interactions. The lower ν_{Fe-His} for the five coordinate heme, the greater the destabilization of the potential will for the corresponding six coordinate species. Thus although the T and R state wells differ in energy they have roughly equal shape near the botton (hence nearly equivalent vibrational frequencies for the heme and ligand associated vibrations). As the iron moves out of the heme plane the repulsive force diminishes resulting in a greater influence of the globin in determining the local geometry. More importantly the energy difference between R and T due to

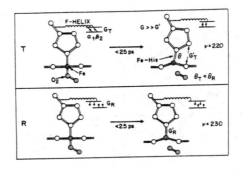

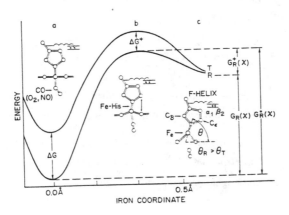

Figure 3. A schematic of both the structural changes and the associated reaction coordinate diagrams accounting for the initial events occurring within picoseconds subsequent to photodissociation in R and T state liganded hemoglobins.

the heme-histidine interaction decreases. To the extent that the transition state has a partially displaced iron, the difference in energy between R and T for example will have been reduced relative to the in plane equilibrium configuration. The more out of plane the iron the greater will be the decrease in the energy difference and hence the larger will be the difference in the dissociation energy barrier. For the slow binding ligands such as CO the transition state is expected to be more in plane than for the fast binding ligands such as O_2. The dissociation energy barrier for CO relative to O_2 is therefore expected to be less responsive to proximal perturbation.

4.2. Geminate Recombination

Geminate recombination (g.r.) in solutions has been observed to occur on two time scales in hemoglobins. There is a ≤ 200 ps rebinding observed for O_2 [28e,40] and NO [28c] and a ~ 100 ns rebinding observed for O_2 and CO [12,44]. The ps and ns processes are thought to correlate respectively with process 1 and the matrix process reported by Frauenfelder and coworkers for cryogenically isolated samples [45]. The yield of geminate rebinding for both processes appears to be sensitive to both the animal source for the Hb and the specific conformational state of a given Hb [40]. For example some Hb's such as HbA and other mammalian Hb's exhibit high yields of g.r. whereas fish and turtle Hb's do not. Distal perturbations appear to contribute to base line differences in the g.r. between certain Hb's such as HbA and HbZ (Zürich, His E7 $\rightarrow$ Arg) [42]. For a given Hb, solution induced perturbations that make the protein more T like, decrease the yield of g.r. These observations can be understood in terms of the same model that we used to explain the influence of protein structure on the rate of spontaneous dissociation.

Solution induced (e.g. pH $\pm$ IHP) perturbations that are known to alter the frequency of ν_{Fe-His} in the ns or ps photoproduct are associated with changes in the geminate yield. A decrease in ν_{Fe-His} is associated with a decrease in the geminate yield. This relationship has been observed directly by monitoring the geminate rebinding on the ps [40,7g] and ns time scales [11]. It has also been indirectly observed by monitoring the relative average photolysis yield over a single 10 ns pulse for solutions that differ only in pH or IHP concentration [7e]. This relationship between $\Delta\nu_{Fe-His}$ and Δ G.Y. includes cases where there is a change in quaternary structure as well as cases where the R structure is only destabilized (e.g. COHbA at pH 6.5 + IHP). The variation in G.Y. among the different Hb's is also at least in part correlated with ν_{Fe-His} [7g]. Those Hb's having the highest ν_{Fe-His} values for the early time photoproduct exhibit the greatest amount of g.r. Those Hb's that do not exhibit g.r. typically have low values for ν_{Fe-His}. The interspecies relationship between ν_{Fe-His} and g.r. is not expected to be a rigorous one since substantial heme pocket modifications from one species to next can be expected to alter the ligand retention time in and the escape time from the heme pocket. Within the vertebrates Hb's this effect may not be serious; however, non-vertebrate Hb's typically exhibit kinetic and Raman spectra indicative of unusual heme environments compared to the vertebrate counterparts.

The induced variations in the geminate process arising from proximal perturbations can be readily explained. Upon dissociation to the dissociated ligand is within the heme pocket. Competition between diffusive(?) escape from the heme pocket and the inner most barrier controlled rebinding determine the yield of the picosecond process (process 1). Rebinding from the "bulk" (non-hemepocket) protein gives rise to the 100 ns process. For both geminate steps, variations in the geminate yield arising from proximal perturbations appear to arise from variation in innermost barrier rather than any change in the diffusional properties of the protein. The model discussed in this work provides a framework for understanding the relationship between variations in ν_{Fe-His} and in the innermost barrier. The ν_{Fe-His} values for given structures indicates the relative resistance to the movement of the iron into the plane of the heme. For the deoxy photoproducts the energy difference due to the heme-histidine interaction will be minimal because the iron is out of the heme plane (see Figure 3). As the ligand starts to rebind the iron moves toward the heme center. The energy difference between different structures (e.g. R vs T) arising from the differences in the proximal environment now begins to increase. Thus to the extend that the transition state has a more in plane iron relative to the starting five coordinate ns or ps photoproduct, the barrier for rebinding will increase with decreasing ν_{Fe-His}. This relationship may directly explain why Mb has a much lower g.y. than HbA (222 cm^{-1} vs 230 cm^{-1} for ν_{Fe-His} of the respective transients). This scheme is supported by the correlation between the 10 ns ν_{Fe-His} values and the actual geminate rebinding rate constants (i.e. k_{21} where k observed $= k_{21} + k_{23}$ and k_{23} is the escape rate) [46]. The escape rate k_{23} exhibits little variation with proximal perturbations. It follows from this and the preceding discussion that a single reaction coordinate diagram can be drawn for a given protein structure in which the relative (vis a vis other similar structures) barrier heights for spontaneous dissociations and geminate rebinding are derived from the ν_{Fe-His} value of the five cocordinate form of the structure. The important question remains as to what extent a single reaction coordinate diagram i.e. a single structure accounts for the observed rebinding kinetics.

There are two categories of dynamical process whose influence on geminate rebinding would necessitate the consideration of more than one reaction coordinate diagrams. The first one we consider is due to thermal fluctuations within a given tertiary structure that take the protein from one substate to another. As envisioned by Frauenfelder and coworkers [45] each of these substates could have a slightly different inner barrier which gives rise to a distribution of barrier heights. At any temperature there is then a characteristic time over which a given molecule samples the full distribution. For rebinding processes that are slow relative to this sampling time, the rebinding process can be described using single average barrier. If the rebinding is fast compared to the sampling time, there there is a transient distribution of substates with an associated distribution of barrier heights. Those molecules that are in the low barrier substrates should exhibit the fastest rebinding rates. The picosecond geminate rebinding of O_2 for HbA in solution is non-exponential over the first few hundred ps subsequent to dissociation [40]. A possible source of this non- exponential behavior is a comparable or slower substate interconversion time. Since ν_{Fe-His} appears to correlate with the barrier height, the line shape of the ν_{Fe-His} Raman band might well be a reasonable place to look for structural manifestations of a distribution substate. Campbell and Friedman have monitored the ν_{Fe-His} Raman line shape associated with the surviving five coordinate population over the time course of the ps geminate rebinding of O_2 in HbA. No change in line shape is observed, only a monotonic decrease in the intensity of the band as the O_2 rebinds. Although this result raises questions regarding the substate description in so far as the distribution maps on to the structural parameters associated with the variation in ν_{Fe-His}, it is nonetheless conceivable that fast dephasing of the excited vibrational levels causes a sufficient line broadening to mask any hole burning or line shape distortions.

The second dynamical process that can modify the simple single reaction coordinate diagram description is structural diffusion. Structural diffusion is the visco-elastic like tertiary relaxation described in earlier sections of this work. From 25 ps to several ns subsequent to dissociation, ν_{Fe-His} does not relax. In contrast to the ps behavior there are over the time course of the 100 ns geminate rebinding, clear demonstrable shifts in ν_{Fe-His} (vide supra) [11]. The relaxations of ν_{Fe-His} increases with conditions that favor the T state (e.g. low pH, IHP).

These results indicate that with respect to $Fe - His$ contributions to the barrier height for rebinding (and dissociations), the averge barrier height is static for the ps rebinding but is increasing with time in a solution dependent fashion for the ns rebinding. Consequently the protein matrix responds to dissociation in manner that may be described as self induced exclusion with respect to the ligand or self induced trapping with respect to the five coordinate heme (Figure 4).

SELF INDUCED TRAPPING

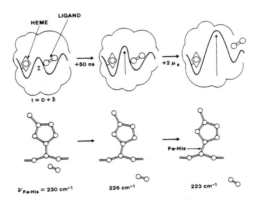

Figure 4. A schematic for self-induced trapping (for the deoxy heme) in hemoglobin based upon time resolved Raman studies.

The longer the ligand remains off, the greater is the barrier for rebinding. Furthermore the kinetics for the standard microsecond bimolecular rebinding will involve structures that are substantially different than those contributing to the faster geminate process. At high pH (> 8) where the relaxation is slowest, the structural differences between the ns and μ sec transients will the smallest whereas at low pH the faster relaxation introduces a substantial structural disparity with time. The fact that the relaxing coordinate is linked to the inner barrier suggests that a full description of rebinding kinetics requires this reaction coordinate in addition to the one used for the diagram shown in Figure 3. An additional reaction coordinate was previously invoked by Hopfield and Agmon [48] to account for the non-exponential kinetics observed for the MbCO system below 200K. Recent transient absorption studies also support the concept of additional degrees of freedom whose evolution in time influences the rebinding kinetics in hemeprotein [12].

References

[1] M.F. Perutz, Proc. R. Soc. London Ser. *B208*, 135 (1980).

[2] R.G. Shulman, J.J. Hopfield and S. Ogawa, Q. Rev. Biophys. *8*, 325 (1975).

[3] J.M. Baldwin and C. Chothia, J. Mol. Biol. *129*, 175 (1979).

[4] J.L. Martin, A. Migus, C. Poyart, Y. Lecarpentier, R. Astier and A. Antonetti, Proc. Nat. Acad. Sci. USA *80*, 173 (1983).

[5] P.V. Argade, M. Sassaroli, D.L. Rousseau, T. Inubushi, M. Ikeda-Saito and A. Lapidot, J. Amer. Chem. Soc. *106*, 6593 (1984).

[6] (a) K. Nagain, T. Kitagawa and H. Morimoto, J. Mol. Biol. *136*, 271 (1980); (b) M.R. Ondrias, D.L. Rousseau, J.A. Shelnutt and S.R. Simon, Biochemistry *21*, 3428 (1982).

[7] (a) J.M. Friedman, R.A. Stepnoski and R.W. Noble, FEBS Letters *146*, 278 (1982); (b) J.M. Friedman, *Time Resolved Vibrational Spectroscopy*, (G. Atkinson, ed.) Academic Press, 307 (1983); (c) J.M. Friedman in *Hemoglobins: Structure and Functions* (A.G. Schneck ed.) Brussels University Press, pp. 269-284 (1984); (d) T.W. Scott, J.M. Friedman, M. Ideka-Saito and T. Yonetani, FEBS Letters *158*, 68; (e) J.M. Friedman, T.W. Scott, R.A. Stepnoski, M. Ikeda-Saito and T. Yonetani, J. Biol. Chem. *258*, 10564 (1983); (f) J. M. Friedman, Science *228*, 1273 (1985); (g) J.M. Friedman, S.R. Simon and T.W. Scott, Copeia *3*, 679 (1985); (h) P. Stein, J. Lerner and T.G. Spiro, J. Phys. Chem. *86*, 168 (1982).

[8] M. Tsubaki and N.-T. Yu, Biochemistry *21*, 1140- 1145 (1982); D.L. Rousseau, S.L. Tan, M.R. Ondrias, S. Ogawa and R.W. Noble, Biochemistry *23*, 2857-2865 (1984); D.L. Rousseau and M.R. Ondrias, Biophys. J. *47*, 726-734 (1985); M. Chance, L. Parkhurst, L. Powers and B. Chance, J. Biol. Chem. *261*, 5689 (1986); S. Pin, P. Valat, R. Cortes, A. Michalowicz and

B. Alpert, Biophys. J. *48*, 997 (1985); A. Bianconi, A. Congin-Castellano, M. Dell'Ariccia and A. Biovanelli, FEBS Letters, in press.

[9] E.W. Findsen, J.M. Friedman, M.R. Ondrias and S.R. Simon, Science *229*, 661 (1985).

[10] B. Campbell, J.M. Friedman and S.R. Simon, to be published.

[11] T.W. Scott and J.M. Friedman, J. Amer. Chem. Soc. *106*, 5677 (1984).

[12] (a) J. Hofrichter, J.H. Sommer, E.R. Henry and W.A. Eaton, Proc. Nat. Acad. Sci. USA *80*, 2235 (1983); (b) J. Hofrichter, E.R. Henry, J. Sommer, R. Deutch, M. Ikeda-Saito, T. Yonetani and W.A. Eaton, Biochemistry *24*, 2267 (1985).

[13] J.A. Shelnutt, D.L. Rousseau, J.M. Friedman and S.R. Simon, Proc. Nat. Acad. Sci. USA *76*, 4409 (1979).

[14] J.M. Friedman, D.L. Rousseau and M.R. Ondrias, Ann. Rev. Phys. Chem. *33*, 471 (1982).

[15] T.G. Spiro, in *Iron Porphyrins* Vol. II, Lever, A.B.P. and Gray, H.B., eds., Addison-Wesley, MA (1983).

[16] D.L. Rousseau and J.M. Friedman, to appear in *Bioloical Applications of Raman Spectroscopy*, (T.G. Spiro, ed.) Vol. 3.

[17] K.B. Lyons and J.M. Friedman, in *Hemoglobin and Oxygen Binding*, C. Ho., Ed. (Elsevier, New York, 1982) pp. 333- 338.

[18] (a) P.V. Argade and D.L. Rousseau, Proc. Natl. Acad. Sci. USA; (b) M. Sassauoli, S. Dargupta and D.L. Rousseau, submitted.

[19] L. Powers, B. Chance, M. Chance, B. Campbell, J.M. Friedman, K.S. Reedy and Y. Zhou, in preparation.

[20] K.B. Lyons, J.M. Friedman and P.A. Fleurey, Nature *275*, 565 (1978).

[21] J. Terner, T.G. Spiro, M. Nagumo, M.F. Nicol and M.A. El Sayed, J. Amer. Soc. *102*, 3238 (1980); J. Terner, J.D. Strong, T.G. Spiro, N. Nagumo, N.F. Nicol and M.A. El Sayed, Proc. Nat. Acad. Sci. USA *78*, 1313 (1981).

[22] E.W. Findsen, T.W. Scott, M.R. Chance, J.M. Friedman and M.R. Ondrias, J. Am. Chem. Soc. *107*, 3355 (1985).

[23] S. Dasgupta, T.G. Spiro, C.K. Johnson, G.A. Dalickas and R.M. Hochstrasser, Biochemistry 24, 5295 (1985).

[24] E.W. Findsen, J.M. Friedman and M.R. Ondrias, unpublished results.

[25] M.R. Ondrias, J.M. Friedman and D.L. Rousseau, Science 220, 615 (1983).

[26] (a) B. Chance, R. Fischetti and L. Powers, J. Am. Chem. Soc. 2, 3820 (1983); (b) L. Powers, J.L. Sessler, G.L. Woolery and B. Chance, Biochemistry 23, 5519 (1984).

[27] Q.H. Gibson, Biochem. J. 71, 293 (1959).

[28] (a) B.I. Greene, R.M. Hochstrasser, R.B. Weisman, W.A. Eaton, Proc. Natl. Acad. Sci. USA 75, 5255 (1978); (b) D.A. Chernoff, R.M. Hochstrasser and A.W. Steele, Proc. Nat. Acad. Sci. USA 77, 5606 (1980); (c) P.A. Cornelius, R.M. Hochstrasser, A.W. Steel, J. Mol. Biol. 163, 119 (1983); (d) C.V. Shank, E.P. Ippen and R. Bersohn, Science 193, 50 (1976); (e) A.H. Reynolds and P.M. Rentzepis, Biophys. J. 38, 15 (1982).

[29] J.M. Friedman, M.R. Ondrias, E.W. Findsen and S.R. Simon, Proceedings SPIE 533, 8 (1985).

[30] J.M. Friedman, in *Time-Resolved Vibrational Sepctroscopy*, p. 204, A. Laubereau and M. Stockbinger, eds. Springer-Verlag, Berlin, Heidelberg (1985).

[31] J.M. Friedman, D.L. Rousseau, M.R. Ondrias and R.A. Stepnoski, Science 218, 1244 (1982).

[32] B.R. Gelin and M. Karplus, Proc. Nat. Acad. Sci. USA 74, 801-805 (1977); B.R. Gelin, A.W-M. Lee and M. Karplus, J. Mol. Biol. 171, 489 (1983).

[33] O. Bancharoenpaurpong, K.T. Schomacker and P.M. Champion, J. Am. Chem. Soc. 106, 5688 (1984).

[34] B.D. Olafson and W.A. Goddard, Proc. Natl. Acad. Sci. USA 74, 1315 (1979); W.A. Goddard and B.D. Olafson, in *Biochemical and Chemical Aspects of Oxygen*, (W. Caughy, ed.) Academic Press, New York, 87-123 (1979); A. Warshel, Proc. Natl. Acad. Sci. USA ait 74, 1789 (1977).

[35] E.R. Henry and W.A. Eaton, Biophysical J. *47*, 208a (1985); E.R. Henry, M. Levitt and W.A. Eaton, Proc. Natl. Acad. Sci. USA *82*, 2034 (1985).

[36] D. Sasgupta, R.A. Copeland and T.G. Spiro, J. Biol. Chem., in press.

[37] H. Hartman, F. Parak, W. Steigmann, G.A. Petsko, D. Ringe, N. Ponzi and H. Frauenfelder, Proc. Nat. Acad. Sci. USA *79*, 4967 (1982).

[38] B. Campbell and J.M. Friedman, unpublished results.

[39] J.O. Alben, D. Beece, S.F. Bowne, W. Doster, L. Eisenstein, H. Frauenfelder, D. Good, J.D. McDonald, M.C. Marden, P.P. Moh, L. Reinisch, A.H. Reynolds, E. Shyamsunder and K.T. Yue, Proc. Natl. Acad. Sci. USA *79*, 3744 (1982); F.G. Fiamingo and J.O. Alben, Biochemistry *24*, 7964 (1985).

[40] J.M. Friedman, T.W. Scott, G.J. Fisanick, S.R. Simon, E.W. Findsen, M.R. Ondrias and V.R. Macdonald, Science *229*, 187 (1985).

[41] S. Matsukawa, K. Mawatari, Y. Yoneyama and T. Kitagawa, J. Am. Chem. Soc. *107*, 1108 (1985).

[42] T.W. Scott, J.M. Friedman and V.W. Macdonald, J. Am. Chem. Soc. *107*, 3702 (1985).

[43] R.J. Morris and Q.H. Gibson, J. Biol. Chem. *259*, 265 (1984); E.R. Henry, J. Hofrichter, J.H. Sommer and W.A. Eaton, in Proceedings of International Conference on Photochemistry and Photobiology (1983), A. Zewail, ed. (Harwood Academic Publishers, New York) pp. 791-809.

[44] (a) D.A. Duddeil, R.J. Morris and J.T. Richards, J.C.S. Chem. Comm., 75-76 (1979); (b) D.A. Duddell, R.J. Morris, N.J. Muttucuman and J.T. Richard, Photochem. Photobiol. *31*, 479-484 (1980); (c) B. Alpert, S. El Moshni, L. Lindquist and F. Tfibel, Chem. Phys. Lett. *64*, 11-16 (1979); (d) J.M. Friedman and K.B. Lyons, Nature *284*, 570-476 (1980).

[45] R.H. Austin, K.W. Beeson, L. Eisenstein, H. Frauenfelder and I.C. Grunsalus, Biochemistry *14*, 5355 (1975); N. Alberding, S.S. Chan, L. Eisenstein, H. Frauenfelder, D. Good, I.C. Gunsalus, T.M. Nordlund, M.F. Perutz, A.H. Reynolds and L.B. Sorenson, Biochemistry *17*, 43 (1978); D.D. Dlott, H. Frauenfelder, P. Langer, H. Roder and E.E. Dilorio, Proc. Natl. Acad. Sci. USA *80*, 623 (1983).

[46] B.F. Campbell, PhD Thesis, University of California at San Diego (1985).

[47] B. Campbell and J.M. Friedman, unpublished results.

[48] N. Agmon and J.J. Hopfield, J. Chem. Phys. *79*, 2042 (1983).

The Connection Between Low-Temperature Kinetics and Life

H. Frauenfelder

Department of Physics

University of Illinois at Urbana-Champaign

1110 West Greenstreet

Urbana, IL 61801

1. The Problem

The title is not my invention, but I will try to do what I was told and describe some of the connections between phenomena observed at low temperatures and reactions taking place near 300 K. We are often asked: Why perform experiments on proteins at temperatures as low as 2 K when life takes place at much higher temperatures? The answer is obvious. Even a simple biological phenomenon such as the binding of O_2 to hemoglobin is in reality a very complex process in which many different parts of the protein participate. Near 300 K all participating reactions are delicately balanced and they occur in such a way that it is difficult to separate the individual components. By lowering the temperature, the component processes can be studied individually. Low temperatures are important to explore the individual steps.

The next question then is: Can low-temperature processes be extrapolated to physiological temperatures? This question is much more difficult and can only be answered by experiment. I will try to show that in some cases the extrapolation works well and gives information that can not be obtained by other means. In other cases, however, the answer is far from clear and more experiments will be needed. Any generalization is dangerous!

The validity of the extrapolation of low-temperature data is not really the most important problem that we try to solve, it is just an essential element. More generally, we are interested in finding the physical principles that govern the structure and reactions of proteins (and possibly also nucleic acids). After a lecture I was once asked by an MD: "Can atoms be sick?" The answer is clearly "NO". All atoms of the same element are identical, proteins can be sick. An understanding of the principles that govern "healthy" and "sick" proteins may teach us a great deal about life. If we are lucky, we will even learn some physics.

In the present lecture I will attempt to describe some of the low-temperature phenomena that we have studied and will indicate how they extrapolate to 300 K. I will concentrate on concepts and results; details can be found in the references.

2. The Approach

Our approach is simple: we use well known proteins (heme proteins such as myoglobin or hemoglobin), study a simple process (rebinding of CO or O_2 after photodissociation) over wide ranges in time and temperature, construct the simplest model we can find, and treat the rebinding process quantitatively [1]. For the Mb-CO system, the reaction can be written as

$$MbCO + photon \rightarrow Mb + CO \rightarrow MbCO.$$

Here we are not interested in the photodissociation, but only consider the last step, the rebinding of CO. Three different aspects of our model are shown in Figure 1. In (a), we sketch *"reality"* and show a schematic cross section through Mb. The heme group with its central iron atom is embedded in the globin.

A CO molecule in the solvent S diffuse toward the protein, enters the protein matrix M, migrates through the matrix into the heme pocket B, and finally binds at the heme iron A. In Figure 1b the process is sketched in a "single-particle model".

The protein is assumed to form a fixed potential in which the CO molecules moves. The form of the potential implies that the motion of the CO is not simple, but a complex random walk in which the molecule may move many times between solvent and pocket before it finally comes to rest covalently bound to the heme iron. The single-particle model is vastly oversimplified. The protein is *not* a rigid

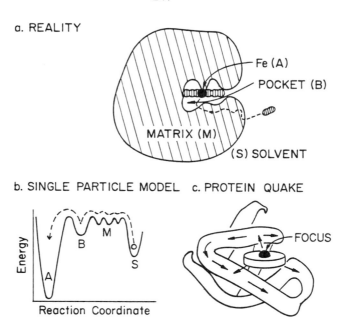

a. REALITY

Fe (A)
POCKET (B)
MATRIX (M)
(S) SOLVENT

b. SINGLE PARTICLE MODEL   c. PROTEIN QUAKE

Energy
M
B
A
S
Reaction Coordinate

FOCUS

Figure 1. Three different aspects of the binding of CO to Mb.

structure but adapts to the moving CO molecule and, on binding the CO, slightly changes its structure [2]. A treatment of the system protein + ligand, with the mutual interaction taken into account, is still too difficult. Nevertheless we can get some information on the protein motion that accompanies ligand dissociation. The way to picture the situation is shown in Figure 1c. After photodissociation the protein rearranges its structure from the liganded to the deoxy form. The rearrangement is similar to an earthquake: Photodissociation releases a strain at the heme iron and the relaxation moves like a quake through the protein. By observing suitable markers, the progress of the quake can be followed and information about the protein motion is obtained. In the following we will sketch some of the main aspects of the single-particle model and the proteinquake and indicate how low- temperature data help elucidate both.

3. Ligand Binding

To study ligand binding, MbCO in a suitable solvent is placed into a cryostat with optical windows. After dissociating the CO with a laserflash, rebinding of CO is followed optically [1]. The result is represented by plotting $\log N(t)$ versus $\log t$, where $N(t)$ is the fraction of proteins that have not rebound a CO at the time t after the flash. The salient features of the behavior of $N(t)$ at low and high temperatures is shown in Figure 2.

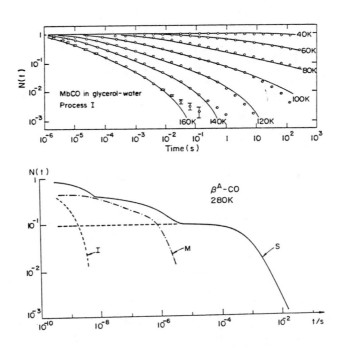

Figure 2. Experimental results at low and high temperatures. The principal features are shown.

At temperature below about 200 K, only a single rebinding process is seen [1] (Figure 2a) which we call process I (For Internal). Process I is independent of the

CO concentration in the solvent and approximately follows a power law in time. Such power-law processes have been observed in many different phenomena, from the discharge of capacitors to the relaxation of glasses [3,4].

Above about 200 K, three different processes can be distinguished as shown in Figure 2b for the binding of CO to the separated beta chain of hemoglobin at 260 K [5]. Processes I and M are independent of the CO concentration in the solvent, S is proportional to it. Processes I and M are not exponential in time, S is.

We interpret the three processes by referring to Figure 1a. In the bound state, MbCO, the CO is covalently bonded to the heme iron, well A in Figure 1b. The laser flash breaks the Fe-CO bond and the CO moves into the heme pocket, well B. Below 200 k, the CO cannot excape from the pocket, but rebinds directly. The direct rebinding is seen as process I. Above 200 K, three different possibilities exist: CO can still rebind directly from the pocket (I), it can move into the protein matrix and rebind from there via the path $M \rightarrow B \rightarrow A(M)$, or it can migrate into the solvent. Any CO molecule in the solvent then can compete for the vacant binding site via the path $S \rightarrow M \rightarrow B \rightarrow A(S)$.

The last process, binding from the solvent, is the one of "interest for life". It is characterized by the association rate coefficient λ_S, indicated in Figure 2b. In the sequential model shown in Figure 1b and described above, λ_S can be written as [6]

$$\lambda_S(T) = \overline{k}_{BA}(T) \; P_B(T) \; N_S(T) \tag{1}$$

where $\overline{k}_{BA}$ is the (average) rate coefficient for the transition $B \rightarrow A$ at temperature T, $P_B(T)$ is the pocket occupation factor or the probability of finding the CO molecule in the pocket in the limit $k_{BA} \rightarrow 0$, and $N_S(T)$ is the fraction of CO molecules that leave the protein after a flash and move into the solvent (Figure 2b).

Experimentally it turns out that in myoglobin, hemoglobin, and also in leghemoglobin [7] $P_B(T)$ and $N_S(T)$ near 300 K vary little from protein whereas $k_{BA}(T)$ can change by orders of magnitude. The reactivity of these proteins is consequently controlled to a large extent by the last binding step, the formation of the covalent bond at the heme iron. This physiologically important conclusion is based on the sequential model. In order to use Eq. (1), k_{BA} must be known at the physiologically interesting temperature. Usually, this value is obtained by extrapolation from low temperatures. Two questions consequently must be asked:

I. Is the sequential model correct?

II. Is the extrapolation of $k_{BA}(T)$ valid?

Before answering these questions, we briefly discuss another problem, the nonexponential time dependence of process I at low temperatures.

We explain the nonexponential time dependence of process I by postulating that the barrier between B and A (Figure 1b) does not have a unique height [1]. If we denote with $g(H_{BA})dH_{BA}$ the probability of finding a protein with barrier height between H_{BA} and $H_{BA} + dH_{BA}$, the rebinding function $N(t)$ at low temperatures can be written as

$$N(t) = \int dH_{BA}\, g(H_{BA})\, \exp\left[-k_{BA}(H_{BA}, T)\, t\right]. \tag{2}$$

Above about 40 K, $k_{BA}(H_{BA}, T)$ is given by an Arrhenius relation,

$$k_{BA}(H_{BA}, T) = A_{BA} \exp\left\{-H_{BA}/RT\right\}. \tag{3}$$

With Eqs. (2) and (3) the observed data can be fitted over a wide temperature range and $g(H_{BA})$ can be determined [1].

Eqs. (2) and (3) imply that the binding step $B \rightarrow A$ is controlled by steric and not by electronic factors. This assumption is in disagreement with at least two calculations [8,9] and consequently leads to the question:

III. Is the binding of CO (and O_2) to Mb controlled by steric or by electronic factors?

This question is again important for "life". The design or evolution of a protein most likely would follow a different path depending on the choice of III. Question III will not be answered here, but we have shown that the experimental data imply dominance of steric factors so that the bond formation $B \rightarrow A$ can be treated adiabatically [10].

The explanation of the nonexponential time dependence of $N(t)$ in Figure 1 in terms of different barrier heights $B \to A$ in different proteins raises the question as to the structural origin of the different barriers. We postulate that a protein cannot be in a unique state of lowest energy, but must have a highly degenerate ground state [1]. Upon folding, the protein does not reach a unique structure but winds up in one of a large number of structurally somewhat different conformational substates. Each substate presumably has a different barrier height. Below about 200 K, each protein freezes into a particular substate and remains there. The result is the observed nonexponential rebinding. At temperatures well above 200 K, each protein molecule moves rapidly from one substate to another and this conformational relaxation leads to the observed exponential rebinding [1]. The rate coefficient for the step $B \to A$ then is an average and is given by

$$\overline{k}_{BA}(T) = \int dH_{BA} k_{BA}(H_{BA}, T) g(H_{BA}) \tag{4}$$

It is essentially this value of $\overline{k}_{BA}$ that we use in Eq. (1). The question of validity of Eq. (4) has already been posed in II above. Another problem is raised by the postulate of conformational substates:

IV. Is the explanation of the nonexponential time dependence through conformational substates correct?

Bialek and Goldstein [11] have proposed an alternate explanaiton for the nonexponential time dependence of process I. They assume that the step $B \to A$ is controlled electronically and that the distribution of rates is caused by the influence of the solvent on the reaction rate. We believe that their explanation is not correct because the experimental evidence speaks against electronic controll [10] and because $g(H_{BA})$ in Mb is solvent independent. Further studies will be required, however, for an unambiguous decision. For the following discussion we assume validity of the substate model.

4. The Single Particle Model

We now return to the question of the validity of the single- particle model.

As shown in Figure 1b, the model assumes *four* wells or regions in sequence, $S \to M \to B \to A$. This assumption leads to the important relation Eq. (1) [6]. Eq. (1) is actually valid even if branching and parallel pathways occur as long as all lead to the same final step $B \to A$ [12]. Doubts concerning the four-well model and the validity of the extrapolation have been expressed [13,14,15]. The answer to questions I and II, posed in Section 3, is consequently important.

The crucial feature of the sequential model, that bond formation is always the last step in the sequence, is supported by three observations [1,5,6,16]: (i) Process I can be seen up to 300 K in β^A, the separated beta chain of human hemoglobin, and in β^{ZH}, the beta chain from the mutant hemoglobin Zurich. (ii) The pH dependence of λ_S at about 300 K in Mb and in β^A is explained entirely by the pH dependence of k_{BA}, measured at low temperatures and extrapolted to 300 K. (iii) The pressure dependence of k_{BA} in the binding of CO to Mb gives approximately the same activation volume at low temperatures [17] and at room temperature [18,19]. While schemes more complicated than the unbranched sequential model may also be able to explain these observations, we assume the simplest one to be correct and use the potential in Figure 1b.

5. The Extrapolation of k_{BA}

In our earliest papers [1] we evaluated the flash photolysis data at temperatures above 200 K by using Eq. (4) with $g(H)_{BA}$ determined below about 160 K. Doubts about this naive extrapolation were raised. Henry and coworkers [15] suggested that process I at room temperature could be as much as 40 times slower than predicted by the extrapolation. Agmon and Hopfield [14] constructed a model in which relaxation of the protein leads to a higher barrier at room temperature and consequently also to a smaller rate for process I. To test the models the extrapolation procedure must be checked.

A test with Mb is difficult. The barrier between B and A is much higher than in other heme proteins. As a result, nearly all CO molecules leave Mb after photodissociation near 300 K. The fraction N_I of CO molecules that rebind directly is very small and process I is extremely difficult to observe at high temperatures.

We consequently have used β^A and β^{ZH} as test proteins. The barriers between B and A in both of these proteins are much smaller than in Mb and process I can still be seen at room temperature [5,16]. By measuring the rate of process I and correcting for the fraction of CO that move from the pocket into the matrix, the average rate coefficient $\overline{k}_{BA}$ is determined as a function of temperature. The result is shown in Figure 3.

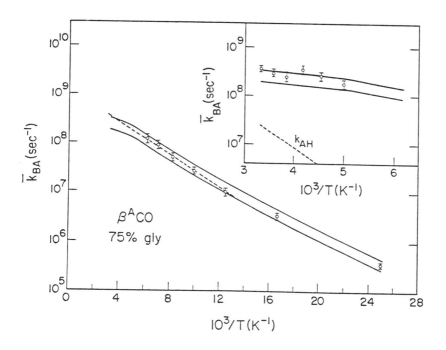

Figure 3. Plot of $\overline{K}_{BA}$ versus $10^3/T$ for β^A CO in 75% glycerolwater. The curves are discussed in the test. After [5].

The open circles in Figure 3 give $\overline{k}_{BA}(T)$ determined directly from the measured data. The solid line is based on the model of Young and Brown [20]. The rate coefficient for the relaxed process I as predicted by Agmon and Hopfield [14] and denoted by k_{AH} is displayed in the insert as dashed line. Figure 3 shows that the agreement between extrapolated and measured rates is very good. The straightforward extrapolation and the model by Young and Bowne yield values that agree within errors. The model of Agmon and Hopfield, however, gives rate coefficients $\overline{k}_{BA}$ that are much smaller than the observed values. We have thus answered question II affirmatively for β^A.

In the case of β^{ZH} the extrapolation does not work quite as well; the measured value is about a factor of three smaller than the extrapolated one [16]. We explain the different behavior of β^A and β^{ZH} as being due to their different structures. In β^A (as in Mb) the solvent does not have direct access to the heme pocket. In β^{ZH}, however, the pocket is open and the solvent can affect the binding process.

The result found here does not imply that the barrier between B and A seen by an incoming ligand can always be obtained by straightforward extrapolation. Eaton and coworkers have pointed out that conformational relaxation may produce a time-dependent rate coefficient [21] which may affect for instance the function in tetrameric Hb [22].

6. Protein Motions

So far we have discussed the binding of CO to heme proteins in terms of the static "single-particle model", Figure 1b. Proteins are, however, dynamic systems that move and their motions are essential to their functions. One main idea can be explained easily with the help of Figure 4. While a skier at rest breathes and moves, it appears at first sight to be difficult to deduce from studies of the resting state how well he or she skis.

To study the motions involved in skiing we must observe the skier in action. While this deduction is correct, careful measurements on the resting state already permit some conclusions concerning the capabilities of the skier. The same argument holds for proteins. To investigate the importance of protein motions for protein function, we must look at a protein in action. Nevertheless, studies of the protein at rest are important. To discuss these questions, we summarize

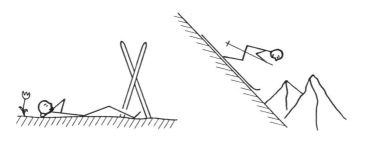

Figure 4. Equilibrium and nonequilibrium motions.

the main concepts.

The binding of CO to Mb involves two *equilibrium states*, deoxyMb and MbCO. Each of these states can exist in a large number of *conformational substates* (CS), briefly discussed in Section 3. Thermodynamics asserts that a small system such as a protein does not possess sharp values of the internal energy, the entropy, and the volume; these quantities fluctuate about their mean values [23]. A resting protein does not remain in one CS but fluctuates from CS to CS. We denote these *equilibrium fluctuations* by EF. The protein reaction, in the case of Mb the transition from MbCO to Mb or from Mb to MbCO, is performed through *functionally important motions* or fims. An understanding of the workings of a protein requires studies of EF and fims. EF and fims are not independent, but are connected by *fluctuation-dissipation* theorems [24,25,26,27,28,29].

EF and fims can be studied by many different techniques. We sketch here only one approach based on photodissociation [30,31].

A protein reaction is similar to an earthquake (Figure 1c): A stress is relieved at the focus. The released strain energy is dissipated in the form of waves and through the propagation of a deformation. In heme proteins, the protein is stressed on binding a ligand, on photodissociation the stress is relieved. In either case the protein finds itself in a state far from equilibrium [32]. Return to equilibrium occurs through a *proteinquake*: the released strain energy is dissipated through waves and through the propagation of a deformation. The experiments

[31] imply that the proteinquake released by the photodissociation of MbCO propagates sequentially:

$$MbCO \overset{h\nu}{\to} Mb_4^* \overset{fim\ 4}{\to} Mb_3^* \overset{fim\ 3}{\to} Mb_2^* \overset{fim\ 2}{\to} Mb_1^* \overset{fim\ 1}{\to} deoxyMb. \qquad (5)$$

$Mb_4^* - Mb_1^*$ are intermediate protein states; fim 4 occurs rapidly even at 3K, fim 3 takes place near 20K, fim 2 starts at about 40K, and fim 1 sets in near 200K. As the proteinquake progresses, substates separated by larger and larger barriers become involved.

X-ray diffraction shows that the difference in structure between MbCO and deoxyMb is observable but small [2]. We can therefore assume that the same substates participate in fluctuations and in the proteinquake. The existence of four fims then implies four tiers of substates, CS^4-CS^1. The arrangement of protein substates, shown schematically in Figure 5, consequently is much more complex than we originally anticipated [1].

The valley at the top of Figure 5a represents one state, say Mb. Mb can exist in a large number of conformational substates of the first tier, CS^1, separated by high barriers. Different CS give rise to different H_{BA}: the valleys are not identical. In spin glasses, such a behavior is called "replica breaking". Each valley in the first tier is structured into substates (CS^2) with smaller barriers. The bifurcation continues through two more tiers, with decreasing barrier heights. The hierarchical arrangement of substates leads to a strong dependence of the dynamics on temperature. Below ≈ 20 K, only, fluctuations among the CS^4 occur.

Above about 20K, the EF4 are so fast that the CS^4 within a given CS^3 are in thermal equilibrium. Fluctuations among the CS^3 then set in, with rates increasing with temperature. The interplay among substates continues to become more complex as the temperature is further increased. At 300K, all substates are involved in the motions of the protein.

At present, the motions involved in the EF and the fims are not very well known. Experiments are difficult and no technique exists that permits us to look at all motions. We will not even discuss what is already known (see for instance

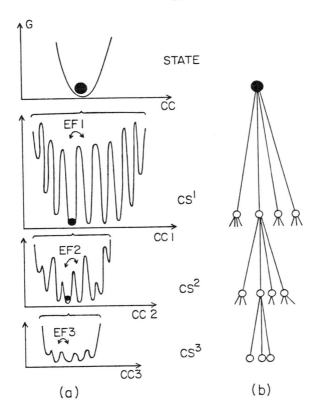

Figure 5. Hierachical arrangement of the conformational substates in myoglobin. (a) Schematized energy surfaces. (b) Tree diagram. G is the Gibbs energy of the protein, CC(1-4) are conformational coordinates. After [31].

[31]), but only outline a number of important properties. Where detailed studies on a particular type of EF or fim have been performed, it has turned out that the motions are not exponential in time. EF and fims consequently must involve distributions. The timescales covered at each temperature by a particular motion can be very large. Figure 6 summarizes the rates of im 1 to fim 4 in Mb. Only one of the motions, fim 2, has so far been explored in some detail.

Fim 2 is markedly nonexponential in time. The range of rates characterizing fim

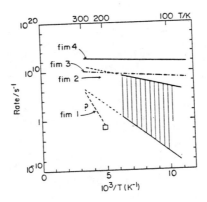

Figure 6. Relaxation rates for fims 1 to 4 as function of $10^3/T$. The shaded area indicates the range of rates for fim 2. The other rates are less well known. (After [31]).

2 are indicated in Figure 6. There is some evidence that the other motions are also nonexponential in time.

The results sketched so far suggest two significant properties of substates and motions in proteins, nonergodicity [33] and ultrametricity [34,35,36]. *Nonergodicity* means that a particular protein does not visit all substates within a given time, for instance the time of a particular experiment. The degree of nonergodicity depends on temperature and time. At low temperatures, each protein is caught in a particular CS^3, but moves from one CS^4 to another. The system is ergodic only within each CS^3. With increasing temperature, higher tiers become ergodic. Even at 300 K, however, there is some evidence of nonergodicity at short times, say ps. *Ultrametricity* describes the structure of the tree in Figure 5b. Simplified, the ultrametric property states that the "distance" between two CS^4 depends on how far up in the hierarchy the two connect. The dynamics in an ultrametric system shows some remarkable similarities to the dynamics of glasses and proteins [37].

Figures 5 and 6 manifest again the importance of low-temperature studies for the exploration of the protein processes that are important for life. If the conformational structure is indeed as complex as shown in Figure 5, it would be extremely difficult to unravel all motions at room temperature. As Figure 6 suggests, the time scale of the different motions overlaps near 300 K. Such an overlap will be particularly confusing if all motions are distributed. It is then difficult to decide the character and assignment of a motion observed near 300 K. By lowering the temperature, the different tiers can be frozen out in steps and the structure and connections become clearer.

The importance of further studies is evident from Figure 6. The extrapolation of the low-temperature behavior is shown as being given by Arrhenius laws. The actual temperature dependence may, however, differ from such a simple law. It is therefore important to investigate the temperature dependence of the various motions over as wide a range as possible to ascertain the validity of the extrapolation.

To sum up, we can state that low-temperature phenomena *are* important for the understanding of life processes because they permit a systematic study of phenomena that are complex and interwoven at physiological temperatures.

Acknowledgements

I thank all my collaborators, Anjum Ansari, Sam Bowne, Joel Berendzen, Ben Cowen, Tim Iben, Todd Sauke, Shyam Shyamsunder, Peter Steinbach, and Bob Young, for their contributions and for many exciting and stimulating discussions. The work was supported by Grant PCM82-09616 from the National Science Foundation and by Grant PHS GM18051 from the Department of Health and Human Services.

References

[1] R.H. Austin, K.W. Beeson, L. Eisenstein, H. Frauenfelder and I.C. Gunsalus (1975). Biochemistry *14*, 5355-5373 (1975).

[2] S.E.V. Phillips, J. Mol. Biol. *142*, 531-554 (1980).

[3] E.W. Montroll and J.T. Bender, J. Stat. Phys. *34*, 129 (1984).

[4] J. Klafter and M.F. Shlesinger, Proc. Natl. Acad. Sci. USA February 1986.

[5] A. Ansari, E.E. DiIorio, D.D. Dlott, H. Frauenfelder, I.E.T. Iben, P. Langer, H. Roder, T.B. Sauke and E. Shyamsunder, Biochemistry, in press.

[6] W. Doster, D. Beece, S.F. Bowne, E.E. DiIorio, L. Eisenstein, H. Frauenfelder, L. Reinisch, E. Shyamsunder, K.W. Winterhalter, and K.T. Yue, Biochemistry *21*, 4831-4839 (1982).

[7] F. Stetzkowski, R. Banerjee, M.C. Marden, D.K. Beece, S.F. Bowne, W. Doster, L. Eisenstein, H. Frauenfelder, L. Reinisch, E. Shyamsunder and C. Jung, J. Biol. Chem. *260*, 8803-8809 (1985).

[8] J. Jortner and J. Ulstrup, J. Am. Chem. Soc. *101*, 3744 (1979).

[9] M.H. Redi, B.S. Gerstman and J.J. Hopfield, Biophys. J. *35*, 471-484 (1981).

[10] H. Frauenfelder and P.G. Wolynes, Science *229*, 337-345 (1985).

[11] W. Bialek and R.F. Goldstein, Biophys. J. *48*, 1027-1044 (1985).

[12] R.D. Young, J. Chem. Phys. *80*, 554-560 (1984).

[13] N. Agmon and J.J. Hopfield, J. Chem. Phys. *78*, 6947-6959 (1983).

[14] N. Agmon and J.J. Hopfield, J. Chem. Phys. *79*, 2042-2053 (1983).

[15] E.R. Henry, J.H. Sommer, J. Hofrichter and W.A. Eaton, J. Mol. Biol. *166*, 443-451 (1983).

[16] D.D. Dlott, H. Frauenfelder, P. Langer, H. Roder and E.E. DiIorio, Proc. Natl. Acad. Sci. USA *80*, 6239-6243 (1983).

[17] L.B. Sorensen, Dissertation, University of Illinois at Urbana-Champaign (1980).

[18] B.B. Hasinoff, Biochemistry *13*, 3111-3117 (1974).

[19] E.F. Cadin and B.B. Hasinoff, J. Chem. Soc. Faraday Trans. I. *71*, 515-527 (1975).

[20] R.D. Young and S.F. Bowne, J. Chem. Phys. *81*, 3730-3737 (1984).

[21] J. Hofrichter, E.R. Henry, J.H. Sommer, R. Deutsch, M. Ikeda-Saito, T. Yonetani, W.A. Eaton, Biochemistry *24*, 2667-2679 (1985).

[22] J.M. Friedman, Science *228*, 1273-1280 (1985).

[23] A. Cooper, Proc. Natl. Acad. Sci. USA *73*, 2740-2741 (1976).

[24] A. Einstein, A. Physik *17*, 549 (1905).

[25] H. Nyquist, Phys. Rev. *32*, 110-113 (1928).

[26] H.B. Callen and T.A. Welton, Phys. Rev. *83*, 34 (1951).

[27] L. Onsager and S. Machlup, Phys. Rev. *91*, 1505-1512 (1953).

[28] R. Kubo, Rep. Progress Phys. *29*, 255-284 (1966).

[29] P. Hänggi, Helv. Phys. Acta *51*, 202-219 (1978).

[30] H. Frauenfelder, in *Structure and Motion: Membranes, Nucleic Acids and Proteins.* Eds. E. Clementi, G. Corongiu, M.H. Sarma and R.H. Sarma, Adenine, Guilderland NY (1985), pp. 205-218.

[31] A. Ansari, J. Berendzen, S.F. Bowne, H. Frauenfelder, I.E.T. Iben, T.B. Sauke, E. Shyamsunder and R.D. Young, Proc. Natl. Acad. Sci. USA *82*, 5000-5004 (1985).

[32] L.A. Blumenfeld, Q. Rev. Biophys. *11*, 251-308 (1978).

[33] R.G. Palmer, Adv. Physics *31*, 669-735 (1982).

[34] G. Toulouse, Helv. Phys. Acta *57*, 459-469 (1984).

[35] M. Mèzard G. Parisi, N. Sourlas, G. Toulouse and M. Virasoro, Phys. Rev. Lett. *52*, 1156-1159 (1984).

[36] R. Rammal, G. Toulouse and M.A. Virasoro, Rev. Modern Phys., in press.

[36] A.T. Ogielski and D.L. Stein, Phys. Rev. Letters *55*, 1634-1637 (1985).

Panel Discussion: The Role of Protein Structure, Dynamics and Ligand Reactivity

Robert Austin, Session Leader

Austin: I think what I'm going to do now because both John, Peter and Hans are up here in front, I'm going to open this to a discussion and direct questions to any of these three characters and I want these to be thoughtful questions, if you can make them that, and try to probe what they are talking about. I want to point out that we've talked about FIMS, and we all know about FUMS. I also want to point out that there's also BUNS which I think John was referring to which are Biologically Unimportant Notions. So keep this in mind and let's find out what's really interesting.

Tom Spiro: I want to support the interest in myoglobin that Hans mentioned, not only is it not boring but it may help understand hemoglobin which is, I think we all agree, an interesting protein. The picosecond Raman experiments which were referred to have been done collaboratively between my group and Robin Hochstrasser's. These experiments show that the photoproduct spectrum generated with 20 picosecond pulses decreases in bands associated with the core size of the porphyrin, which have the same shifts which we saw earlier in short time hemoglobin photolysis spectra. Assuming that these come from the same coupling to the protein, it is then of great interest that in hemoglobin the shifts persist beyond 10 nanoseconds, in fact, we think its likely that they relax with a 50 nanosecond or so time constant. But in myoglobin, they are definitely gone at 10 nanoseconds and experiments are currently underway to try to pin down the relaxation time. It looks as if 1 nanosecond already is also too long, so that there appears to be some parallel process which is much faster for myoglobin than the hemoglobin. We interpret the core size relations as suggesting that this involves a resistance to motion out of the plane of the hemes which was the subject we heard about yesterday and it's still undefined – somewhat controversial. But in any event, it occurs to me to wonder about the connection. You classified this as possibly FIN 3 but how do we know whether this is really FIN 3 or FIN 2? If you extrapolate your infrared data to high temperatures, I realize this is dangerous, is it possible that we are in fact looking at the same thing, and secondly does this infrared band likewise show a shift in hemoglobin photorecombination experiments and what is the time constant for that?

Frauenfelder: We haven't measured hemoglobin. We will. It's somewhat more difficult because CO rebinding is faster. Whether FIN 2 and FIN 3 are identical or not, I doubt it at the moment from looking at the extrapolation but that obviously is something that one has to look at now and do so carefully and in detail, and I think both you and Denis and we are all working on that and trying to figure it out. It is clear that the present scheme is just a scheme to organize knowledge, or maybe the actions of knowledge, and that it produces the experiments and the ideas on certain parts and obviously a year from now it may look totally different, but at least it gives us something which makes the questions clearer.

Joel Friedman: I would just like to comment, and I'll talk about this in greater detail in my talk later. There appears there be no correlation whatsoever between that process 1 and this core size marker. Instead we've looked at the iron histidine stretching motion and there's a dramatic correlation between that frequency and very fast geminate rebinding. Indeed, in myoglobin at room temperature, there's minimal geminate rebinding on the time scale of the relaxation of that core size marker. In addition, and I don't have a firm conclusion yet in the terms of the origin of this effect, we see a power dependent shift in the frequency of a particular raman line that is sensitive to the core size, whereas the ion proximal histidine stretching mode is essentially fixed within 20 picoseconds of photolysis. It's at its equilibrium value. So, I would just add the caveat that it may be that at room temperature, the core size is not all that critical. What is critical is whether the fluctuation and substates are relevant to the geminate rebinding, and we're in the midst of an experiment right now which would test it, and I'd just like to briefly describe it because it actually is the politicians experiments because it could show that both John is right and you are right and maybe I'll be right too.

This is a Raman profile of the iron histidine stretch and this is an increase in frequency based on the correlation, the highest frequency should be associated with the lowest barrier if in fact this is inhomogeneously broadened on a time scale of the geminate rebinding. That is, if the conversion rate is slow or comparable to the geminate rebinding then those proteins having this configuration will undergo the fastest rebinding and this the slowest. So as the ligand rebinds after a picosecond photolysis, one should rebind preferentially to this population and one should in effect get Raman hole burning or curve distortion and actually see the rebinding demonstrate the relevance of the substate to room temperature rebinding.

Linda Powers: Hans, you ignored the CO in your very nice scheme of the intermediates, while at least at low temperature, there's Mossbauer data, and there's infrared data, x-ray absorption data, that consider this to be important. Do you have anything to say about that?

Frauenfelder: Yes. First of all, we don't know really quite as completely our experiments in this area as in my talk because I agree with Joel that it seems that what we call FIN 2 does not affect the rebinding. That is, the rate of rebinding goes regardless of how far this relaxation has occurred. I don't understand that. Neither do you, I think, but that's an experimental result about which I cannot find an explanation. The fact that all the other things are different seem to have nothing to do, in a first approximation, with the fact that the CO is still there. So, that gives us some confidence that we can disregard CO in a first approximation, not in the next, but in a first approximation. Now I may be sorry again a year from now for having said that, but at the moment it looks quite safe: that all these other data, Mossbauer, the position of the 756 is really not connected to the CO or to the oxygen but is connected to the relaxation of the protein and that you can forget the CO. That's simply our present opinion.

Powers: I would disagree. The Mossbauer data is rather significant, in fact, it senses, at least at low temperature, the presence of the CO in the pocket, as does the x-ray absorption spectroscopy. Because, if you'll remember we see the iron to porphyrin nitrogen distance to be something certainly compatible with what has also been observed both in the room temperatures, the very fast Raman stuff that Tom with his core size marker and the data which you quoted from Denis. So we don't see an FeN distance as long as the fully deoxy form, while we still know that there is some presence of the CO in the pocket.

Frauenfelder: I wouldn't believe that. I think there is a very slight effect, in the second approximation. I think the presence of either CO or O_2 changes the Mossbauer spectrum very slightly. But, I think in the first approximation, the Mossbauer spectrum is already so close to the deoxy immediately after the photodissociation and the good agreement that we get with fitting both FIN 2 data and the Mossbauer effect would tell me that the CO simply plays no role. I know we disagree, and I think we'll just have to have more data to figure that out.

Watt Webb: I want to ask a naive question and I think I want to ask it of John Hopfield. Hans always makes everything so clear with his beautiful graphics and the nice names that he applies to these processes. So I can usually follow everything that he says. But John, I think, said that most of the processes that Hans was talking about at low temperature aren't relevant to the room temperature kinetics. And what I would like to hear is the same kind of language describing what's controlling what goes on at room temperature. Is that a fair question?

John Hopfield: That's a fair question. First of all one of the things that is a fairly fundamental disagreement between Hans and me about the importance of myoglobin comes in part from our own particular perversions. When he's hanging on a cliff at 19,000 feet by one hand in the snow wondering whether his heart is going to keep beating, myoglobin may have a relevance. When I'm sitting there waiting to receive the serve, thinking, using fast twitch molecules, the brain doesn't have myoglobin in it, fast twitch molecules don't need myoglobin. It's unimportant to tennis. In this kind of description, I would have described things as saying you photolyze off, you come back at low temperatures, this coordinate has been frozen in the distribution. At high temperature, I would have insisted you come off, photolyze off, you are able to move, you slide down hill, and the high temperature recombination is from this regime, not from that one, and so the barrier distribution which you see here is not directly relevant to high temperatures. In every talk I've ever heard Hans give before, he was actually trying to use the low temperature barrier distribution as the thing which you also were involved with at high temperatures. You saw on the other hand, in the beta hemoglobin data a change in which actually what he was describing was, for the first time, evidence that actually as you go up in temperature, you lose the distribution of barriers here. The barrier effect at high temperatures increases, in short, you can actually see what appears to be the beginning of the falling down toward this high temperature regime. It is a degree apart.

Frauenfelder: You know, I won't say my opinion about tennis, that's maybe why I am not playing tennis any longer, but the point is that the moving up towards high temperatures actually checks totally with the fact that the second channel is open which goes out into the matrix, also we have independent measurements from the intensity of the various components. My feeling is that while we may over- complicate, you have too simple a model. That is, if I had to state it about your figure, you're missing one important configuration or coordinate, that with two you cannot do it, because in two you are forced to ski down into the lowest valley and then at room temperature, climb over the barrier. In my model, it's very different because the energy isn't so strictly tied to what you do, but it is a different coordinate which determines the barrier height - I think you call it Θ, and so I think it's really very different. The basic ideas are similar, but we have at least one additional coordinate and all I can say is that my data seem to support our model and not yours.

Now in your model, as you go to high temperature, the direct binding from the pocket dramatically slows down which means you have to get a larger probability of being in the pocket to explain the binding. Now we have calculated in our model, the probability of being there corresponds to a pocket volume of about 30 $\mathring{A}^3$, which is quite reasonable. In your model, it would correspond to 3,000 $\mathring{A}^3$ and the x-ray data simply do not give a pocket of that size within the protein. So this is a second independent evidence which would throw doubt on the over-simplified map that you draw.

Hopfield: I don't doubt for a moment that our map is over- simplified, but I don't agree with the notion that the fluctuations and barrier height in the protein just happen to see all that's been photolyzed which you examine when you are doing geminate recombination, and before the protein has had a chance to readjust, and the notion that fluctuation from barrier height are the same as if the CO has been off for 10 seconds, and was at room temperature coming back into that same location. I think that is just probably false.

Now of course you do have one tremendous advantage. I have never seen an experimentalist yet whose data did not fit his theoretical explanation.

Frauenfelder: I think that's a very good point, however, I think there was a famous discussion in which, I forgot whether it was Pauli to Einstein or the other way, I think it was Pauli to Einstein who said "Einstein stop telling God what to do." And it seems to me, the theorists are trying to do that. We are trying to listen to the protein.

Peter Wolynes: I was going to say that this discussion of what is exactly transferrable in protein kinetics reminds me a little bit of quantum chemistry where of course there was the early sort of valence bond picture which is something like Hans' theory, and then people moved on to Huckel theory and said the transfer integrals were transferable and now finally no one believes any of that except in some qualitative way. And, we have in a sense now the ab initio theory of protein dynamics in the computer simulations. I think one of the problems that we see from the experiments is that, for example in John's and Agmon-Hopfield's theory, they have to introduce things like diffusion constants for this protein coordinate. Those diffusion constants are clearly involved in other activated processes and whether at the same time we're hearing that there's all kinds of activated processes going on in this system. So it may be that this is intrinsically complicated and you can't have just two- degrees of freedom.

Buhks: I just want to follow upon on the question of Linda Powers and about CO participation in the binding process. The question was whether indeed what happens with the ligand itself matters. There are lots of experimental data where a change in CO distance was observed, that data I believe was measured by Linda herself and Britton Chance. They also observed a change in the angle of the iron-CO bond. Ok, there are some conformational changes involved in the CO molecule while the process occurs. Theoretically, this problem was also treated by Joshua Jortner and Lundstrom some time ago, and they showed that probably mostly conformation in the protein would determine the rate and its temperature dependence. We also did, with Joshua, isotope effect calculations and that data was by the way demonstrated very nicely by Frauenfelder's group; that if you change carbon - go from carbon 12 to carbon 14, you observed isotopic effects of 1.5 at 20 K which decreases to 1.2 at 60 K and so forth. And if you plug in the data related to conformational changes of CO molecules, we come up with an uncoupling parameter of the order of 10 units while in order to explain the entire temperature dependence, we need something about 180. So in other words, the isotopic effects studies really demonstrate that structural changes in CO very widely affect the entire rate behavior of the protein rebinding process, as exactly as Frauenfelder mentioned that protein quakes probably indeed play a major role.

Burt Bronk: I have a low brow experimental question. In studying the relaxation from bound carbon monoxide to release, at the low temperatures you get various low release rates isn't that right? Well, if one is studying a large population of states, when one is part way through at a low temperature releasing the carbon monoxide, shouldn't you simplify the population distribution and be able to show this up on a temperature shift experiment to a higher temperature?

Frauenfelder: This is experimentally very difficult to do.

Bronk: But is it the correct data though? That you would simplify the distribution?

Frauenfelder: It's basically a hole burning experiment and we have done it – we call multiple flash. We had done that many years ago with Bob Austin and the group. We have done that and you can simplify it. You get additional information. But it's very difficult. The experiments, at some point, simply become so difficult. There is also another problem in this field which I've never encountered before. It is the following – anytime you start writing down all the parameters you should look at, and then you write down the parameters that your friend suggested you should look at, then you calculate the time to do it. It always exceeds essentially infinity. So you always have to make a selection of things to do. It simply so complex than anything you do gets you into the problem of infinite number of experiments.

Ed Stern: I wanted to ask a question of Hans, I guess. I find it actually amazing that you can extrapolate at low temperature to high temperature and this may be related to a misunderstanding that I have and I want to clarify that. It seems to me that at low temperature, you're dominated by energy surfaces but at room temperature, the barriers that you should talk about seem to be should be free energy as opposed to just ordinary potential energy surfaces, unless I have a fundamental misunderstanding, and the fact that your results say that you can extrapolate to room temperature, would imply to me that entropy of the protein is not important. Do I have a fundamental misunderstanding here?

Frauenfelder: No, you are correct. Entropy plays a major role, but the entropy in any rate relation goes into the pre- exponential factor and while somebody made a nasty remark about the pre-exponential, I forgot who it was, it could have been by my friend Peter Wolynes, the fact is that the pre-exponentials, as you find them, are absolutely essential. Let me just point that out, because it's a photobiological argument. Now on the last binding step, we find that the pre-exponential from going to the pocket to the island is 10^9. Now, there are two ways of explaining 10^9: One is nonadiabaticity and the other one is an entropy change. If it's nonadibaticity, then the other one from here to here should also be 10^9. If it is entropy change, then obviously from here to here is much larger, because then it is essentially $10^{12} + 3$ instead of -3 so it should be 10^{15}. Now while the latter rate agrees with our theory, I don't think we could have cheated by a fact of 10^6 to be in agreement with the theorists. It came out of the experiment. Now, that's an absolutely crucial number because if the pre-exponential were 10^9 we couldn't exist. Once hemoglobin binds an oxygen molecule, it will be staying there for hours and hours, and our whole system wouldn't work. So entropy plays an absolutely crucial role. But it is of course also in the process of extrapolation.

Ed Stern: Yes, but what I don't understand is how the entropy – you say it's only in the prefactor, but ...

Frauenfelder: In the reaction, it is in the prefactor.

Stern: And therefore, it's something you can even see at low temperature.

Frauenfelder: Yes, so it is there. Entropy plays an absolutely crucial role. No question.

Stern: And it wouldn't effect the barrier except for only as a prefactor?

Frauenfelder: Well, the way we extrapolate them, it's in the barrier.

Feher: I'd just like to make a short comment on this violent disagreement between Hans and John. Of course, it makes for a lively and interesting discussion, but I was just questioning whether it's not a case of whether this is not just a BUD: Biologically Unimportant Disagreement. After all, both of you agree that you have a distribution of barriers and the CO travels across the barrier.

Is it really biologically that important, what the details of the distribution are, whether they are the same or not and whether the pass is the same or not?

Hopfield: The question is whether you can get from low temperature to room temperature in any useful way and it's only for that reason that the details really matter. If you look for example at this beautiful hierarchy that Hans is putting up, that kind of thing is characteristic of glasses. You go well above the glass temperature, that's our hierarchy – normally disappears and what you have left is very simply dynamics. To the extent that proteins don't have something more complicated left at higher temperatures, and you know that they do, it really is very likely a structure of this kind of thing. So the question of trying to extrapolate from below where you have these complications to above where you don't, really is a difficult one and I don't think it can be easily glossed over. I don't know how to do it well. I wouldn't admit that Hans knows how to do it better. One comment on entropy, Hans. Entropy of course, things freeze out as you go to low temperature. You can't use the same entropy at high temperatures. Entropy itself is going to be temperature dependent. You can't use a fixed prefactor and say: ah, that's an entropy term.

Frauenfelder: I want to add one point to the last argument. I have to think longer. I think I also agree with John that it is an important thing. Of course, you realize that this disagreement is a friendly one, and because we both are interested in finding the right model and I've never claimed that I make better theories, all I claim is that I look at the data. And so, I think we need more data in that field to really get the complete understanding and then we will understand which of the problems are minor and which are major. Some of them are clearly minor, but the extrapolation, I think, is an important aspect. Now we should realize that while in a glass, above the glass temperature much disappears. A protein doesn't have such a luxury because each atom is still covalently bonded and some of the states that are accessible at lower temperature on the slowest scale, it still has to be restricted to freeze. So there's truly fundamental differences between glasses and proteins which I think we're just beginning to understand.

Rousseau: Since the topic of this round table discussion is supposed to be the relationship between dynamics and protein function, I think we have the ideal situation where we've been talking about both hemoglobin and myoglobin, to consider what their functions are, and see what experiments have been done that might be related directly to the functional differences between these two proteins. Hemoglobin is the oxygen transport protein which binds oxygen such that it has four binding sites, when the oxygen partial pressure is high the affinity of each of these binding sites is high, when the partial pressure of the oxygen is low the affinity becomes low. So there has to be communication within the protein between one of the binding sites and other binding site. This is very different than myoglobin where there is just a single subunit, and it only has one binding site, so there is no need for the protein that surrounds the heme to know whether or not a ligand, like oxygen or CO, has actually bound to the heme. One can conceive of doing experiments to test this and these relate to the protein quake kind of experiments that Hans was talking about. Namely, the experiments that we have talked about today where you bind CO to the heme, you photodissociate it and you look at the resulting properties of the photodissociated heme. There are a variety of ways of looking at this as we've all heard. But one wants to do the comparison between a deoxy, relaxed, fully relaxed heme and a photodissociated heme to see what these differences are.

In hemoglobin, what is found by optical absorption, by low temperature Raman scattering and room temperature Raman scattering looking at transient experiments, we find that there are very large differences between a fully relaxed deoxy heme and the photodissociated heme. What this means then is that after you take a ligand off the heme, that resulting deoxy heme on a very short time scale does not fit into the heme pocket. It was really energy minimized for having the ligand bound – the CO bound. So what this means, this is in relation to the protein quake, that there has to be a disturbance then that's going to propagate from there through the rest of the protein until the protein can accommodate that deoxy heme.

In myoglobin, the differences between the photodissociated heme and the fully relaxed deoxy heme are very small. Thus, you can make the analogy then between the protein quake in terms of an East Coast earthquake and a West Coast earthquake, where the West Coast earthquake is much more like hemoglobin where buildings fall down and so forth; and the East Coast with much smaller earthquakes are much more closely related to myoglobin.

I want to make one more comment in terms of the relationship between low temperature experiments and room temperature transient experiments in terms of the dynamics. A few years ago, Joel Friedman and I did a comparison between 10 nanosecond photodissociated hemoglobin and hemoglobin photodissociated at 77 K. We found that the resulting structure, the perturbation on the heme resulting from the photodissociation was the same in these two cases. So I think as far as that experiment goes, the constraints imposed by the protein on the heme are the same when you do that kind of a comparison at low temperature as they are at high temperature.

Ali Naqui: I would like to add a piece of evidence because we are discussing and what happens at low temperature and at a higher temperature. There are people who are truly biologists and truly biochemists and try to understand how things are happening in a more complicated system, not so simple as myoglobin, molecular weight 18,000, but rather things like cytochrome oxidase which has two hemes and two coppers and maybe some zinc and something else has been reported. Britton Chance did his low temperature experiments starting in the early '70's with the stop flow at -30°, starting with fully reduced cytochrome oxidase and how it reacts with oxygen. And later in '75 he went down to -135, -140 K and then slowly raised the temperature and saw some reaction going on. And he found different compounds A, B, and starting from something else C and so on and so forth. Now the question was, what does that mean. Ok, it's a beautiful experiment but what does that mean? At that time, technology as far as optical spectroscopy was concerned, technology was not available to do the room temperature experiments. But ten years later I think in the late '84, there are three papers from different groups and they actually repeated those experiments done in 1967 by Gibson and Greenwood using modern day technology and they find small differences so that the reaction sequence are very much similar. And now, there are people again who are again going back to the low temperature to look at things more carefully. So I think there is, perhaps not myoglobin or hemoglobin, but there are systems in other biological systems that work more or less similarly at low temperature and room temperature and as we believe that if you lower the temperature, you've just lowered the motion and see the film in a slow motion that was shown yesterday. That's another piece of evidence that everybody should take into account.

David Kleinfeld: This is more of a pedestrian question on Hans' data. You showed data for beta zurich and you showed sort of a recombination rate vs.

temperature and from that you said that there was sort of break points in the data at around 200 K. The interpretation was that the solvent comes in and the solvent is freezing out around 200 K, so I guess it's like 50% glycerol. If you had to change the nature of the solvent, does the nature of its break temperature change.

Frauenfelder: Yes.

Barry Honig: I just want to the extend the point that there are other proteins other than hemoglobin and myoglobin and I wanted to give you all some evidence about visual pigments which accomplish something that we all agree is biologically relevant to help us see. It's become clear that the photochemistry of these systems is identical at room temperature, and vise versa. On the other hand, these proteins have been designed to do something very important and very interesting. So, by freezing out what they do, the fact that you see it even at low temperature may result from the fact that the protein has worked so hard to do one particular thing. It could well turn out to be the case that in proteins such as myoglobin, and here I have to come down on the other side, and well maybe because what it does isn't so interesting, if that turns out to be the case then you're going to see differences at room and low temperature simply because the protein doesn't have to work very hard to do anything.

Kleinfeld: I just have a question over this notion of a glass temperature. If I understand it right, the correct analogy is that at room temperature things are moving around very rapidly. Protein motion is rapid compared to any oxygen binding/unbinding. Well, maybe the first thing you can do is, tell me what's a glass temperature? In a magnetic system, it's a well defined temperature where if you take a glass and put a magnetic field on, cool it, it will finally reach the temperature where you can take off the magnetic field and magnetization stays. But what's the analogy for a protein?

Frauenfelder: I have it on good authority that the theorists don't know what the glass transition is in glasses. In spin glasses, I think they agree; in a conformational glass I think they disagree.

Kleinfeld: Experimentally, it's a well defined glass temperature? But I guess the question in the proteins is if you experimentally find a sort of a break temperature between when you see exponential behavior and this geminate recombination.

Frauenfelder: No.

Kleinfeld: Not sharp break temperature, but at room temperature things are exponential?

Frauenfelder: In a glass you always have to specify two parameters: time and the temperature. One alone is not enough because even above the glass temperature relaxation phenomena occur at a certain rate and if you have something that faster than the particular relaxation time, you see more glass like behavior, if it's slower, you see a liquid-like behavior. And since in proteins, if our model is right, there are 44 processes taking place at all temperatures. You have a whole spectrum of rates and unless you tell me exactly what you are looking at, at which temperature, and which time, I am in principle not able to answer. In the other case, I'm only unable to answer because we don't know.

Kleinfeld: Let me get the experimental question then. Does the cooling rate matter?

Frauenfelder: Only logarithmically. That's also known from glasses. I assumed, you know Vitali showed the pressure titration. If you do it under pressure, Vitali showed the data, then there is an enormous difference? Everyone who works with glasses knows that cooling rates are very infinitesimal, you have to measure very accurately to get dependence of cooling rates, and that's difficult in proteins, but in principle yes.

Wolynes: In many glasses, although not all, in particular pure silicone glass is not one of these materials, have unusual non-Arrhenius temperature dependence and it's not just having nonexponential behavior and distributed barriers and things like that. There's also a behavior that at a fairly well defined temperature, the rates tend to vary much more rapidly with temperature than you would expect from an Arrhenius law. I believe that Vitali Goldanskii has found a little bit of evidence for some thing unusual happening as a function of temperature, but I don't think that particular sign of a glass transition has been seen clearly.

Goldanskii: I think I just gave several examples and several arguments which speak in favor of the glass like type transition, but certainly doesn't mean that it is the final evidence of exactly the glass type transition. And, as soon as I have the possibility, I would like to continue the discussion on the connection between the low temperature, high temperature experiments. The investigations of rhodopsin and transformations of rhodopsin – the low temperature studies of the transformations of rhodopsin gave the possibility to obtain all intermediates after the absorption of the primary quantum, and there are four intermediates. That was the key to this illusion of this problem long before it was possible to obtain these results with the high time resolution as it was done very recently and recent results confirm the results which were obtained at room temperature. So this is just a good example that shows that low temperature data really have many applications to the high temperature function.

Charge Exchange Between Localized Sites

J. Jortner and M. Bixon

Department of Chemistry

Tel Aviv University

69978 Tel Aviv Israel

Abstract

We examine two distinct classes of quantum mechanical effects on the dynamics of electron transfer (ET) processes between a spatially fixed donor-acceptor pair. First, we review the conventional ET processes, which occur from a thermally equilibrated initial vibrational manifold and consider the nature of the electronic coupling, i.e., direct exchange and superexchange, as well as nuclear overlap contributions. Conventional ET with some quantitative modifications is applicable for the description of the later charge separation events in bacterial photosynthesis, which occur on the time scale $t > 10$ psec. Second, we consider the implications of the competition between ET and medium-induced vibrational relaxation.

Such a modified ET theory may be applicable for the description of the primary charge separation in bacterial photosynthesis, which occurs on the psec (or subpsec) time scale.

1. Introduction

The description of electron transfer (ET) processes between molecular ions in solution, in glasses, in crystals and in biophysical systems rests on the conceptual framework developed for the description of radiationless electronic transitions in condensed phases and in biomolecules, which span a broad spectrum of intermolecular reactions. Relevant examples are:

(1) Thermal ionization of impurity states in semiconductors [1].

(2) Electron-hole recombination in crystalline and amorphous semiconductors [2,3].

(3) Small polaron motion, involving the hopping of an electron in a molecular solid [4].

(4) Intermolecular electronic energy transfer between molecules or ions [5,6].

(5) Intramolecular electronic relaxation in large molecules [7].

(6) Nonradiative electronic relaxation in ionic centers [8] and in impurity states in insulators [9].

(7) Localization of an excess electron in a polar fluid [10].

(8) Low spin to high spin interconversion in transition metal complexes [11].

(9) Electron transfer between ionic or molecular pairs in solutions, glasses and solids [12].

(10) Electron transfer in biophysical systems [13,14].

(11) Recombination of carbon monoxide or dioxygen with hemoproteins [15-17].

All these radiationless phenomena are described in terms of relaxation processes between two distinct zero-order electronic configurations, being induced by a (weak) residual electronic interaction. The ET process [12-14, 18-27], which is the subject matter of the present discussion, involves nonradiative relaxation within a "supermolecule" consisting of the donor(D)-acceptor(A) pair, together with the entire medium. The relevant electronic states are DA and D^+A^-. The ingredients of this relaxation process are [13,14,19-27]:

(A) The electronic coupling V, which involves the two-center transfer integral between D and A.

(B) The modification of the nuclear vibrational states of the "supermolecule" by the change in the electronic states. These involve: (i) The changes in the intramolecular vibrational equilibrium configurations (and vibrational frequencies), which are specified by the coordinates q_c of the D and A centers. These are characterized by the nuclear reorganization energy E_c. (ii) The changes in the equilibrium configurations (and frequencies) of the exterior medium modes, which are characterized by the coordinates q_m. These are specified by the reorganization energy E_s.

The microscopic rate constant W_I for ET from a single vibrational state $\chi_I(\mathbf{Q})$, where $\mathbf{Q} = \mathbf{q}_c, \mathbf{q}_m$, of DA with the energies E_I to the manifold $\chi_J(\mathbf{Q})$ of vibrational states of D^+A^- characterized by the energies E_J can usually be described in terms of a nonadiabatic formalism

$$W_I = \frac{2\pi}{\hbar}|V|^2 F_I \tag{1.1}$$

being given in terms of a product of an electronic term $|V|^2$ and a nuclear Franck-Condon factor

$$F_I = \sum_J |\langle \chi_I(\mathbf{Q})|\chi_J(\mathbf{Q})\rangle|^2 \, \delta\left(E_J - E_I\right). \tag{1.2}$$

The basic implicit assumption underlying the conventional ET theory is that medium-induced vibrational relaxation and vibrational excitation processes in the initial $\{|I\rangle\}$ manifold are fast on the time scale of the electronic-vibrational process [25], i.e.,

$$W_I \ll \gamma_{VR} \tag{1.3}$$

where $\gamma_{VR} \approx 10^{11} - 10^{12} sec^{-1}$ is a typical vibrational relaxation rate [25,26]. Condition (1.3) allows for a separation of the time scales for "fast" vibrational relaxation and the "slow" electronic-vibrational process. The conventional rate constant k for ET is given by the thermal average of the microscopic rate

$$k = \frac{2\pi}{\hbar}|V|^2 F \tag{1.4}$$

where

$$F = Z^{-1} \sum_I \exp(-E_I/kT) F_i \tag{1.5a}$$

is the thermally averaged nuclear vibrational overlap Franck-Condon factor and

$$Z = \sum_I \exp(-E_I/kT) \tag{1.5b}$$

is the partition function for the initial manifold.

The conventional nonadiabatic multiphonon rate constant, eq. (1.4), is expressed as a product of the electronic coupling $|V|^2$ and the thermally averaged nuclear vibrational overlap factor. The electronic and nuclear contributions manifest the implication of "intermolecular engineering" and "intramolecular engineering" of the D and A centers, respectively. The spatial spacing and orientation of D and A determines the electronic term V, while the nuclear contribution F is governed by the distortions of the nuclear equilibrium configurations accompanying the ET. We shall now proceed to examine the contribution of the electronic and nuclear ingredients to ET between spatially fixed simple molecular D and A centers. We shall subsequently proceed to describe the application of the ET theory to the charge separation events in bacterial photosynthesis. We shall argue that, as long as condition (1.3) is satisfied, the ET dynamics within the reaction center (RC) of photosynthetic bacteria is amenable for the description in terms of the conventional ET theory, eq. (1.4). The major difference between the mechanism of ET between simple ions or molecules in solutions as well as in glasses on the one hand, and in a protein medium of the RC on the other hand, is then a matter of details rather than of principles. In particular, the major contribution to the reorganization in the RC originates from intramolecular nuclear modes, rather than from the medium modes, while the protein medium exerts a large effect on the energetics of the D^+A^- system. The conventional ET theory may break down for the description of the primary processes of charge separation in bacterial photosynthesis, which occurs on the psec (or subpsec) time scale when ultrafast ET competes with vibrational relaxation. The entire ET theory has then to be modified. Our discussion is aimed towards the establishment of a conceptual framework for the description of the mechanisms of useful disposal of energy in the RC of photosynthetic bacteria, as explored from the microscopic point of view.

2. Nuclear and Electronic Effects on Conventional Electron Transfer Rates

Quantum effects on conventional ET processes occurring from a thermally equilibrated initial vibrational manifold fall into two categories [28]:

(a) Nuclear tunneling effects, which originate from the contribution of the F term.

(b) Electron coupling for nonadiabatic processes, which is manifested by the $|V|^2$ term.

2.1 Nuclear Contributions

Two limiting forms of the nuclear vibrational overlap factor can be considered:

(a) The low-temperature limit is exhibited when the thermal energy kT is low relative to all the characteristic relevant vibrational energies $\{\hbar\omega_k\}$. The low-temperature Franck-Condon factor is $F_{I=0}$, while the low T rate is $W_{I=0}$. This low-temperature limit corresponds to nuclear tunneling from the vibrationless nuclear state of the initial-state potential surface to the final vibronic states, which are quasidegenerate with it [12-14, 26-28].

(b) The high-temperature limit occurs when the thermal energy kT exceeds all the characteristic frequencies. A classical treatment of F can be utilized which amounts to the replacement of the discrete sums in eq. (1.5) by multidimensional configurational integrals [1, 28]

$$F^c = \int d\mathbf{Q} \exp\left[-U_I(\mathbf{Q})/kT\right] \delta\left[U_F(\mathbf{Q}) - U_I(\mathbf{Q})\right] / \int d\mathbf{Q} \exp\left[-U_I(\mathbf{Q})/kT\right],$$
(2.1)

where $U_I(\mathbf{Q})$ and $U_F(\mathbf{Q})$ are the nuclear potential surfaces of the initial and final states, respectively. The high-temperature rate, eq. (1.4), is $k = 2\pi|V|^2 F^c/\hbar$. The classical nuclear contribution F^c to the high-temperature ET rate within the framework of the harmonic approximation [13,14,18-25] is a Gaussian function of the Marcus form [12,29,30]

$$F^c(\Delta E; E_r) = (4\pi E_r kT)^{-\frac{1}{2}} \exp\left[-(\Delta E + E_r)^2/4E_r kT\right]$$
(2.2)

where ΔE is the electronic energy gap [18-24, 28], while $E_r = E_s + E_c$.

The effects of nuclear tunneling on the ET rates originate from the contribution of the intramolecular high-frequency vibrational modes to F, and are manifested by the following phenomena:

(1) Enhancement of absolute rates [37,38]: Nuclear tunneling effects involving high-frequency intramolecular modes are expected to increase the absolute rate of ET relative to the classical value.

(2) Temperature dependent activation energy [19]. The contribution of the high-frequency modes results in the increase of the activation energy with increasing temperature.

(3) Kinetic isotope effects [39]. Isotopic substitution of the molecular donor or/and acceptor will result in frequency changes and in the distortions of the (reduced) equilibrium configurations to totally symmetric modes leading to a (normal) isotope effect on the ET rate.

(4) Final-state vibrational and electronic excitations [22-24, 29, 30, 40]. Excitations of internal quantum states of the donor and the acceptor for strongly exoenergic reactions exhibit large quantum corrections on the ΔE dependence of the rates, which result in appreciable deviations from the classical relation.

(5) Transition from nuclear tunneling to an activated rate equation with increasing temperature [13,14,31,32]. The temperature range, where the rate is temperature independent $kT \leq \eta\hbar\omega_\kappa$ can be specified [31,32] by the condition $\eta = 0.2 - 0.3$ for all the relevant vibrational frequencies, while for higher values of kT the high-temperature limit (2.1) is approached.

2.2 Electronic Contribution

Three electronic coupling schemes can be considered: (1) Spin-allowed direct exchange [13,14,26,33,34]. (2) Spin-forbidden direct exchange [35]. (3) Spin-allowed superexchange [36]. The magnitude of $|V|$ is determined in all cases by the intermolecular organization, i.e., relative spacing and orientation of the D and A centers.

Consider first V for direct exchange. The simplest picture involves the one-electron impurity model [13,14]. The electron coupling exhibits an exponential dependence on the D-A separation R, i.e.,

$$V = A \exp[-\alpha R] \qquad (2.4)$$

with the exponent α being given by a one-electron contribution of the asymptotic form

$$\alpha = \hbar^{-1}(2m^*B)^{\frac{1}{2}}, \qquad (2.5)$$

where m^* is the electron effective mass and B is its binding energy. This expression seems to be grossly oversimplified to provide a quantitative description of the electronic term.

In order to obtain semiquantitative information on the transfer integral, a proper quantum mechanical calculation has to be conducted transcending the naive one-electron picture. Such an approach has to utilize many-electron wavefunctions for the D and A centers. The absolute magnitude and the distance dependence of the transfer integral depends crucially on the "tails" of the electronic wavefunctions. Such information regarding the functional form of the electronic wavefunctions at large distances from the nuclei is not easy to come by from electronic structure calculations, which usually focus on the energetics and short-range charge distribution of molecules. The "tails" of electronic wavefunctions determine several interesting solid state and molecular observables. In the area of organic solid state, these involve the band structure of triplet excitons and triplet exciton dynamics in molecular solids [42], excess electron and hold band structure, as well as electron and hole mobility [43-45] in organic solids. In the area of molecular physics, relevant phenomena involve triplet electronic energy transfer between pairs of molecules in solution, as well as Penning ionization in the gas phase, $M + A^* \rightarrow M^+ + A + e$, where A is a rare-gas atom and M is a molecule. The cross section for this process is determined by the excess electron density (EED) located beyond the molecular van der Waals radius, e.g., $EED \simeq 0.075$ for the NH_3 molecule [46].

Detailed many-electron calculations for the two-center one-electron transfer integrals between large aromatic molecules were conducted about twenty years ago in the context of electron and hole band structure of organic solids [43-45]. These calculations incorporated both (one-electron and two-electron) Coulomb contributions, which were summed into an effective Coppert-Meyer-Sklar potential, and (two-electron) exchange terms [43-45]. Extreme care was exerted to account for the "tails" of the electronic wave-functions of the π electrons, which were represented in terms of antisymmetrized products of molecular orbitals. The molecular orbitals were taken as linear combinations of carbon SCF functions, which properly account for the asymptotic behavior of the wavefunctions [48].

It will be instructive to examine the distance dependence of V, which is relevant for ET between aromatic molecules in model systems [45] and for the elucidation of ET dynamics in biophysical systems. Figure 1 shows the distance dependence of the electron (anthracene$^-$-anthracene) transfer integral and the hole (anthracene$^+$-anthracene) transfer integral between a pair of parallel anthracene molecules, while Figure 2 shows some of the orientation effects of the ET integral. From these results [45], we conclude that

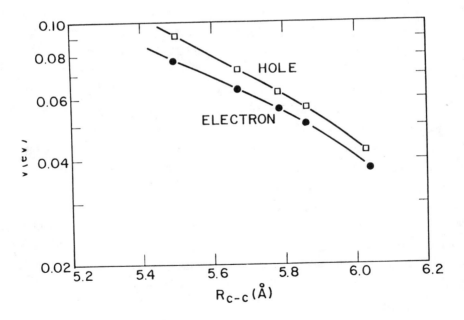

Figure 1. The distance dependence of the electron and hole transfer integral between a pair of parallel anthracene molecules. R_{c-c} corresponds to the center-to-center distance between anthracene-anthracene and anthracene$^+$-anthracene data from reference 45.

(A) The electron and hole transfer integrals exhibit, qualitatively and quantitatively, a similar distance dependence.

(B) The distance dependence of both the electron and hole transfer integrals between parallel molecules can be reasonably well accounted for by the exponential distance dependence, eq. (2.4), with the following parameters: $\alpha = 1.35 \mathring{A}^{-1}$ and $A = 13eV$ for the electron; $\alpha = 1.45 \mathring{A}^{-1}$ and $A = 14eV$ for the hole.

(C) There is a large difference in the absolute value of V at different orientations. However, the functional exponential dependence of the transfer integral on the distance is invariant with respect to the relative orientation.

(D) The effects of the intermolecular π overlap effects, which provide the basis for semiempirical estimates of V, cannot be worked for the numerical estimates of the transfer integral. For example, for the pair of perpendicular molecules the π overlap is practically zero, while the detailed calculations (Figure 2) result in a finite and quite large value of V. Reliable numerical data have to rest on many-electron calculations.

The experimental data for the distance dependence of the rate constants, $k(R)$, were fit in the form $k(R) \propto \exp(-\gamma R)$ so that, according to the relations $k(R) \propto |V|^2$, eq. (1.4) and $V \propto \exp(-\alpha R)$, eq. (2.4), we get $\gamma = 2\alpha$. Table I presents the experimental estimates for the exponents α extracted from ET between aromatic molecules in glasses. The experimental data fall in the range $\alpha = 0.5 - 0.6 \mathring{A}^{-1}$, which is considerably lower than the theoretical estimates of $\alpha = 1.3 - 1.5 \mathring{A}^{-1}$ based on direct transfer, all of which were incorporated in Table I. This discrepancy may originate from the following causes:

(i) Poor electronic wavefunctions used in the calculations. We are reluctant to accept this easy way out, as the same wavefunctions result in quantitative estimates of band structure of triplet excitons [42,47], which are also dominated by the "tails" of the electronic wavefunctions.

(ii) The effects of medium dielectric screening effects on the electron transfer integrals. We do not expect that these effects will exert a gross modification of the distance dependence of V.

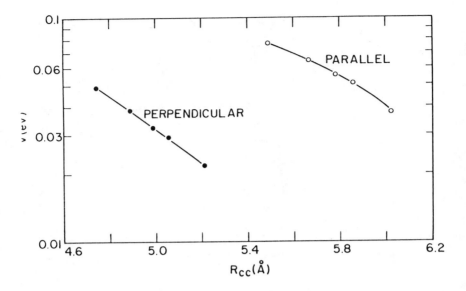

Figure 2. Orientational dependence of the electron transfer integral for anthracene⁻-anthracene. Data for parallel and perpendicular molecules are presented. R_{c-c} corresponds to the center-to-center intermolecular distance. Data from reference 45.

(*iii*) Nonorthogonality corrections for the electronic wavefunctions of D and A, which are due to their overlap with the intervening medium molecules. Such nonorthogonality effects, which are imposed by the Pauli exclusion principle, were estimated for two-center, two-electron, exchange integrals between aromatic molecules and were found to be small ($\sim$ 10%) [42,46]. We expect that a similar state of affairs will prevail for the transfer integrals.

(*iv*) The breakdown of the direct transfer scheme. The contribution of higher order contributions due to superexchange interactions may dominate the electronic coupling between D and A in a medium. The role and implication of superexchange interactions will now be considered.

Table I: Experimental and Theoretical Values of $2\alpha^{2)}$ for energy Transfer between Aromatic Ions and Molecules

SYSTEM	2α $(\mathring{A}^{-1})$	REFERENCE
Biphenyl$^-$ - aromatics	1.2	b)
Biphenyl$^-$ - naphthalene	1.2	c,d)
Biphenyl$^-$ - phenylethylene	1.0	c,d)
Biphenyl$^-$ - fluoranthene	1.43	e)
Pyrene$^+$ - TMPD	1.15	f)
Biphenyl$^+$ - TMPD	1.15	f)
Anthracene$^-$ - anthracene	2.70	g)
Anthracene$^+$ - anthracene	2.90	g)

a) The distance dependence of the ET rate constant is $k(R) = k_0 \exp(-2\alpha R)$.

b) J.R. Miller, J.V. Beitz and R.K. Huddelston, J. Am. Chem. Soc. 106 (1983) 5057.

c) I.V. Alexandrov, R.F. Khairutdinov and K.I. Zamaraev, Chem. Phys. 32 (1978) 123.

d) J.R. Miller, Science 189 (1975) 221.

e) S. Strauch, G. McLendon, M. McGuire and T. Guarr, J. Phys. Chem. 87 (1983) 3579.

f) J.R. Miller and J.V. Beitz, J. Chem. Phys. 74 (1981) 6746.

g) Reference 45 and present work.

Up to this point we were concerned with direct transfer. An additional electronic coupling mechanism involves the superexchange interaction, which was originally invoked by McConnel [36]. Subsequently, it was noted [14] that this electronic coupling mechanism may be important for biophysical systems. The super exchange mechanism involves D-A electronic coupling via mediating electronic states of the solvent molecules. The superexchange coupling prevails in parallel to the direct coupling. This is evident from Figure 3, where we schematically present the quasicontinua of vibronic levels for the initial and final states, which are coupled directly and via the off-resonance interactions with the solvent states. This physical picture can be expressed formally in terms of the scattering matrix, T, between the continua $T = \hat{V} + \hat{V}\hat{G}\hat{V} = \hat{V} + \hat{V}\hat{G}^o\hat{V} + \hat{V}\hat{G}^o\hat{V}\hat{G}^o\hat{V} + \ldots$, where $\hat{V}$ is the interaction, $\hat{G}$ is the Green's function for the system and $\hat{G}^o$ is the zero-order Green's function, which corresponds to a system with $\hat{V} = 0$. The transition probability is proportional to $|T|^2$, which replaces $|V|^2$ in the first-order perturbation expansion, eq. (1.4). From these relations it is apparent that the electronic coupling consists of two additive contributions, that is the direct coupling $\hat{V}$ and the superexchange terms, which contain the series $\hat{V}\hat{G}^o\hat{V} + \hat{V}\hat{G}^o\hat{V}\hat{G}^o\hat{V} + \hat{V}\hat{G}^o\hat{V}\hat{G}^o\hat{V}\hat{G}^o\hat{V} + \ldots$. The dominant contribution to the superexchange consists of nearest-neighbor interactions. The relevant energy parameters, which characterize superexchange, are the nearest-neighbor transfer integrals β_D, β and β_A, the average energy gap Δ between the mediating levels and the donor and acceptor (D-A) state (Figure 3). In the limit $|\beta|, |\beta_A|, |\beta_D| \ll \Delta$, the superexchange contribution to the transfer integral assumes the form [36]

$$V = (\beta_D\beta_A/\Delta)\,(\beta/\Delta)^{n-1} \tag{2.6}$$

where n is the number of mediating states. This result can be recast in the form

$$V = \frac{\beta_D\beta_A}{\beta}\,\exp[-n\,ln(\beta/\Delta)] \tag{2.7}$$

utilizing the relation $n = R/a$, where a is the (average) value of the diameter of a mediating molecule, the interaction assumes the form

$$V \simeq \bar{A}\exp(-\bar{\alpha}R) \tag{2.8}$$

with

$$\bar{\alpha} = a^{-1}ln(\Delta/\beta) \tag{2.9a}$$

and

$$\bar{A} = \beta_A\beta_D/\beta. \tag{2.9b}$$

The superexchange contribution, eq. (2.8), exhibits the same distance dependence as the direct transfer, eq. (2.3). However, the exponential parameter in

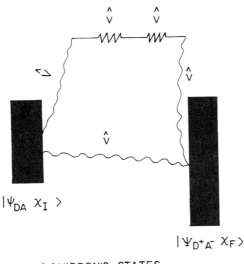

$|\Psi_{DA}\,\chi_I\rangle$

$|\Psi_{D^+A^-}\,\chi_F\rangle$

(a) VIBRONIC STATES

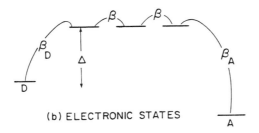

(b) ELECTRONIC STATES

Figure 3. Energy levels scheme for superexchange.

eq. (2.3) is now replaced by $\bar{\alpha}$, eq. (2.9a), whereupon the superexchange and the direct contributions reveal a quantitatively different distance dependence.

It is pertinent to establish the crossing over from the direct transfer to the superexchange mechanism. As both coupling mechanisms occur in series then, according to eqs. (2.3) and (2.8), the superexchange will dominate when

$$\bar{A}\exp(-\bar{\alpha}R) > A\exp(-\alpha R) \qquad (2.10)$$

so that

$$\bar{\alpha} < \alpha - \frac{1}{R} ln(A/\bar{A}). \tag{2.11}$$

Taking typical values of $\bar{A} \approx \beta \approx 0.1 eV$ for mean-neighbor solvent molecules (figure 1) $\alpha = 1.35 \mathring{A}^{-1}$ and $A \approx 10 eV$, together with $a \simeq 4 \mathring{A}$, we estimate from eqs. (2.11) and (2.9a) the characteristic parameters for the dominance of superexchange

$$R = 10 \mathring{A}; \bar{\alpha} < 0.9 \mathring{A}^{-1}; \Delta/\beta < 40$$
$$R = 20 \mathring{A}; \bar{\alpha} < 1.1 \mathring{A}^{-1}; \Delta/\beta < 80 \tag{2.12}$$

Equation (2.12) provides a diagnostic tool for the identification of the superexchange interaction. Accordingly, the ET rate constants between aromatic molecules in glasses (Table I), which are characterized by $\alpha \approx .05 - 0.6 \mathring{A}^{-1}$, do not correspond to direct transfer but are rather induced by the superexchange electronic coupling via the mediating (empty) levels of the intervening solvent molecules. This conclusion concurs with a recent analysis of Miller [48]. We can also estimate the distance R_0, beyond which the ET proceeds via superexchange interactions. Utilizing again eqs. (2.11) and (2.9a), we get

$$R_0 > \frac{1}{\alpha - \bar{\alpha}} ln(A/\bar{A}) \tag{2.13}$$

which, with the characteristic values $\alpha = 1.35 \mathring{A}^{-1}, \bar{\alpha} = 0.6 \mathring{A}^{-1}, A = 10 eV$ and $\bar{A} = 0.1 eV$, results in $R_0 > 6 \mathring{A}$ for the dominance of superexchange coupling.

2.3 Electron Transfer Through Molecular Bridges

Four interesting experimental contributions were reported recently [49-52], which explored experimentally the mechanism of ET through rigid bridges in model compounds. In particular, the studies of the Caltech group [5] and the Munich group [52] have provided quantitative information on the ET rates at variable fixed distances. These novel data provide interesting information regarding the mechanism and distance dependence of the ET rates, where R is now presented in terms of the edge-to-edge distance. The rate constants were fit (figure 4) by the relation $k \propto \exp(-2\alpha R)$.

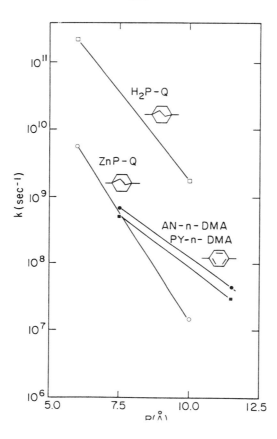

Figure 4. Distance dependence of ET rates through molecular bridges.

From these results it is apparent that:

(1) For the aliphatic bicyclo (2,2,2) octane bridge [50] $\alpha = 0.6 - 0.7 \mathring{A}^{-1}$, while for the aromatic benzene bridge [52] $\alpha = 0.35 \mathring{A}^{-1}$. These numerical values of α imply, according to relation (2.12), that the electronic coupling for both the aliphatic and aromatic bridge corresponds to superexchange.

(2) The ratio of the ET rate constants $k(1)$ and $k(2)$ for a bridge with $n = 1$ and $n = 2$, respectively, is given according to eq. (2.5) by $k(2)/k(1) = (\beta/\Delta)^2$. From the data of Table II, we assert that for the aliphatic bridge [51] $\Delta/\beta \simeq 10 - 17$, and for the aromatic benzene bridge [52] $\Delta/\beta = 4$.

(3) The appreciably lower values of α and of Δ/β for the benzene bridge,

as compared to the bicyclo (2,2,2) octane bridge, manifest the dramatic enhancement of the efficiency of the superexchange coupling via aromatic bridges.

Table II

MOLECULE	k sec^{-1}	α Å^{-1}	(Δ/β)
H_2 Porphyrin-Benzene-Quinone[a]	$5.8x10^9$		
		0.7	19.6
H_2 Porphyrin-Benzene-Bibicyclo (222) Octane-Quinone[a]	1.5×10^7		
Zn Porphyrin-Benzene-Quinone[a]	2.2×10^{11}		
		0.6	11.1
Zn Porphyrin-Benzene-Bibicyclo (222) Octane-Quinone[a]	1.8×10^9		
Pyrene-Benzene-DMA[b]	5.2×10^8		
		0.35	4.2
Pyrene-(Benzene)$_2$-DMA[b]	3.0×10^7		
Anthracene-Benzene-DMA[b]	7.0×10^8		
		0.35	3.9
Anthracene-(Benzene)$_2$-DMA[b]	4.5×10^7		

a) Reference 51.

b) Reference 52.

The dominating role of the superexchange in model systems, e.g., for ET between aromatic molecules in glasses (section 2.1) and through rigid bridges implies that this electronic coupling mechanism is central for ET in biophysical systems, as was suggested a few years ago [14].

3. Structure and Electron Transfer Dynamics in Reaction Centers of Photosynthetic Bacteria

The crystallographic data of the Munich group [53] on the reaction center (RC) of Viridis (Figure 5) resulted in a wealth of structural information, which provides the basis for the understanding of the "intermolecular engineering" that governs the ET dynamics in the primary charge separation events in bacterial photosynthesis.

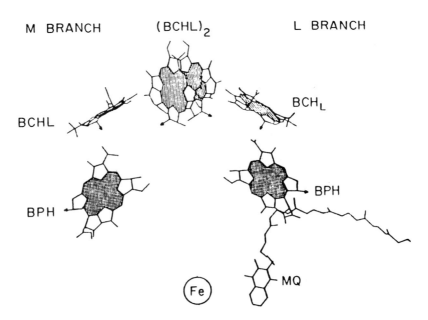

Figure 5. The structure of the reaction center of Rhodopseudomonas viridis. Reprinted by permission of J. of Molecular Biology, Vol. 180 (1984), p. 385. (reference 53).

The existence of quasisymmetric branching of the components in the RC raises two questions. First, what are the structural implications of the gross unidirectional charge separation in the RC? The directionality control of the charge separation may be accomplished by charged or/and polar groups in the protein, which energetically stabilize the ion pair, i.e., charge transfer state(s) along one branch. Second, what is the reason for the structural redundancy of the RC, which effectively utilizes only a single branch in the charge separation process? One may conjecture that this redundancy originates from the biosynthesis of the RC. Proceeding to mechanistic issues, the detailed structural information raises the important issue regarding the role of the auxiliary bacteriochlorophyll (BChl) monomer, which mediates structurally and dynamically between the bacteriochlorophyll dimer $(BChl)_2$ (also to be denoted as P) and the bacteriopheophytin (Bph). The energy of the $P^+(BChl)^-(Bph)Q$ state is not known (Figure 6) and it is an open question whether this state constitutes a real intermediate state or, alternatively, a virtual mediating state (Figure 6) in the ET process. We shall return to this problem in section 3.2.

A central feature of ET dynamics in the RC involves the possible role of the protein, as manifested by the energy levels, vibrational excitations and nuclear motion in this unique medium. The following features are relevant in this context.

(A) Distribution of local conformational substates [15] with each distinct local configuration performing the same function with a different rate [16]. Such dynamic inhomogeneous broadening effects may prevail for ET processes, being manifested by differences in medium reorganization energies and by a spread in the D-A distances.

(B) The occurrence of low-frequency vibrational motion in proteins [17], which may couple in the electronic states.

(C) The possible existence of tunneling states in proteins [54,55], which are analogous to those postulated for glasses and which may couple to the electronic states.

Effects (A), (B) and (C) may be important for controlling the ET dynamics in the RC only providing that the coupling of the electronic states to the protein modes is substantial, i.e., the medium rearrangement energy, E_m, is large [26].

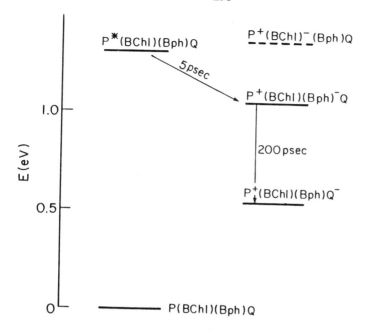

Figure 6. A schematic energy levels scheme for ET in the reaction center of a photosynthetic bacteria. The energy of the $P^+(BChl)^-(Bph)Q$ state is unknown.

In section 3.2, we shall argue that in the RC $E_m/E_c < 1$ and that the nuclear coupling is dominated by the intramolecular modes.

(D) We now consider the effects of charges and polar groups and their motion with respect to the components of the RC. Electrical charges on the protein (Figure 7) modify the energy levels of the D^+A^- pair. The most straightforward effects pertains to the modification of the energy gap, which in the presence of the external charges $\{Z_j\}$ is

$$\Delta E(Z) = \Delta E + \sum_j \frac{Z_j e^2}{\varepsilon} \left(\frac{1}{R_{D_j}} - \frac{1}{R_{A_j}} \right) \qquad (3.1)$$

where ΔE is the gap in the absence of charges (Figure 7), while ε is an effective dielectric constant. The modification of the energy gap will effect the ET dynamics. Of particular interest in this context is the energetic stabilization of some ion-pair states exerted by electrostatic interaction with the protein charges or/and polar groups. In fact, the low energy of the well-known ion pair (charge transfer) state $P^+(BChl)(Bph)^-$, which is located at the range 1.20-1.35 eV above the ground state of $P(BChl)(Bph)$, presumably originates from electrostatic stabilization by the proteins medium. Invoking the well-known relationship for the energy E_{CT} of the charge transfer state [56]

$$E_{CT} = I - EA + C + PO \qquad (3.2)$$

where I is the ionization potential of the donor, EA is the electron affinity of the acceptor, C is the Coulomb interaction energy between D^+ and A^- and PO is the polarization energy of the medium by the D^+A^- pair. Taking reasonable values of $I = 6.5eV$ [57], $EA = 1eV$ and $C = -0.4eV$, one then requires the value of $PO = -3.9eV$ to obtain the experimental value of $E_{CT} = 1eV$. From model calculations on molecular crystals [36], it appears that this value of PO is exceedingly low for a neutral medium. Obviously, such a low value of PO can be accomplished by electrostatic interactions of D^+ and A^- with charged and polar groups within the protein medium.

We now proceed to discuss the mechanism of ET in the RC. The sequence of the primary electron transfer processes in the RC (Figure 6) can conveniently be subdivided into two sets of processes, that is, (a) early ET, i.e., the first 10 psec, and (b) later stages of ET, corresponding to the time scale exceeding 10 psec. These two distinct time domains will now be discussed.

3.1 Later Electron Transfer

The two major ET processes, which occur in vivo and which correspond to the later processes involve ET from Bph^- to Q [27,60]

$$P^+(BChl)(Bph)^-Q \rightarrow P^+(BChl)(Bph)Q^- \qquad (3.3)$$

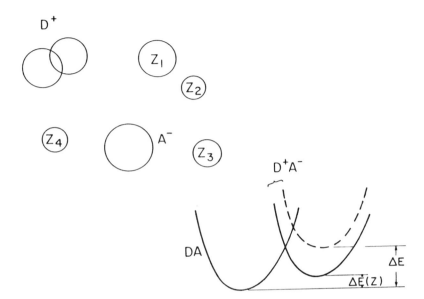

Figure 7. The effect of electrostatic interactions with protein charges on the energy gap in ET.

and ET from cytochrome c to the dimer cation $(Bph)_2^+$ [13,14,27]

$$(\text{ferrocytochrome c})P^+(BChl)(Bph)Q^- \rightarrow (\text{ferricytochrome c}))P(BChl)(Bph)Q^-$$
$$(3.4)$$

Both processes occur on a sufficiently long time scale to satisfy the validity condition, eq. (1.3), for the applicability of the conventional theory of ET [26,27]. Accordingly, the conventional nonadiabatic ET theory, eq. (1.4), is applicable to describe the ET processes occurring in a protein medium. The major difference between the mechanism of ET in solutions as well as in glasses on the one hand, and in a protein medium on the other hand, is just a matter of technical details rather than of principles. Following the outline of section 1, two types of intramolecular vibrational modes contribute to the Franck-Condon factor, i.e., (i) intramolecular vibrational modes of the donor and acceptor centers, and (ii) intermolecular vibrational polar modes of the exterior proteins medium. The available experimental data for reactions (3.3) and (3.4) strongly indicate that its major contribution to the nuclear reorganization for the later ET reactions within the RC originates from the intramolecular vibrational modes of the donor and the acceptor centers (type (i)), rather than from the exterior medium modes (type (ii)). This evidence rests on the temperature dependence of the ET rates for reaction (3.3) and (3.4), which are portrayed in Figure 8. Reaction (3.4) reveals the well-known transition from nuclear tunneling at low temperatures to an activated rate process at high temperature [13,14]. The transition temperature, T_0, from the tunneling to the activated domain is $T_0 \simeq 100K$ (Figure 8) which, according to eq. (2.3), implies that the relevant nuclear frequency is $\langle \omega \rangle \simeq 300 - 400 cm^{-1}$. The unique temperature dependence of reaction (3.3) (Figure 8) again provides evidence for the dominating role of intramolecular vibrational modes in determining the ET dynamics. Reaction (3.3) exhibits a temperature independent rate below 100 K followed by a decrease of the rate with increasing temperature at $T \geq 100K$. This rather uncommon pattern was attributed [26,27,58,59] to an activationless ET. Activationless (or barrierless) ET, which involves the crossing of the potential surfaces at the minimum of the initial DA state is characterized by the following features [26,27]:

(1) The rate constant is temperature independent over the range determined by eq. (2.3).

(2) The low temperature rate is

$$\left(2\pi |V|^2 / \hbar^2 \langle \omega \rangle\right) \left(\hbar \langle \omega \rangle / 2\pi \Delta E\right)^{\frac{1}{2}} \tag{3.5}$$

where $\langle \omega \rangle$ is the characteristic relevant nuclear frequency.

(3) At high temperature

$$W \propto 1/\sqrt{T} \tag{3.6}$$

exhibiting negative apparent activation energy.

The recent experimental results of Parson [60] on the (weak) temperature dependence of reaction (3.3) are in semiquantitative agreement with these predictions.

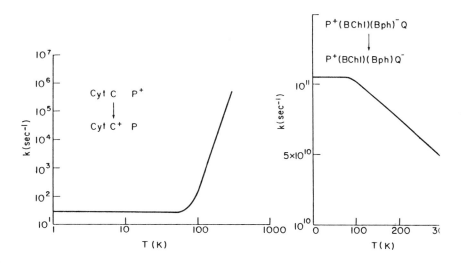

Figure 8. Temperature dependence of the reaction rates for cytochrome ox-idation [D. DeVault and B. Chance, Biophys. J. 6 (1966) 825. Reprinted by permission of the Biophysical Journal] and ET from $(Bph)^-$ to Q [W.W. Parson, reference 60].

The rate constant in the temperature range 100-300 K is expected on the basis of eq. (2=3.6) to decrease by a numerical factor of 1.7, while the experimental data (Figure 8) exhibit a decrease by a numerical factor of 2.3 over this temperature range. What is more important for the establishment of the general conceptual framework for the elucidation of the later ET processes is that the crossing over from the region of nuclear tunneling to the $1/\sqrt{T}$ temperature domain occurs at $T_0 \simeq 100K$, which again implies a high value $\langle\omega\rangle \simeq 300 - 400cm^{-1}$ of the relevant vibrational frequency. We thus conclude that both later ET processes (3.3) and (3.4) are dominated by coupling high intramolecular vibrational modes with $\langle\omega\rangle = 300-400cm^{-1}$, which are attributed to the intramolecular vibrational motion of the porphyrin ring. This analysis demonstrates a major contribution of nuclear reorganization for later ET in the RC originating from the high frequency intramolecular vibrational modes, i.e., $E_c/E_m \gg 1$. This state of affairs for ET in the RC is qualitatively different from that encountered for ET between organic molecules held by rigid bridges in a polar solution, where $E_c/E_m < 1$ and a major contribution to the nuclear reorganization energy originates from the exterior medium modes. We conclude this analysis by pointing out that:

(a) The dynamics of later ET processes in understood.

(b) These ET processes in the RC differ qualitatively from a variety of ET reactions in view of the dominating role of intramolecular vibrational mode(s) in the dynamics.

(c) The important role of activationless ET in the first step of the later ET has been noted. This involves the optimization of intramolecular distortions to ensure directionality and specificity of charge separation on the time scale $t > 10psec.$

Obviously, details of the conventional theory have to be filled in. The available information regarding the details of the relevant intramolecular vibrational modes of the donor and acceptor is meager. Also, the nature of the electronic coupling, which on the basis of the discussion in section (2.2) presumably proceeds via superexchange mechanism, has to be further elucidated. Finally, additional correction terms have to be incorporated in the description of activationless processes. An important ingredient in this context involves thermal expansion effects, which result in the decrease of $|V|$ and of k with increasing temperature. The contribution to the temperature dependence of the ET rate due to thermal expansion is [27] d $ln(k)/dT = -(2/3)\alpha R\kappa_T$, where κ_T is the thermal expansion coefficient of the protein. This contribution may well account for the slight discrepancy between the estimated and experimental decrease of the ET rate of reaction (3.4) over the temperature range of 100-300 K. It is also worth while noting that thermal expansion effects are large for those activationless reactions which are characterized by a large D-A separation and, consequently, a low value of $|V|$, i.e., slow ET processes. The slow (msec) recombination reaction between Q^- and D^+ [61] is an activationless process [27,28,58] and the rate of this reaction is expected to be very sensitive to thermal expansion effects and to structural modifications of the RC.

3.2 Early Electron Transfer

The understanding of the first 10 psec of the ET events in the RC requires the elucidation of the following three ingredients:

(a) The electronic states involved.

(b) The mechanism of the primary ET events.

(c) The proper description of ultrafast ET processes.

Regarding point (a), we note that an interesting feature of the electronic states pertains to the role of ion-pair, charge-transfer states. We have already dwelt on the charge transfer state $(BChl)_2^+ (BChl)(Bph)^- Q$, which is important for later ET. Within the context of early ET two types of charge transfer states are relevant:

(1) Dimer charge transfer states,

$$\alpha | (BChl^+ BChl^-) \rangle + \beta | (BChl^- BChl^+) \rangle, \qquad (3.7)$$

where $(BChlBChl)$ corresponds to the two components of the dimer.

(2) Dimer-monomer charge-transfer states,

$$\psi_{CT} = \alpha | (BChl^+ BChl)(BChl)^- \rangle + \beta | (BChlBChl)^+ \rangle (BChl)^- \rangle. \quad (3.8)$$

The location of states (3.7) and (3.8) is not known. As we have noted, the energy of these states may be crucially determined by the protein environment. Stark spectroscopy of the reaction center of the sphaeroides bacteria [62] has provided strong evidence for the extensive admixture of charge transfer states, which are characterized by a high permanent dipole moment μ into the dimer excitation, while the higher energy excitations of the $(BChl)$ and of the (Bph) are not contaminated. Provided that the protein environment of the two components of the dimer is nearly identical, the state (3.7) will be devoid of a permanent dipole moment. A likely candidate for mixing, which is characterized by a high value of μ, is the state (3.8). The scrambling of the lowest energy neutral exciton state

$$\psi_N = |(BChl^*BChl)\rangle - |(BChl)BChl^*)\rangle \qquad (3.9)$$

of the dimer with the charge transfer state (3.8) results in the mixed state

$$\psi = a\psi_N + b\psi_{CT} \qquad (3.10)$$

whose ψ_{CT} components do not contribute to the integrated intensity of the dimer band, but rather appreciably to the Stark enhancement [62] and to the redistribution of intensity in the dimer band. In the absence of quantitative intensity data one cannot provide an estimate of the charge transfer mixing coefficient $|b|^2$. Furthermore, the location of the zero-order of estimates indicate that the mixing coefficient of the charge transfer states is

$$b \simeq \langle \psi_N | H | \psi_{CT} \rangle / \Delta E \qquad (3.11)$$

where the matrix element is the two-electron and one-electron coupling, while ΔE corresponds to the energy gap between the zero-order states. Model calculations for charge transfer interactions in organic molecular crystals [64], i.e., naphthalene and anthracene, resulted in $|\langle \psi_N | H | \psi_{CT} \rangle| \simeq 100 cm^{-1}$, so that reasonable mixing of charge transfer states, i.e., $|b|^2 \sim 0.1$, implies that $|\Delta E| \simeq 200 - 500 cm^{-1}$ for the energy gap between the zero-order neutral and the dimer-monomer charge transfer excitation ψ_{CT} in the RC. On the basis of this crude estimate, one cannot decide whether the zero-order charge transfer state is located above or below the neutral excitation. In any case, the close proximity of the neutral dimer excitation and the charge transfer state implies extensive mixing of these two electronic configurations.

The mechanism of the primary ET events in the RC is determined by the energetics of the dimer-monomer charge transfer state P^+BChl^- (where P denotes the $BChl$ dimer) relative to the lowest electronic excitation P^*BChl of the dimer. Provided that the energy of the electronic excitation P^*BChl is located above the equilibrium charge transfer state, then the P^+BChl^- state, which contains a high admixture of ψ_N, constitutes a genuine intermediate state in the ET sequence. On the other hand, when the $P^*(Bchl)$ excitation is located somewhat below the P^+BChl^- state, but mixes heavily with it, then the scrambled initial excited state, eq. (3.10), transfers an electron to the Bph. The mechanism of ET is schematically represented in terms of the initial state $a\psi_N + b\psi_{CT}$ and the final state $|P^+(BChl)(Bph^-)\rangle$.

The electronic coupling

$$V \simeq b\langle P^+(BChl^-)(Bph)|H|P^+(BChl)(Bph^-)\rangle, \qquad (3.12)$$

which induces this process, is essentially determined by the transfer integral between the dimer-monomer charge transfer state and the separated $P^+(Bph^-)$ ion pair. The amplitude b, eq. (3.12), is $|b| \sim 0.2 - 0.3$, being sufficiently large to provide efficient coupling. In this mechanism a distinct charge transfer state is not involved as a genuine intermediate state. Rather, the charge transfer state, $P^+(BChl^-)$, mediates the primary ET process.

In spite of the current mechanistic uncertainty regarding the primary charge separation in the RC, it is evident that the formation of the $P^+(BChl)(Bph^-)$ separated ion pair occurs on the time scale of 4-5 psec [27, 60], and that some additional ET processes may be even faster ($\sim 1psec$) [63]. These ET rates occur on the time scale of medium-induced vibrational relaxation rates of the donor and acceptor molecules. There is a distinct possibility that for early ET events within the RC the assumption of fast vibrational relaxation, eq. (1.3), cannot be taken for granted and the conventional ET theory may be inapplicable. A treatment of ET during vibrational relaxation has already been provided [28, 29]. The major physical aspect of this process (Figure 9) is that when the donor-acceptor system is excited at the energy E_e it will start to lose energy via vibrational relaxation. If the initially excited configuration E_e is located below the crossing point E_x of the potential surfaces, the nuclear Franck-Condon contributions to the relevant microscopic ET rates are sufficiently low to allow for the system to relax vibrationally to a thermally equilibrated configuration from which "conventional" ET will occur. However, if the initially excited state, E_e, is located above the crossing point, E_x, the system will pass on its way downward through E_x. At the vicinity of the crossing point, the nuclear contribution to the microscopic ET rates is large, i.e., Franck- Condon factors being $\sim 10^{-1} - 10^{-2}$ [25,26]. Thus, provided that the electronic coupling is large due to spatial proximity of the donor-acceptor pair and/or efficient mediating of the electronic coupling, efficient ET will occur at the crossing point. Thus, the necessary condition for ET during vibrational relaxation is $E_e \geq E_x$, which ensures efficient ET. There is a distinct possibility that the primary ET events in the RC compete with medium-induced vibrational relaxation occurring in a system which does not reach thermal equilibrium. A possible experimental test of this conjecture will involve an interrogation of wavelength dependence of the quantum yield for charge separation in the RC following photoselective excitation in the dimer absorption band. At low excitation energies a drop of the quantum yield will then be exhibited. The proposal that ultrafast ET [26] and energy transfer [65] electronic processes in the RC overwhelm vibrational relaxation may provide a general mechanism for the efficient utilization of energy in primary photobiological processes.

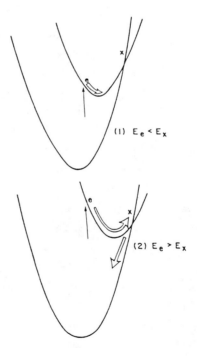

$(1) \; E_e < E_x$

$(2) \; E_e > E_x$

Figure 9. A schematic description of potential surfaces $(DA$ and $D^+a^-)$ for ET initiated by initial excitation and followed by vibrational relaxation.

305

Acknowledgement

We are indebted to Professor M.E. Michel-Beyerle for prepublication information and for interesting discussions.

References

[1] R. Kubo and Y. Toyozawa, Prog. Theoret. Phys. 13 (1966) 160.

[2] C.H. Henry and D.V. Lang, Phys. Rev. B15 (1977) 989.

[3] N.F. Mott, E.A. Davis and R.A. Street, Phil. Mag. 32 (1975) 961.

[4] T. Holstein, Ann. Phys. 8 (1959) 343.

[5] T. Forster, Naturwissenschaften 33 (1946) 166.

[6] D.L. Dexter, J. Chem. Phys. 21 (9153) 836.

[7] R. Englman and J. Jortner, Mol. Phys. 18 (1970) 145.

[8] R.H. Bartram and A.N. Stoneham, Solid State Commun. 17 (1975) 1593.

[9] N.D. Sturge, Phys. Rev. B8 (1983) 6.

[10] P.M. Rentzepis, R.P. Jones and J. Jortner, J. Chem. Phys. 59 (1973) 766.

[11] E. Buhks, J. Jortner, G. Navon and M. Bixon, JACS 102 (1980) 2918.

[12] R.A. Marcus, J. Chem. Phys. 24 (1956) 966.

[13] J.J. Hopfield, Proc. Nat'l. Acad. Sci. USA 71 (1974) 3640.

[14] J. Jortner, J. Chem. Phys. 64 (1976) 4860.

[15] R.H. Austin, K.W. Beeson, L. Eisenstein, H. Frauenfelder and I.G. Gunsalus, Biochemistry 14 (1975) 5355.

[16] H. Frauenfelder, *Structure and Dynamics: Nucleic Acids and Proteins*, eds. E. Clementi and R.H. Sarma (Adenine Press, New York, 1983) p. 364.

[17] J. Jortner and J. Ulstrup, J. Am. Chem. Soc. 101 (1979) 3744.

[18] V.G. Levich, *Physical Chemistry: An Advanced Treatise*, eds. H. Eyring, D. Henderson and W. Jost (Academic Press, New York, 1970), Vol. 9B.

[19] N.R. Kestner, J. Logan and J. Jortner, J. Phys. Chem. 78 (1974) 2148.

[20] R.R. Dogonadze and A.M. Kuznetsov, Elektrokhimiya 3 (1967) 1324.

[21] R.R. Dogonadze, A.M. Kuznetsov and M.A. Vorotyntsev, Phys. Status Solidi B. 54 (1972) 125; 425.

[22] R. Van Dyne and S. Fischer, Chem. Phys. 5 (1974) 183.

[23] S. Efrima and M. Bixon, Chem. Phys. Lett. 25 (1974) 34; Chem. Phys. 13 (1976) 447.

[24] J. Ulstrup and J. Jortner, J. Chem. Phys. 63 (1975) 4358.

[25] J. Jortner, Phil. Mag. Ser. B14 (1979) 317.

[26] J. Jortner, J. Am. Chem. Soc. 102 (1980) 6676.

[27] J. Jortner, Biochimica Biophysica Acta 594 (1980) 193.

[28] M. Bixon and J. Jortner, Discuss. Farraday Soc. (London) 74 (1982) 17.

[29] R.A. Marcus, Discuss. Faraday Soc. (London) 29 (1960) 21.

[30] R.A Marcus, Annu. Rev. Phys. Chem. 15 (1964) 155.

[31] V.I. Goldanskii, Dokl. Acad. Nauk. USSR. 124 (1959) 1261; 127 (1959) 1037.

[32] E. Buhks and J. Jortner, J. Phys. Chem. 84 (1980) 3370.

[33] J. Ulstrup, *Charge Transfer Processes in Condensed Media*, (Springer-Verlag, New York, 1979).

[34] M.D. Newton, Int. J. Quantum Chem. 14 (1980) 363.

[35] E. Buhks, M. Bixon, J. Jortner and G. Navon, Inorg. Chem. 18 (1979) 2014.

[36] H.M. McConnell, J. Chem. Phys. 35 (1961) 508.

[37] P. Siders and R.A. Marcus, J. Am. Chem. Soc. 103 (1981) 741.

[38] E. Buhks, M. Bixon, G. Navon and J. Jortner, J. Phys. Chem. 85 (1981) 3759.

[39] E. Buhks, M. Bixon and J. Jortner, J. Phys. Chem. Soc. 103 (1981) 748.

[40] J.B. Beitz and J.R. Miller, J. Chem. Phys. 71 (1979) 4579.

[42] J. Jortner, J.L. Katz, S.I. Choi and S.A. Rice, J. Chem. Phys. 42 (1965) 309.

[43] J.L. Katz, S.A. Rice, S.I. Choi and J. Jortner, J. Chem. Phys. 39 (1963) 1638.

[44] R. Silbey, S.A. Rice, M.T. Vala and J. Jortner, J. Chem. Phys. 42 (1965) 733.

[45] S.A. Rice and J. Jortner, *Physics of Solids under High Pressure*, eds. T. Tonizuka and R.M. Emrick (Academic Press, New York, 1965), p. 65.

[46] K. Ohno, S. Matsumoto and Y. Harada, J. Chem. Phys. 81 (1984) 2183.

[47] R. Silbey, N.R. Kestner, J. Jortner and S.A. Rice, J. Chem. Phys. 42 (1965) 444.

[48] J.R. Miller, "Paper presented at the Workshop on Antennas and Raction Centres in Photosynthetic Bacteria", Falderfing, W. Germany (1985).

[49] J.R. Miller, L.T. Calcaterra and G. Closs, J. Am. Chem. Soc. 106 (1984) 3047.

[50] M.R. Wasielewski, M.P. Niemczyk and E.P. Pewitt, J. Am. Chem. Soc. 107 (1985) 1080.

[51] A.D. Joran, B.A. Leland, G.G. Geller and J.J. Hopfield, J. Am. Chem. Soc. 106 (1980) 6090.

[52] H. Heitele and M.E. Michel-Beyerle, (submitted).

[53] J. Deisenhofer, O. Epp, K. Miki, R. Huber and H. Michel, J. Mol. Biol. 180 (1984) 385.

[54] V.I. Goldanskii, Yu. F. Krupyanskii and V.N. Fleurov, Doklady A.N. USSR 272 (1983) 978.

[55] G.A. Saigh, H.J. Shink, H.J. Lohneysen, F. Parak and S. Honklinger, Z. Phys. B55 (1984) 23.

[56] R.S. Berry, J. Jortner, J.C. Mackie, E.S. Pysh and S.A. Rice, J. Chem. Phys. 42 (165) 1535.

[57] P.R. Dupius, R. Roberge and C. Sandorfy, Chem. Phys. Lett. 75 (1980) 75.

[58] J.J. Hopfield, *Electrical Phenomena at the Biological Membrane Level,* ed. E. Roux (Elsevier, Amsterdam, 1977) p. 471-490.

[59] J.J. Hopfield, *Tunneling in Biological Systems,* (Academic Press, New York, 1977), pp. 417-432.

[60] W.W. Parson, Paper presented at the Workshop on Antennas and Reaction Centres of Photosynthetic Bacteria, Falderfing, W. Germany (1985).

[61] W.W. Parson, R.K. Clayton and R.K. Cogdell, Biochem. Biophys. ACta 387 (1975) 265.

[62] D. de Leeuv, M. Malley, G. Butterian, M.Y. Okamura and G. Feher, Biophys. J. 37 (1982) 111a.

[63] W. Kaiser, Paper presented at the Workshop on Antennas and Reaction Centres of Photosynthetic Bacteria, Falderfing, W. Germany (1985).

[64] S.I. Choi, J. Jortner, S.A. Rice and R. Silbey, J. Chem. Phys. 41 (1964) 3294.

[65] M.E. Michel-Beyerle and J. Jortner, (to be published.).

The Effect of Conformational Dynamics and Phase Transitions on Electron and Atom Group Transfer Processes

A.M. Kuznetsov

The A.N. Frumkin Institute of Electrochemistry of the
Academy of Sciences of the USSR
Leninskij Prospect 31
Moscow V-71, USSR

and

J. Ulstrup
Chemistry Department A
Building 207
The Technical University of Denmark
2800 Lyngby, Denmark

1. Introduction

Large-amplitude motion of molecular fragments in widely different time ranges in protein molecules is reflected in a variety of molecular and environmental features presently being uncovered. Techniques which have provided information about protein dynamics include x-ray crystallographic [1-3] and Mössbauer spectral properties [4], NMR relaxation of suitable probe nuclei [5,6], fluorescence quenching [7], and time resolved resonance Raman spectroscopy [8].

The physical nature of the different kinds of protein motion can be substantiated by molecular dynamics, stochastic dynamics, and other theoretical approaches [9]. Motion of individual surface groups is sufficiently flexible to acquire diffusional character, in anharmonic multiple-barrier potentials [10]. Internal motion can either be elastic, lattice dynamic-like, it can involve local vibrational motion, or it can be "collective" diffusive motion along conformational multiple-barrier modes. The latter can involve larger-scale "hinge-bending" of linked globular domains in which case the character of the motion approaches that of a phase transition.

Temperature variation can induce rearrangements in the conformational system of both solute and membrane-bound proteins.

In some cases this occurs over narrow temperature ranges, as revealed by protein mobility [11,12] and by sharp transitions in the Arrhenius relation for membrane-bound enzymes [13-16]. In such cases the transitions are likely to originate from phase transitions in the membrane, which is also corroborated by computer simulation [17,18]. In other cases transitions from a "frozen" to a mobile state occurs at lower temperatures [11,12,19-22] and over wider ranges, and reflect thermal freezing of conformational states.

Large conformational fluctuations are finally often seen to be associated with crucial elementary steps in biological processes.

One example is conformational changes induced by photoelectron transfer [19,22]. Another example is that the "resting" conformation of myoglobin leaves no access to the binding site of small ligand molecules, and a gate to the heme pocket must be provided by conformational dynamics [24]. Substantial group displacements also accompany substrate binding to enzymes such as carboxypeptidase [25] and the group transfer steps themselves such as proton transfer in serine proteases [26].

Biological electron and atom group transfer processes have been subjected to analysis in terms of quantum mechanical theories of electron transfer. While low-frequency conformational nuclear modes have been recognized in such approaches [27-31], the possible 'triggering" function of such modes does not seem to have received explicit attention. It is the purpose of the present work to point to ways in which these aspects of conformational and phase transitions can be incorporated in electron transfer theory.

2. Adiabatic Separation of a System of Slow Conformational Modes

Electron transfer theory often rests on the concept of a set of fluctuating environmental nuclear modes, the vibrational free enthalpy of which compensates for the donor-acceptor electronic energy difference [32-34]. In the "linear" and continuum limits this leads to a representation of the medium Hamiltonian as an assembly of harmonic oscillators subject to equilibrium coordinate shifts only, and the kinetic parameters depend weakly, by their bulk properties on the temperature. This approach needs modification when applied to conformational transitions in a macromolecular medium. In certain temperature ranges these modes undergo structural changes which affect both the electronic interstate coupling and the nuclear reorganization parameters. In molecular terms these features must be represented by anharmonic, multi-barrier potentials as in stochastic solvent representations [35]. Such large-amplitude motion is also encountered when ferroelectrical materials undergo second order phase transitions.

A set of slow conformational nuclear modes can be handled by separating adiabatically this mode set $\{r\}$ from the total set, $\{R\}$, leaving the local molecular modes $\{Q\}$. For each electronic state, s, this corresponds to the total potential energy surfaces

$$U_s(Q;r) = U_s^{cl}(r) + U_s^{mol}(Q) + V_s(Q;r) \tag{2.1}$$

"cl" and "mol" denoting the classical environmental and the molecular contributions in the initial (i) and final (f) state, and $V_s(Q;r)$ the coupling between the two sets. A fully classical approach can furthermore be regarded as adequate for the set $\{r\}$. Denoting the local mode vibrational levels in the initial and final states by ν and μ, respectively, the transition probability becomes

$$W = \frac{2\pi}{\hbar} Z_i^{-1} \sum_{\nu,\mu} \int dr \exp\left[-\frac{U_{i\nu}(r)}{k_B T}\right] \left(\chi_f^\mu(Q)|v_{fi}(Q)|\chi_i^\nu(Q)\right)^2 \delta\left[U_{i\nu}(r) - U_{f\mu}(r)\right] \tag{2.2}$$

$$U_{s\lambda}(r) = U_s^{cl}(r) + \varepsilon_{s\lambda}(r); Z_i = \sum_\nu \int dr \exp\left[-\frac{U_{i\nu}(r)}{k_B T}\right] (\lambda = \nu, \mu) \tag{2.3}$$

$\varepsilon_{s\lambda}$ and $\chi_s^\lambda(Q)$ are the vibrational energy spectrum and nuclear wave functions of the local modes in the electronic state "s", k_B Boltzmann's constant, T the absolute temperature, $\hbar$ Planck's constant divided by 2π, and $v_{fi}(Q)$ is the electronic transition matrix element leading to reaction.

In order that the role of the conformationally "active" modes becomes explicitly apparent we next invoke two steps. We first assume that all reaction properties depend only on *some* conformational coordinates, but insignificantly on the remaining ones. We can then average eq. (2.2) with respect to the latter to obtain

$$W = \frac{2\pi}{\hbar} \tilde{Z}_i^{-1} \sum_{\nu,\mu} \int dq \exp\left[-\frac{F_{i\nu}(q)}{k_B T}\right] \left(\chi_f^{\mu}(Q;q)|v_{fi}(Q;q)|\chi_i^{\nu}(Q;q)\right)^2$$
$$\delta\left[F_{i\nu}(q) - F_{f\mu}(q)\right] \tag{2.4}$$

where $F_{s\lambda}(q)$ are now the configurational free enthalpies with respect to the conformationally "active" modes for which we have also used the symbol q. Z_i is a configuration integral with respect to the coordinates q

$$F_{s\lambda}(q) = F_s(q) + \varepsilon_{s\lambda}; \tilde{Z}_i = \sum_{\nu} \int dq \exp\left[-\frac{F_{i\nu}(q)}{k_B T}\right]. \tag{2.5}$$

The second step can be appreciated by introducing an explicit representation for the high-frequency modes such as the harmonic approximation

$$\varepsilon_{s\lambda}(q) = \hbar\Omega_s(q)\left(\lambda + \frac{1}{2}\right) \tag{2.6}$$

$\Omega_s(q)$ being the vibrational frequency at given q. Eqs. (2.4) and (2.5) then take the form

$$W = \tilde{Z}_i^{-1} \int dq \exp\left[-\frac{F_i(q)}{k_B T}\right] W_{fi}(q); \tilde{Z}_i = \int dq \exp\left[-\frac{F_i(q)}{k_B T}\right] \tag{2.7}$$

where $W_{fi}(q)$ is the transition probability at given q, and $F_i(q)$ includes the conformational system part only.

Electron transfer theory provides detailed prescription for $W_{fi}(q)$. In the simplest case where local mode frequencies are constant, the Condon approximation with respect to $\{Q\}$ valid, and the harmonic approximation adequate for all the other system modes, $W_{fi}(q)$ takes the form [32,36]

$$W_{fi}(q) = (v_{fi}(q))^2 A \exp -H[\Delta F(q)]$$
$$H[\Delta F(q)] = \beta\alpha(q)\Delta f(q) + \beta\alpha(q)[1 - \alpha(q)]E_r \tag{2.8}$$
$$+ \sum_{\rho} m_{\rho}\Omega_{\rho}(q)\left[Q_{\rho o}^f(q) - Q_{\rho o}^i(q)\right]^2$$

$$\frac{sh\left\{\frac{1}{2}\beta\hbar\Omega_\rho(q)[1-\alpha(q)]\right\}sh\;\frac{1}{2}\beta\hbar\Omega_\rho(q)\alpha(q)}{sh\left[\frac{1}{2}\beta\hbar\Omega_\rho(q)\right]}$$

where $\Delta F(q) = F_f(q) - F_i(q)$ is the reaction free enthalpy at given q, E_r the reorganization free enthalpy in all other modes than Q and q, $\{\rho\}$ the set of local modes with effective masses m, and initial and final state equilibrium coordinates $Q^i_{\rho o}(q)$ and $Q^f_{\rho o}(q), \beta = (k_B T)^{-1}$, and A a constant which depends relatively weakly on the topology of the potential surfaces. $\alpha(q)$ is finally the transfer coefficient determined by the extremum condition $\partial H / \partial \alpha = 0$.

3. Specification of Conformational System Modes

With the transition probability at hand we now specify the conformational modes. We do so with reference to the strongly simplified two-state model shown in Figure 1, which, however, still represents some crucial features of conformational relaxation. A conformational state is represented by a local minimum at the coordinate values q_{10} and q_{20} on the free enthalpy surface spanned by q, and separated from other minima by free enthalpy barriers.

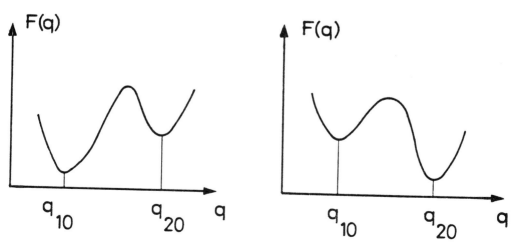

Figure 1

A rather crucial point is now that a change in the external conditions (temperature, pressure, electric field etc.) can modify the free enthalpy surfaces in such a way that a new conformational state, say q_{20} in Figure 1b after the change is the lower. If the change is fast, the system relaxes diffusively to its new overall equilibrium state, provided that the diffusive energy barrier exceeds the thermal energy $k_B T$, which we shall assume in the following. Otherwise the motion is rather to be regarded as anharmonic but ballistic in the sense that the system cannot be accommodated in given conformational states.

Considering first temperature effects only, we can put this in quantitative terms by the following one-dimensional free enthalpy surface for the "pure" conformational medium

$$F(q) = -\frac{1}{2}aq^2 + \frac{1}{4}bq^4 + C(T)q; a, b > 0 \qquad (3.1)$$

$C(T) \gtrless 0$ for $T \gtrless T_o$, and $C(T) = 0$ for $T = T_o, T_o$ being a temperature where conformational transitions occur. As a first approximation, in the following we shall give $C(T)$ the same form as the free enthalpy close to a second order phase transition, i.e. $C(T) \approx ?(T - T_o)$ [37].

Eq. (3.1) is phenomenological but still represents some important features of conformational relaxation within the two-state scheme. In particular, eq. (3.1) is symmetric with respect to $T = T_o$, and conformational transitions can only occur here. At other temperatures either the "left" $(T > T_o$, fig. 1a) or the "right" $(T < T_o$, fig. 1b) minimum is the lower, and in the absence of solute molecular interactions no transformation occurs.

In the presence of solute molecules the surfaces are modified by electronic-conformational interaction terms in much the same way as the polarization state of a polar solvent. These terms display a smooth functional dependence of q. We shall illustrate this by the simplest, linear form for which the reactant and product surfaces then take the form

$$\begin{aligned}
F_i(q) &= -\frac{1}{2}aq^2 + \frac{1}{4}bq^4 + C(T)q - \gamma_i q \\
F_f(q) &= -\frac{1}{2}aq^2 + \frac{1}{4}bq^4 + C(T)q - \gamma_f q
\end{aligned} \qquad (3.2)$$

where γ_i and γ_f are appropriate "coupling constants".

The interplay between the chemical process and conformational transitions is now reflected in both the terms $C(T)q$ and $\gamma_s q$. Conformational relaxation can only accompany the process at given T when $\gamma_i \neq \gamma_f$, but whether it does so depends crucially on the magnitude of $C(T)$ and γ_s, and in particular on the sign of $C(T) - \gamma_i$ and $C(T) - \gamma_f$. When the signs coincide, i.e. when $C(T) > \gamma_i, \gamma_f$, or $C(T) < \gamma_i, \gamma_f$, then the free enthalpy surfaces are only distorted by a charge transfer process, while no significant conformational state repopulation occurs, as either the inequalities $F_i(q_{10}^i) > F_i(q_{20}^i)$ and $F_f(q_{10}^f) > F_f(q_{20}^f)$ or the same, inverted inequalities remain valid throughout the process. On the other hand, if

$$\gamma_f > C(T) > \gamma)i, \quad \text{or} \quad \gamma_f < C(T) < \gamma_i \tag{3.3}$$

i.e.

$$\begin{aligned} F_i(q_{10}^i) &< F_i(q_{20}^i)) \text{ and } F_f(q_{10}^f) > F_f(q_{20}^f), \text{ or} \\ F_i(q_{10}^i) &> f_i(q_{20}^i) \text{ and } F_f(q_{10}^f) < F_f(q_{20}^f) \end{aligned} \tag{3.4}$$

(Fig. 2), then electronic-conformational coupling is a crucial element of the process. Due to temperature dependence of $C(T)$ this can be expected to be important in the temperature interval between T_{up} and T_{low} where

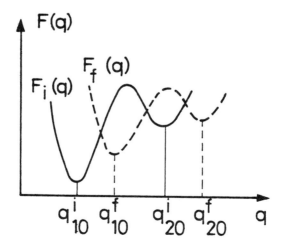

Figure 2

$$T_{\mathrm{up}} = T_o + \max(\gamma_s/mu); T_{\mathrm{low}} = T_o + \min(\gamma_s/mu) \qquad (3.5)$$

if $C(T)$ is given the linear form $\mu(T - T_o)$. "max" and "min" here refer to the value of γ_i and γ_f which is the larger and the smaller, respectively.

4. Charge Transfer in Conformationally Dynamic Systems

We consider separately charge transfer systems with weak and strong electronic-conformational coupling in the sense of eqs. (3.2)-(3.4).

4.1. Absence of transfer-induced conformational transitions

When $\gamma = \gamma_i = \gamma_f$, the effect of the conformationally "active" coordinate q appears formally in the q-dependence of the kinetic parameters in eq. (2.8), while the reaction free enthalpy is independent of q. The dependence is, however, the strongest for both the electronic factor and for nuclear factors representing atom group transfer when the latter involves tunnelling and thermal motion. Nuclear tunnelling is numerically the simpler, as the q-dependence of $\alpha(q)$ can then be approximately disregarded. Presently we consider this case only, representative for example of proton transfer in hydrolytic enzyme systems. We can then combine the high-frequency limit of the $\{\rho\}$-modes with, for example an exponential decay of the electronic factor to give

$$v_{fi}(q) \approx v_{fi}^o \exp\left(-\frac{\sigma}{2}\right) \exp(-\xi q); \quad \sigma = \frac{1}{\hbar} \sum_\rho m_\rho \Omega_{\rho o} \left(Q_{\rho o}^{io} - Q_{\rho o}^{fo}\right)^2 \qquad (4.1)$$

where ξ is a decay parameter, and the whole q-dependence of $\Omega_\rho(q), Q_{\rho o}^i(q)$ and $Q_{\rho o}^f(q)$ is contained in the factor $\exp(-\xi q)$.

Together with eqs. (2.8) and (3.2), eq. (4.1) provides the necessary basis for the transition probability in the absence of conformational transitions. Considerable simplification arises if, close to the conformational minima, we replace eq.

(3.2) by the harmonic approximation

$$F_s(q) \approx \frac{1}{2} \sum_{\ell=1,2} \left[\tilde{\gamma}_\ell (q - q_{\ell 0})^2 + F_s(q_{\ell 0}) \right] + F_{s0} \tag{4.2}$$

where (cf. eq. (3.2))

$$\tilde{\gamma}_\ell = 3bq_{\ell 0}^2 - a; \quad -aq_{\ell 0} + bq_{\ell 0}^3 + C(T) - \gamma = 0. \tag{4.3}$$

This is a valid approximation when the potentials are not too shallow. The transition probability, eqs. (2.7) and (2.8) then becomes:

$$W = W_o \left[\tilde{\gamma}_1^{-1/2} + \tilde{\gamma}_2^{-1/2} \exp\left\{ -\beta \left[F_i(q_{20}) - F_i(q_{10}) \right] \right\} \right]^{-1}$$

$$\left\{ \tilde{\gamma}_1^{-1/2} \exp\left[(\xi^2/\beta\tilde{\gamma}_1) - 2\xi q_{10} \right] + \tilde{\gamma}_2^{-1/2} \exp\left\{ -\beta \left[F_i(q_{20}) - F_i(q_{10}) \right] \right\} \right.$$

$$\left. \cdot \exp\left[(\xi^2/\beta\tilde{\gamma}_2) - 2\xi q_{20} \right] \right\} \tag{4.4}$$

where W_o is the transition probability in the absence of a conformational system, i.e. $W_o \approx A(v_{fi}^o)^2 \exp[-H(\Delta F_o)]$, and $q_{\ell 0}$ are to be regarded as the particle transfer distance at the bottom of the conformational potential wells.

Eq. (4.4) shows that the conformational system modifies the rate by the population of different conformational states and by the decay parameter. The effect of the latter is caused by temperature dependence of the particle transfer distance, being approximately $q^* \approx q_{\ell 0} - 2k_B T \xi / \tilde{\gamma}_\ell$ in the conformational state ℓ. The equation also shows that when conformational transitions are induced by temperature variation, sharp, phase transition-like changes in the Arrhenius slope towards either larger or smaller values can emerge.

When the coupling terms γ_i and γ_f differ but conformational transitions can still be disregarded, then conformational effects are no longer restricted to the electronic and nuclear tunnelling factors, but now also reflected in the energetics of the process. The system is still located in the same conformational well throughout the reaction, but the well is shifted vertically and horizontally in the final state relative to the initial state. As the system, however, never approaches the barrier top, motion along this mode is no different from any local mode and can be represented (still close to the bottom of the wells) by suitable bound potentials such as

$$F_s(q) = \frac{1}{2}\tilde{\gamma}_\ell(q - q_{\ell 0})^2 + F_s(q_{\ell 0}^s); \quad s = i, f; \ell = 1, 2. \tag{4.5}$$

This gives for W_{fi} (cf. eq. (2.8))

$$W \approx A v_{fi}(q^*)^2 \exp\left[-\sigma(q^*)\right]\exp\left[-H(\Delta F_o)\right]$$
$$H(\Delta F_o) = \beta\left[\alpha F_o + \alpha(1 - \alpha)(E_r^{con} + E_r)\right] \tag{4.6}$$

where $\Delta F_o = F_f(q_{10}^f) - F_i(q_{10}^i)$, E_r^{con} is the reorganization free enthalpy in the conformational system, i.e. $E_r^{con} = \frac{1}{2}\gamma_1\left(q_{10}^f - q_{20}^i\right)^2$, and $q^* \approx (1 - \alpha)q_{10}^i + \alpha q_{10}^f$.

In the absence of transfer-induced conformational transitions the linear electronic-conformational coupling terms can thus (a) induce equilibrium shifts in the conformational coordinates leading to conformational reorganization free enthalpies, and (b) invoke conformational coordinate dependence on the electronic and nuclear tunnel factors. Both effects can be envisaged to display fairly abrupt changes when conformational transitions are induced thermally.

4.2. Effects of transfer-induced transitions

Depending on the location and topology of conformational potential surfaces, different effects can appear when the charge transfer process itself is accompanied by conformational transitions and repopulation at a given temperature. Consider first the surface in Figure 3. This location implies the charge transfer occurs *prior* to the conformational transition, while the system is still in the most stable initial conformational well. Conformational relaxation only occurs afterwards and - within the two-state approximation - inside a single well. Conformational relaxation is therefore fast and reduces to statistical averaging with respect to the surface location

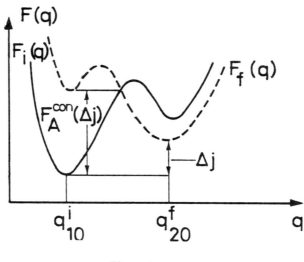

Figure 3

$$W = (Z_i^{\mathrm{con}})^{-1} \int d\Delta j \exp\left[-F_{A1}^{\mathrm{con}}(\Delta j)/k_B T\right] W_{fi}[q(\Delta j)]$$

$$Z_i^{\mathrm{con}} = \int d\Delta j \exp\left[-F_{A1}^{\mathrm{con}}(\Delta j)/k_B T\right]$$

(4.7)

where $F_{A1}^{\mathrm{con}}(\Delta j)$ is the activation free enthalpy along q for given "local" reaction free enthalpy Δj (Figure 3). As long as the dominating contributions to eq. (4.7) are provided by values of Δj which do not lead to significant deviations of the surface locations from that in figure 3, the transition probability is therefore in basic respects no different from that of processes in molecular solvents.

A different pattern emerges for surfaces such the ones in Figure 4. Charge transfer is now either preceded by a conformational transition from q_{10}^i to a metastable configuration q_{20}^i, or charge transfer occurs first leading from q_{10}^i to the metastable configuration q_{10}^f which then relaxes to the configuration q_{20}^f. In either case the overall rate pattern depends on the rates of both the particle transfer itself and of conformational relaxation, and the process can thus either belong to the statistical equilibrium regime or to the "stochastic" regime. We shall show this for the second of the two cases represented by Figure 4.

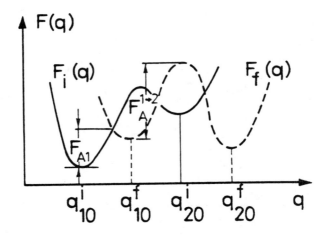

Figure 4

The two limiting cases correspond to charge transfer or conformational relaxation being the slower step. In the former case the rate is determined by charge transfer in the first conformational state, and the transition probability is approximately

$$W \approx \left(\tilde{Z}_i^{con} \right)^{-1} \exp\left[-F_{A1}^{con}(\Delta j^*)/k_B T \right] W_{fi}[q(\Delta j^*)]$$

$$\tilde{Z}_i^{con} = \int dq \exp\left\{ -[F_i(q) - F_i(q_{10}^i)]/k_B T \right\}$$

$$\approx (2\pi k_B T)^{-1} \left\{ \tilde{\gamma}_1^{-1/2} + \tilde{\gamma}_2^{-1/2} \exp\left\{ -[F_i(q_{20}^i) - F_i(q_{10}^i)]/k_B T \right\} \right\} \quad (4.8)$$

(cf. eq. (4.4)) where the activation free enthalpy $F_{A1}^{con}(\Delta J^*) = F_i[q(\Delta j^*)] - F_i(q_{10}^i)$. The temperature effects arise from $F_{A1}^{con}(\Delta j^*)$, from $W_{fi}[q(\Delta j^*)]$, and from $\left(\tilde{Z}_i^{con} \right)^{-1}$, the latter effect being associated with a decreasing initial state equilibrium population and could lead to "inverse" temperature dependence of the overall rate when the other activation free enthalpy terms are small.

In the second case the rate determining step is conformational relaxation from q_{10}^f to q_{20}^f, q_{10}^f being pre-equilibrated with q_{10}^i by the chemical process. W

now takes the approximate form

$$W \approx \exp\left\{-\left[F_f(q^f_{10}) - F_i(q^i_{10})\right]/k_B T\right\} W^{con}_{rel}; \quad W^{con}_{rel} = \frac{\omega^{con}_{eff}}{2} \exp\left[-F_A^{1\to2}/k_B T\right]$$

(4.9)

where W^{con}_{rel} is the conformational relaxation probability determined by the diffusional barrier $F_A^{1\to2}$ and the frequency for conformational relaxation ω^{con}_{eff}.

5. Charge Transfer Close to a Second Order Phase Transition Temperature

Conformational transitions bear some resemblance to second order phase transitions in the sense that the transitions occur in a narrow temperature range in the absence of external fields and can be regarded as accompanied by small changes in certain "ordering" parameters $\vec{\eta}(\vec{r})$ [37]. As the solute-solvent interaction, V_s^{int}, is determined by $\vec{\eta}(\vec{r})$ and the appropriate (electric, pressure, etc.) fields, $\vec{\phi}_s(\vec{r})$

$$V_s^{int} = -\int \vec{\phi}_s(\vec{r})\vec{\eta}(\vec{r})d\vec{r}$$

(5.1)

and is different for different electronic states, this can lead to "abrupt" changes in the pattern for the chemical process.

We consider briefly the effect of a macroscopic phase transition on the basis of the free enthalpy functional for a continuous medium [37]

$$F[\vec{\eta}(\vec{r}), T] = F_o(T) + \frac{1}{2}a(T)\int d\vec{r}[\vec{\eta}(\vec{r})]^2 + \frac{1}{2}\psi(T)\int d\vec{r}[\text{div }\vec{\eta}(\vec{r})]^2.$$

(5.2)

The second term contains a quantity $a(T)$, the properties of which reflect the phase transition, i.e. $a(T) = a_o(T - T_o)$ close to T_o (cf. eq. (3.1)). The third term is only important when the ordering variation is comparable to the spatial correlation length of the local environmental structure, and for the sake of simplicity we disregard this term in the following [38].

If $\vec{\phi}_s(\vec{r})$ and $\vec{\eta}(\vec{r})$ represented the electric fields, $\vec{E}_s(\vec{r})$, and the polarization,

$\vec{P}(\vec{r})$, then eq. (5.2) would take the form

$$F\left[\vec{P}(\vec{r}),T\right] - F_o(T) \approx \frac{2\pi}{c} \int d\vec{r} \left[\vec{P}(\vec{r})\right]^2 . \tag{5.3}$$

Formally $c = \varepsilon_o^{-1} - \varepsilon_s^{-1}$, ε_o and ε_s being the optical and static dielectric constant. Comparison between eqs. (5.2) and (5.3) then shows

$$a(T) = \frac{4\pi}{c(T)}, \text{ or } c(T) \approx \frac{4\pi}{\eta(T - T_o)} \to \infty \quad \text{for } T \to T_o. \tag{5.4}$$

The direct association of the dielectric parameter with such a behavior is, however, not obvious except perhaps for ferroelectric materials, and we shall therefore let $\vec{\phi}_s(\vec{r})$ and $\vec{\eta}(\vec{r})$ represent more general field and ordering parameters.

In the absence of spatial and vibrational frequency dispersion the reaction and environmental reorganization free enthalpies become

$$\Delta F_o = \Delta F_o^o + \frac{1}{2} \frac{1}{a(T)} \int d\vec{r} \left\{ \left[\vec{\phi}_{fo}(\vec{r})\right]^2 - \left[\vec{\phi}_{io}(\vec{r})\right]^2 \right\} \equiv \Delta F_o^o + \frac{F_o^{\text{env}}}{a(T)} \tag{5.5}$$

$$E_s^{\text{env}} = \frac{1}{2} \frac{1}{a(T)} \int d\vec{r} \left[\vec{\phi}_{fo}(\vec{r}) - \vec{\phi}_{io}(\vec{r})\right]^2 \equiv \frac{E_{so}^{\text{env}}}{a(T)} \tag{5.6}$$

where $\vec{\phi}_{io}(\vec{r})$ and $\vec{\phi}_{fo}(\vec{r})$ are the initial and final state fields at equilibrium and we have exploited the local [38] relation $\vec{\eta}(\vec{r}) = \frac{1}{a(T)}\vec{\phi}(\vec{r})$.

These equations show that critical behavior of the chemical process near the phase transition can be expected. Close to this point of activation free enthalpy is thus approximately $(E_{so}^{\text{env}} + \Delta F_o^{\text{env}})^2 / [4a_o(T - T_o)E_{so}^{\text{env}}]$ giving formally an infinitely small rate constant as $T \to T_o$, unless $\Delta F_o^{\text{env}} \to -E_{so}^{\text{env}}$. This result formally originates from the extremum values of thermodynamic quantities close to second order phase transitions [39].

6. Concluding Remarks

Two features in particular have emerged from our analysis. First, we have incorporated conformational relaxation in the form of a two-state model and shown that this can give rise to conformational coordinate and temperature dependence of the electronic and nuclear tunnelling factors, the reaction and reorganization free enthalpies, and the effective frequency for motion on the free enthalpy surfaces. Whether the effects do appear, depends on the location and topology of the surfaces. The pre-exponential factors may thus be strongly affected by conformational motion, also when electronic-conformational coupling is weak, and strong electronic-conformational coupling may have a quite insignificant impact on the reaction pattern. Secondly, we have shown that a second order environmental phase transition can lead to critical behavior of the rate constants and activation parameters.

Reported data for conformational relaxation and phase transitions do not in general provide sufficient details to point to particular cases among the ones discussed above. Ferroelectric and ferromagnetic materials are known to exhibit "extremal" behavior close to phase transitions [39-42]. For example, the activation free enthalpy for surface oxidation of iron to FeO exhibits a sharp maximum close to the transition temperature [40,41], and "anomalously" low rates are also found for dihydrogen desorption from thin nickel films close to the transition. No such effects appear for biological materials. The enzyme "activity" of membrane-bound nitrogenase and several phosphatases are found to exhibit breaks in the Arrhenius relation around room temperature [11-16]. In view of the pressure sensitivity the effects have been ascribed to membrane gel-liquid phase transitions which possibly induce conformational transitions in the proteins, but the elementary reaction steps are not sufficiently disentangled to warrant closer analysis in terms of the present formalism.

Photoinduced electron transfer in protein systems sometimes go from a low-temperature activationless region to a high-temperature Arrhenius relation. The low transition temperature of about $100°K$ for the cytochrome c/bacteriochlorophyll system suggests that the molecular origin of the activationless branch for this system is nuclear tunnelling [28,29-31,43]. For several other related processes, such as electron transfer from the primary to the secondary quinone acceptor in the photosynthetic reaction centers of purple bacteria and several electron transfer processes in plant photosynthesis the transition temperature is much higher $(150-200°k)$. The temperature dependence of some the processes is furthermore quite similar to that of the conformational mobility as reflected by Mössbauer absorption [22,23] or rotational relaxation of attached spin labels [21] and strongly suggests that electronic-conformational coupling occurs.

In terms of the formalism above such a process would then appear to be dominated by energetic effects of the conformational system, the effects on the electronic factor and from conformational state repopulation being of minor importance. Since conformational relaxation in single wells would presumably occur also at lower temperatures (eq. (4.7)) these systems are likely to correspond to the strong electronic-conformational coupling limit. From the temperature dependence alone we cannot, however, distinguish whether the process is dominated by conformational relaxation (eq. (4.10)) such as concluded in ref. 22, or by the electron transfer step itself (eq. (4.9)).

In one case, viz. the back reaction from the primary acceptor to the primary donor, an "inverse" temperature effect is found, i.e. the rate decreases by a factor of 5-10 when the temperature is lowered from 300 to 150°K and then remains almost constant on further lowering [19,43]. This effect is understandable if the reaction belongs to the activationless region [31,44], but conformational relaxation could in fact give a similar effect. Inverse temperature dependence could arise in the weak electronic-conformational coupling limit largely from the temperature dependence of the electronic factor (eqs. (4.4) and (4.5)), giving a smaller electron transfer distance at lower temperatures (cf. ref. 19). In the strong-coupling limit it could arise either from the same origin, from repopulation of conformational states (eq. (4.9)), or from a gradual shift from the statistical equilibrium to the conformationally "stochastic" regime (eqs. (4.9) and (4.10)).

References

1. H. Frauenfelder, G.A. Petsko and D. Tsernoglou, Nature, 280 (1979) 558.

2. M.J.E. Sternberg, D.E.P. Grance and D.C. Phillips, J. Mol. Biol., 130 (1979) 231.

3. P.G. Debrunner and H. Frauenfelder, Ann. Rev. Phys. Chem., 33 (1982) 283

4. F. Parak, E.N. Frolov, R.L. Mössbauer and V.I. Gol'danskii, J. Mol. Biol., 145 (1981) 825.

5. F.R.N. Gurd and T.M. Rothgeb, Adv. Protein Chem., 33 (1979) 74.

6. O. Jardetzky, Acc. Chem. Res., 14 (1981) 291.

7. J.R. Lakowicz and G. Weber, Biochemistry, 12 (1973) 4171.

8. J.M. Friedman, D.L. Rousseau and M.R. Ondrias, Ann. Rev. Phys. Chem., 33 (1982) 471.

9. J.A. McCammon and M. Karplus, Ann. Rev. Phys. Chem., 31 (1980) 29.

10. B. Gavish, Proc. Natl. Acad. Sci. USA 78 (1981) 6868

11. D.D. Thomas and C. Hidalgo, Proc. Nat. Acad. Sci. USA, 75 (1978) 5488.

12. J.H. Davis, M. Bloom, K.W. Butler and I.C.P. Smith, Biochim. Biophys. Acta, 597 (1980) 477.

13. F. Ceuterick, J. Peeters, K. Heremans, H. de Smedt and H. Olbrechts, Eur. J. Biochem., 87 (1978) 401.

14. W.L. Dean and C. Tanford, Biochemistry, 17 (1978) 1683.

15. H. de Smedt, R. Borghgraef, F. Ceuterick and K. Heremans, Biochim. Biophys. Acta, 556 (1979) 479.

16. a. K. Heremans and F. Wuytack, FEBS Lett., 117 (1980) 161; b. K. Heremans, D. de Smedt and F. Wuytack, Biophys. J., 37 (1982) 74.

17. J.F. Nagle, Ann. Rev. Phys. Chem., 31 (1980) 157.

18. O.G. Mouritsen, A. Boothroyd, R. Harris, N. Jan, T. Lookman, L. Mac-Donald, D.A. Pink and M.J. Zuckerman, J. Chem. Phys., 79 (1983) 2027.

19. B. Hales, Biophys. J., 16 (1976) 471.

20. L.A. Blumenfel'd, D.S. Burbaev, R.M. Davidov, L.N. Kubrina, A.F. Vanin and R.O. Vilu, Biochim. Biophys. Acta, 379 (1975) 512.

21. A.I. Berg, P.P. Noks, A.A. Kononenko, E.N. Frolov, I.N. Krymova, A.B. Rubin, G.I. Likhtenshtein, V.I. Gol'danskij, F. Parak. M. Bukl and R. Mössbauer, Molekulyarnaya Biologiya, 13 (1979) 81.

22. F. Parak, E.N. Frolov, A.A. Kononenko, R.L. Mössbauer, V.I. Gol'danskij and A.B. Rubin, FEBS Lett., 117 (1980) 368.

23. F. Parak, E.W. Knapp and D. Kucheida, J. Mol. Biol., 161 (1982) 177.

24. D. Beece, L. Eisenstein, H. Frauenfelder, D. Good, M. Marden, L. Reinisch, A.H. Reynolds, L.B. Sorensen and K.T. Yue, Biochemistry, 15 (1979) 3421.

25. F.A. Quiocho and W.N. Kipscomb, Adv. Protein Chem., 25 (1971) 1.

26. H. Sumi and J. Ulstrup, in preparation.

27. R.R. Dogonadze, A.M. Kuznetsov and J. Ulstrup, J. Theor. Biol., 69 (1977) 239.

28. A.M. Kuznetsov, N.C. Sondergaard and J. Ulstrup, Chem. Phys., 29 (1978) 383.

29. A. Sarai, Biochim. Biophys. Acta, 589 (1980) 71.

30. E. Buhks, M. Bixon and J. Jortner, Chem. Phys., 55 (1981) 41.

31. A.M. Kuznetsov and J. Ulstrup, Biochim. Biophys. Acta, 636 (1981) 50.

32. a. R.R. Dogonadze and A.M. Kuznetsov, Physical Chemistry. Kinetics, VINITI, Moscow, 1973; b. Progr. Surf. Sci., 6 (1975) 1; c. R.R. Dogonadze, A.M. Kuznetsov and T.A. Marsagishvili, Electrochim. Acta, 25 (1980) 1.

33. N.R. Kestner, J. Logan and J. Jortner, J. Phys. Chem., 78 (1974) 2148.

34. J. Ulstrup, Charge Transfer Processes in Condensed Media, Springer-Verlag, Berlin, 1979.

35. H. Frauenfelder and P.G. Wolynes, Science, in press.

36. R.R. Dogonadze, A.M. Kuznetsov and M.A. Vorotyntsev, Phys. Stat. Sol., 52b (1972) 125, 425.

37. L.D. Landau and E.M. Lifshitz, Statistical Physics, Pergamon, Oxford (1970).

38. A.A. Kornyshev, J. Chem. Soc. Faraday Trans. II, 79 (1983) 651.

39. N.V. Grabinskij, A.P. Levanyuk and A.S. Sigov, Fiz. Tverd. Tela, 24 (1982) 1936.

40. G.C. Measor and K.K. Afzulpurkar, Phil. Mag., 10 (1964) 817.

41. V.D. Borman, A.N. Pivovarov and V.I. Troyan, JETP Lett., 39 (1984) 556.

42. M.R. Shanabarger, Phys. Rev. Lett., 43 (1979) 1964.

43. D. DeVault, Quart. Rev. Biophys., 13 (1980) 387.

44. E. Buhks and J. Jortner, FEBS Lett., 109 (1980) 117.

Effects of Distance, Energy and Molecular
Structure on Electron Transfer Rates

J. Miller

Chemistry Division, Argonne National Laboratory

Argonne, IL 60439

Work performed under the auspices of

the Office of Basic Energy Sciences

Division of Chemical Science

US-DOE contract number W-31-109-ENG-38.

Electron transfer reactions are known to occur at long distances in a number of proteins. Among these are the photosynthetic reaction centers which have recently been crystallized to yield an exciting picture of an arrangement of chromophores designed to facilitate charge separation in photosynthesis. We do not yet know in detail how this structure carries out its function. We know a few of the basic principles, but some of the others still need to be learned. At Argonne National Laboratory, our research has focused on the question, "How do long distance electron transfer processes occur; at what distances can they occur; and how are they controlled by energy, distance and molecular structure?" Here, in this extended abstract, I present in a brief summary form some of our findings. I will also review some of what is known about how charged molecules interact with their surrounding medium and the way in which the dynamics of the medium can control the rates of electron transfer processes. It is plausible that similar effects occur to couple the protein dynamics to electron transfer processes occurring within proteins.

In our laboratory we've used two methods to determine electron transfer rates at fixed distances between the electron donors (D) and acceptors (A). In the first of these methods, D and A molecules are dispersed in a rigid, glassy matrix [1-11]. Then a pulse of radiation [1-5, 8-11] or light [6-7] adds or takes away an electron from a few of the D and A molecules or excites a few of them to an electronically excited state. The kinetics of the subsequent electron transfer processes are followed, usually with optical absorption measurements. If the D and A molecules do not aggregate, then the random distribution of distances between D and A can be calculated. From these experiments, one can obtain the rate of an electron transfer process as a function of distance, or conversely, the distance of electron transfer as a function of time. Table 1 summarizes results obtained in our laboratory, giving electron transfer distances found in various kinds of reactions.

EXPERIMENTALLY MEASURED TUNNELING DISTANCES
(Center to Center)

$$D^{\bar{\cdot}} + A \longrightarrow D + A^{\bar{\cdot}} \quad \text{or} \quad D_h^+ + A_h \longrightarrow D_h + A_h^+$$

$$15 \, \overset{o}{A} \quad \text{in} \ 10^{-8} s \, ; \qquad 35 \, \overset{o}{A} \quad \text{in} \ 10^2 \, s$$

$$
\left.
\begin{array}{l}
D^* + A \\[1em]
D + A^*
\end{array}
\right\} \longrightarrow D^+ + A^{\bar{\cdot}}
\left\{
\begin{array}{l}
\text{singlets} = 15 \, \overset{o}{A} \ \text{in} \ 10^{-8} \, s \\[1em]
\text{triplets} = 31 \, \overset{o}{A} \ \text{in} \ 1 \, s
\end{array}
\right.
$$

These are for optimally exothermic reactions. Reactions either too weakly or too strongly exothermic are slower.

Table 1

Distances listed in Table 1 are for optimally exothermic reactions. For reactions with less or more exothermicity, the distances or the rates at a particular distance would be smaller.

A remarkable feature of Table 1 is that the distances obtained are almost independent of the type of reaction. Long distance transfer of positive charge between two molecules occurs almost as effectively as transfer of negative charge. The creation of an ion pair from two neutral molecules, one of which is electronically excited, occurs almost as effectively as do the transfer reactions involving a charged molecule and a neutral molecule. This seems surprising, because the electrons being transferred are at very different energy levels in the different types of reactions, so that, considered as electron tunneling processes, the simple electron tunneling barrier heights are much larger for the hole transfers and the excited state reactions than for the transfer of an electron from a radical anion. Table 1 tells us that electrons may be transferred over about 15 Å in time of about 10^{-8} seconds or over more than 30 Å in a time of about a minute.

These measured values of electron transfer distances are subject to one additional assumption. That assumption is that the broad distribution of rates observed in a rigid glassy matrix is due predominantly to the distribution of distances. Two types of experiments have been performed [9-11] which provide a check on this assumption and the assumption of a random distribution of distances. Both types of experiments supported the assumptions, indicating a valid measurement of distances of electron transfer.

With these experiments, it has been possible to establish by experimental measurements that electron transfer rates decrease exponentially with distance:

$$k(r) = \nu \exp(-\alpha r).$$

(1)

According to the simples models of electron tunneling, the parameter α in equation 1 would vary as $B^{1/2}$, where B is the binding energy on the electron donor, or the electron tunneling barrier height. This simple form should be valid if there is a smooth potential barrier between the donor and acceptor. Experiments designed to test this notion by varying the binding energy of the electron donor [5] have instead been more consistent with another form, $\alpha \sim lnB$ (see Table 2).

This form is consistent with the idea that the region between the donor and acceptor contains a complicated molecular structure, and that the electron being transferred propagates through states of the medium as in Figure 1b. That idea also includes in a natural way the possibility that electronic coupling between the donor and acceptor propagates through filled orbitals of the medium [3]. This "hole tunneling" can explain why hole transfer [3] occurs at distances similar to those for electron transfer (see Table 1).

Another important question is the role of energy in electron transfer. For decades, theories have predicted a relationship between the kinetics and the thermodynamics of electron transfer.

Distance Dependence of Electron Transfer Rate Constants

$$k(r) = \nu \, exp[-\alpha(r - R_0)]$$

The Dependence of α on the Tunneling Barrier Height B

D	B (eV)	α (Å^{-1})	$2(2mB)^{1/2}/\hbar$ [a]	$(2/d)ln(B/\beta)$ [b]
Triptycene	1.81	1.10	1.38	1.10
Biphenyl	2.50	1.22	1.62	1.20
Biphenylene	2.84	1.26	1.73	1.23
Pyrene	3.00	1.13	1.77	1.25
Fluoranthene	3.34	1.37	1.87	1.28
Acridine	3.49	1.22	1.91	1.30
Fluorenone	3.78	1.24	1.99	1.32
2-EtAQ	4.00	1.43	2.05	1.34

[a] Calculation of α as the splitting between levels of two potential wells separated by a smooth potential barrier.

[b] Calculation of the superexchange model with the separation between sites in the medium, d = 6.7Å, and the electron exchange coupling between neighboring sites, β = 0.0466 eV.

Table 2

This relationship in which the electron transfer rate increases with more thermodynamic driving force until it reaches a maximum at the total reorganization energy of the reaction and then decreases with further driving force is illustrated for the case of a reaction in a protein [16] in Figure 2.

For these intraprotein electron transfers, the relationship was not definitely verified because there were too few data points, and quantities other than the thermodynamic driving force changed from one reaction to another. The relationship has been clearly established, however, for fixed distance electron transfer, both in glasses (Figure 3) and in liquids [2,4,14], where the bifunctional molecules of the type illustrated in Figure 4 were used [14]. For many electron transfer reactions in fluids, most of the decrease in the highly exothermic region and a substantial part of the increase in the weakly exothermic region are obscured by the diffusion controlled limit. For many years, this led people to question whether the so-called inverted region actually exists.

In systems designed to separate charges, such as photosynthetic reaction centers or artificial photochemical energy storage devices, it is desirable to make forward charge separating reactions fast with little free energy change, but to have backward energy wasting reactions slow, despite their very large free energy changes. This is contrary to the usual correlation between rate and energy in chemistry.

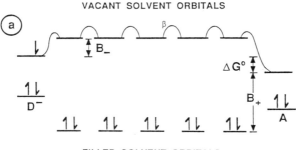

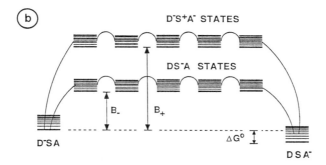

Figure 1. A molecular orbital diagram (top) and a system state diagram for the superexchange mechanism for the electron transfer reaction $D^- + A \to D + A^-$. In this mechanism, the electron exchange interaction between the reactants $(D^- S A)$ and products $(D S A^-)$ states occurs indirectly via virtual states which place an electron $(D S^- A)$ or a hole $(D^- S^+ A^-)$ on sites S^- in the medium (or solvent) between D and A.

Usually, more driving force means a faster rate – the opposite of what is desired. Electron transfer reactions in the inverted region, the opposite can be true, and this is much more marked in nonpolar solvents. This is clearly seen by the comparison of a weakly and strongly exothermic reaction in a moderately and very weakly polar solvent, given in Table 3. These data strongly suggest to optimize a charge separation system, getting fast forwards rates and slow

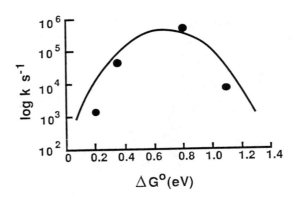

Figure 2. Electron transfer as a function of free energy change for protein/protein complexes of native and modified cytochrome c with cytochrome b₅ (data from McLendon et al., ref. 16).

backward rates, the reactions should occur in a medium of very low polarity. It is not necessary that the medium have no dipoles, but only that they are unable to respond to the movement of charge in an electron transfer reaction, that is, there is little "reorganization energy" for the electron transfer reaction. An important job of the protein environment in photosynthetic reaction centers and other biological electron transfer reactions may be to provide a medium having low reorganization energy surrounding the donor and acceptor.

Table 3: Rate constants for electron transfer in moderately polar (THF) and nonpolar (isooctane) solvents. Electron transfer is from biphenyl⁻ across the rigid androstane spacers to the acceptor.

Acceptor	$-\Delta G°$(eV)	$k(s^{-1})$ THF	Isooctane
naphthalene	0.05	1.5×10^6	1.5×10^9
benzoquinone	2.1	2.5×10^8	3.6×10^6

There is another important effect seen in glasses which may also occur in

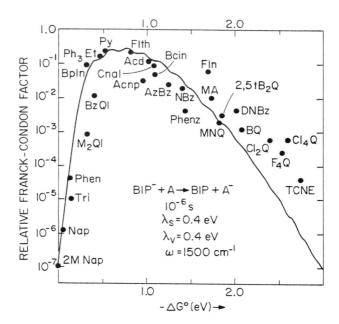

Figure 3. Relative rates of electron transfer as a function of free energy change,
$\Delta G°$, *for electron transfer at fixed distance in MTHF glass at 77K. The donor*
is biphenyl radical anion (BIP⁻). *Various acceptors are used to vary* $\Delta G°$.
Reprinted with permission from (4). Copyright (1984) American Chemical Soci-
ety

proteins. That is, that the reorganization energy of the medium for glass is a
dynamic quantity. It depends on time, because there are relaxation times of the
rigid glassy medium which are as slow as or slower than the times of the electron
transfer process. The relaxations of glasses around charged species have been observed indirectly through time dependent spectral shifts of radical ions created
in the glasses [17]. Radical ions which show such spectral shifts are those for
which the charge distribution is significantly different in the ground and excited
electronic states. In rigid glasses, these spectral shifts span at least eight decades
in time [17]. The time dependent spectral shifts imply a time dependent reorganization energy for electron transfer reactions of these transient charged species.
Unfortunately, no experiment has been performed which quantitatively measures
these reorganization energies. The kinetics of electron transfer reactions in these

Figure 4. A molecule containing two aromatic groups, biphenyl and naphthalene, separated at fixed distance by the rigid steroid spacer androstane. The nearest edges of the aromatics are 10 Å apart, and the centers are 17 Å apart. Addition of an electron to one of the groups allows measurement of the intramolecular electron transfer rate constant by optical spectroscopy.

rigid glasses for reactions having weak exothermicities show complications which can be readily understood [4] with the idea of the time dependent reorganization energy. At the same time, no transfer theories to correctly account for these time dependent relaxation processes are available. It seems reasonable to conclude that the idea of a time dependent reorganization energy is qualitatively correct, but no quantitative description is presently possible.

Is such an effect important in proteins? There is really no information to shed light on that question, but it is plausible that it may be so. Since the relaxations of a medium are temperature dependent, that would mean that time dependence of the reorganizational energy will also be temperature dependent. This can lead to unusual effects such as decreased, zero or even negative apparent activation energies for an electron transfer process. This can occur even though the activation energy for the electron process at constant reorganization energy may be substantial and positive. Even more so than in rigid glasses, in proteins we have great difficulty determining what the relaxations around transient charge species are. Therefore, the time dependence of reorganization energies, if it is substantial, will be even less well known. Looking at the situation in a positive light, we can understand that extremely interesting dynamical effects can occur for electron transfer processes in proteins, but the more negative view is that the factors giving temperature dependence of electron transfer processes in proteins may be very complicated and the information needed to understand them may be very difficult to obtain. Thus, temperature dependence of electron transfer reactions and proteins, while very interesting may be very difficult to use to reach definite conclusions about the nature of the process going on.

Not long ago, very little was known about long distance electron transfer processes. One important thing that was now known was whether they even occurred to a substantial extent. Progress has been made. We know, in many cases, what distances are possible and what the dependence is on distance, and something about how that is affected by energetic parameters. We also know something about how the rates at a fixed distance are affected by the free energy of the overall reaction as a whole, and about the polarity of the medium. In each case, however, the details are very sketchy. We only know the answers for a small number of media and electron donors and acceptors. We know some of the principles that would go into constructing an efficient photochemical charge separation device, such as the bacterial photosynthetic reaction center, but we don't know enough to evaluate the known structures of these reaction centers, which have been described in this meeting.

References

This abstract summarizes and cites only our own work, but the references given here will guide the reader to extensive related work by others.

1. J.R. Miller, Science *189*, 221-222 (1975).

2. J.V. Beitz and J.R. Miller, J. Chem. Phys. *71* (11), 4579-4595 (1979).

3. J.R. Miller and J.V. Beitz, J. Chem. Phys. *74* (12), 6746-6756 (1981).

4. J.R. Miller, J.V. Beitz and R.K. Huddleston, J. Am. Chem. Soc. *106* (18), 5057-5068 (1984).

5. V. Krongauz, R.K. Huddleston and J.R. Miller, to be published.

6. J.R. Miller, K.W. Hartman and S. Abrash, J. Am. Chem. Soc. *104* (15), 4296-4298 (1982).

7. J.R. Miller, J.A. Peeples, M.J. Schmitt and G.L. Closs, J. Am. Chem. Soc. *104* (24), 6488-6493 (1982).

8. R.K. Huddleston and J.R. Miller, J. Phys. Chem. *86* (2), 200-203 (1982).

9. R.K. Huddleston and J.R. Miller, J. Phys. Chem. *86* (8), 1347-1350 (1982).

10. R.K. Huddleston and J.R. Miller, J. Phys. Chem. *87* (24), 4867-4872 (1983).

11. R.K. Huddleston and J.R. Miller, J. Phys. Chem. *85* (15), 2292-2298 (1981).

12. L.T. Calcaterra, G.L. Closs and J.R. Miller, J. Am. Chem. Soc. *105* (3), 670-671 (1983).

13. R.K. Huddleston and J.R. Miller, J. Chem. Phys. *79* (11), 5337-5344 (1983).

14. J.R. Miller, L.T. Calcaterra and G.L. Closs, J. Am. Chem. Soc. *106* (10), 3047-3049 (1984).

15. G.L. Closs, L.T. Calcaterra, N.J. Green, K.W. Penfield and J.R. Miller, J. Phys. Chem., in press.

16. G. McLendon and J.R. Miller, J. Am. Chem. Soc. *107*, 7811 (1985).

17. R.K. Huddleston and J.R. Miller, J. Phys. Chem. *86* (13), 2410-2415 (1982).

Electron Exchange in Cytochrome c

S. Isied, R. Bechtold, A. Vassilian

Dept. of Chemistry, Rutgers University

New Brunswick 08903

Although the title of my talk is on cytochrome c, I would like to start out by introducing some of the results from my laboratory on *intramolecular* electron transfer reactions across polypeptides. Over the last several years we have synthesized and studied three series of binuclear transition metal complexes bridged by polypeptides (Table I) [1-3]. The general experiment involves the following series of compounds:

$$M = [(H_2O)(NH_3)_4Ru^{II}-], \ [(NH_3)_5Os^{II}-] \ ; \ n = 0\text{-}4; \ iso =$$

$$M' = [(NH_3)_5Co^{III}-] \ , \ [(NH_3)_5Ru^{III}-]$$

where the complexes are prepared with two metal centers in the oxidized form. One electron per complex is then added to these systems using chemical or pulse radiolytic techniques to generate the kinetic donor-acceptor intermediate which transfers an electron from the metal donor M to the metal acceptor M in a unimolecular process. For specific members of the series I where n is held constant, variation of the rates of electron transfer can be related to peptide structure and conformation.

In this talk I will limit myself to the series of oligoprolines that we have studied. Proline peptides are among the most rigid of all peptides. Under the conditions chosen for these experiments (aqueous acidic media), proline polypeptides exist mainly in the trans-configuration [3] and there is a defined distance between the C and N terminals. The five membered proline ring limits the conformational flexibility of these peptides, such that there is only one major slow motion, that is, trans to cis proline isomerization (timescale - 1-2 minutes) [4,5]. Thus, oligoprolines with all trans-configuration have the longest distance between the N and C terminals. When one or more of the trans bonds isomerize to the cis configurations, the N to C terminal distance generally decreases.

cis proline   trans proline

Table I shows the variation in rates of electron transfer for a series of oligoproline complexes ($n = $ 0-4 prolines) with a $Ru(II)$ donor and a $Co(III)$ acceptor. In this series of $Ru - Co$ complexes, electron transfer becomes so slow with increasing distance, that the prolines have ample time for trans to cis isomerization to occur prior to electron transfer, especially for the $n = 3$ and $n = 4$ proline isomers. Thus for $N = 0, 1, 2$ the expected decrease in rate with increasing number of prolines is observed, while for $n = 3, 4$, the rate of electron transfer actually increases as the number of proline rings increases. This is presumably because of the trans to cis proline isomerization, bringing the donor and acceptor to closer proximity.

Table I also shows the same series of oligoproline complexes with $(NH_3)_5Os^{II}-$ donor replacing the analogous ruthenium donor.

With this substitution the driving force for the electron transfer increases by over 600 mV, resulting in a decrease of the timescale of reaction from hours to fractions of seconds. For this series of Os-Co oligoproline complexes (Table I) the rate of electron transfer is much more rapid than proline isomerization. In the Os-Co oligoproline complexes ($n = 0, 1, 2$) where electron transfer occurs in timescales ranging from microsecond to seconds, an exponential decrease of rate with distance is observed. Note that this decrease in rate in these Os-Co complexes is in the range of that predicted by earlier theories [6,7]. However, when the rate of electron transfer approaches the time of trans to cis isomerization, changes in the polyproline structure result in the bending of the curve observed (figure 1).

Table I

n = 0, 1, 2, 3, 4

n	Complex	$k_{Ru\text{-}Co}$	$k_{Os\text{-}Co}$	$k_{Os\text{-}Ru}$
0	(M)-iso-(M')	1.2×10^{-2}	2.4×10^{5}	–
1	(M)- iso(Pro)-(M')	1.0×10^{-4}	2.9×10^{2}	3×10^{6}
2	(M)- iso(Pro)$_2$-(M')	0.64×10^{-5}	0.6	7×10^{4}
3	(M)- iso(Pro)$_3$-(M')	5.6×10^{-5}	0.4×10^{-1}	–
4	(M)- iso(Pro)$_4$-(M')	14×10^{-5}	0.1×10^{-1}	–

$(M) = (H_2O)(NH_3)_4 Ru^{II}$, $(NH_3)_5 Os^{II}$

$(M') = (NH_3)_5 Co^{III}$, $(NH_3)_5 Ru^{III}$

In order to achieve a faster electron transfer process, we have used a $Ru(III)$ acceptor instead of a $Co(III)$ acceptor. In our preliminary data for the Os-Ru oligoproline complexes for $n = 1, 2$ prolines (Table I), the variation of the rate of electron transfer with distance is much smaller than that for the analogous Os-Co series. Because the structure of the bridging ligand is identical, the rates observed allow us to compare the efficiency of two different metal acceptor orbitals in long range electron transfer, i.e. compare the properties of $Co(III)$ (d^6) with that of $Ru(III)(d^5)$ as acceptors. We are continuing our work on this Os-Ru oligoproline series to study how the conformation of the peptide bridging ligand responds to the fast rates of electron transfer expected in this series.

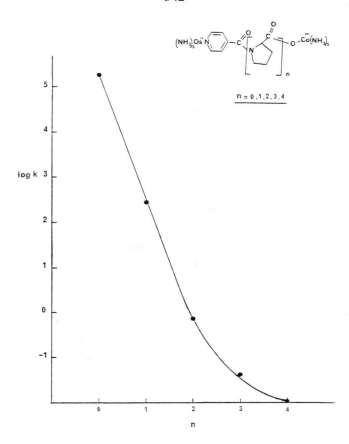

Figure 1. Plot of log k, the rate constant for intramolecular electron transfer, vs. the number of proline residues in the series $[(NH_3)_5Os^{II} - L - Co(NH_3)_5]$ $L = iso(Pro)_n, n = 0 - 4.$

Let me now turn to cytochrome derivatives. Over the last several years we have prepared derivatives of cytochrome c with different ruthenium complexes, $[(NH_3)_5Ru-]$ and $(NH_3)_4Ru(isn)-]$, isn $= N$ $C_{\|} -NH_2$, covalently attached to His-33 [8-11]. We have isolated the derivatives where a single ruthenium complex is bound per cytochrome c molecule (Figure 2). These species were characterized by a number of physical chemical and biological techniques, including Fe-Ru analysis, visible spectral, CD, gel electrophoresis, HPLC, peptide mapping, and reaction with cytochrome c oxidase and cytochrome c peroxidase. All these techniques indicate that the new molecules generated maintain their native conformation and biological activity [8-11].

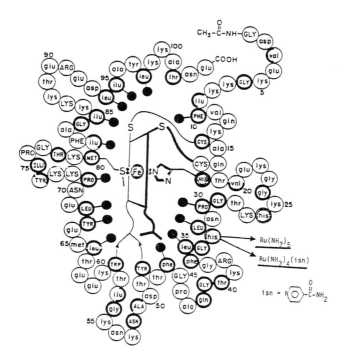

Figure 2. Two cytochrome c derivatives: $[(NH_3)_5 Ru - cyt\ c]$ *and* $[(isn)(NH_3)_4 Ru - cyt\ c]$ *where the ruthenium complex is covalently attached to* $His - 33$.

The most important experiments we conducted with these derivatives are the studies on the kinetics and thermodynamics of electron transfer. Figure 3 shows the electrochemistry for both of these molecules. Note that the reduction potential of the heme site remains unchanged in the two ruthenium derivatives (Figure 3).

The two ruthenium sites differ in reduction potential, one, the $(NH_3)_5 Ru-$, is a better reductant and the other, $(NH_3)_4 (isn) Ru-$, is a better oxidant than the heme site in cytochrome c. For the $(NH_3)_5 Ru - cyt\ c$ derivative, studies on the rate of intramolecular electron transfer with different radicals (generated in situ

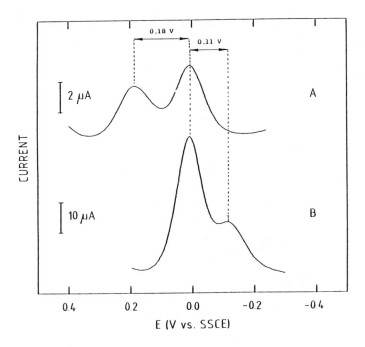

0.18 V

0.11 V

2 μA

A

CURRENT

10 μA

B

0.4   0.2   0.0   -0.2   -0.4

E (V vs. SSCE)

Figure 3. Differential pulse polarograms of $[(isn)(NH_3)_4Ru - cyt\ c](A)$ *and* $[(NH_3)_5Ru-cyt\ c](B)$: *Volts vs. standard saturated calomel electrode (SSCE); 2 mm gold disk electrode, scan rate = 2mV/s, pulse amplitude 25 mV; approximately 0.25 μmol* $[Ru^{III} - cyt\ c^{III}]$ *in 0.1 NaC10₄, 0.01 M bipyridine, 0.08 M phosphate buffer, pH 7.*

by pulse radiolysis) (Figure 4) show that intramolecular electron transfer from the $Ru(II)$ site to the heme (III) site occurs with a rate constant of $k \sim 52s^{-1}$. The activation parameters for this process are $\Delta H^{\neq} = 3.5$ Kcal/mol and $\Delta S^{\neq} = -39$ e.u.

This rate of electron transfer occurs on the same timescale as many conformational changes in the cytochrome c molecule [12,13]. Therefore, it was a challenge for us to demonstrate that electron transfer, rather than conformational motion, is limiting in this long range electron transfer process. We felt that the use of a second derivative, the $Ru(NH_3)_4(isn) - cyt\ c$, would clarify this issue

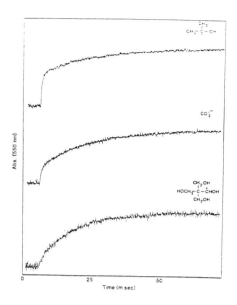

Figure 4. Pulse radiolytic reduction of $[(NH_3)_5 Ru^{III} - cyt\ c^{III}]$ *by* $(CH_3)_2 \overset{\bullet}{C} OH$, CO_2-, *and* $(CH_2OH)_3 \overset{\bullet}{C} HOH$ *radicals.*

if we demonstrate that electron transfer can proceed in the reverse direction, that is an electron can be transferred from the reduced heme site to the $Ru(III)$ derivative on the surface of the protein. Pulse radiolysis studies (Figure 5) for the $(NH_3)_4(isn)Ru - cyt\ c$ derivative showed that while the heme is rapidly reduced by CO_2^- or $(CH_3)_2 \overset{\bullet}{C} -OH$ radicals; the electron is *not* rapidly transferred between the heme(II) and the $Ru(III)$ site (over the timescale of several seconds).

The combined results of our experiments constitute strong evidence for *directional electron transfer* in horseheart cytochrome *c*. There can be many explanations for the origin of this directionality. One of the most obvious explanations would be a conformational change between the oxidized and reduced proteins, resulting in a change in the electron coupling between the donor and acceptor. In other words, the barrier to heme oxidation is different than that for heme reduction in cytochrome c. This observation is in agreement with other work on cytochrome c by Tabushi et.al [14] who showed that intermediates formed during the reduction of cytochrome c are not similar to those formed during reoxidation. Further work on other systems currently going on in our laboratory will shed some light as to the importance and the generality of this observation.

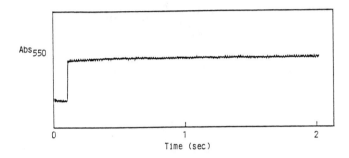

Time (sec)

Figure 5. Pulse radiolytic reduction of $[(NH_3)_4(isn)Ru^{III} - cyt\ c^{III}]$ *with isopropanol radical.*

I would like to conclude now by saying that for small molecule electron transfer reactions reasonable agreement is found between theory and experiment. However, for protein electron transfer, the electron transfer process may be a multistep process which includes more than single electron transfer elementary step. Thus, caution should be exercised when one tries to apply the formalisms and models developed for small molecules to electron transfer proteins. **Acknowledgements**

The authors wish to thank Dr. Harold Schwarz of Brookhaven National Laboratories for his help in the pulse radiolysis experiments. The authors gratefully acknowledge the support of the National Institutes of Health Grant (GM 26324) and the National Science Foundation Grant 840552. S.I. is the recipient of a

National Institutes of Health Career Development Award (AM 00732) (1980-85) and a Camille and Henry Dreyful Teacher Scholar Award (1981-1985).

References

[1] S. Isied, A. Vassilian, J. Am. Chem. Soc. *106*, 1726-1732 (1984).

[2] S. Isied, A. Vassilian, J. Am. Chem. Soc. *106*, 1726-1732 (1984).

[3] S. Isied, A. Vassilian, R. Magnuson, H. Schwarz, J. Am. Chem. Soc. *107*, 000 (1985).

[4a] L. Lin, J.F. Brandt, Biochem. *22*, 553-0 (1983).

[4b] J.F. Brandt, H.R. Halvorson, M. Brennan, ibid, 1985 *14*, 4953.

[5] H.N. Cheng, F.A. Bovey, Biopolymers *16*, 1465-72 (1977).

[6] J. Jortner, J. Chem. Phys. *64*, 4860 (1976).

[7] J. Hopfield, Proc. Nat. Acad. Sci. USA, 3640-3644 (1974).

[8] S. Isied, G. Worosila, S.J. Atherton, J. Am. Chem. Soc. *104*, 7659-61 (1982).

[9] S. Isied, C. Kuehn, G. Worosila, J. Am. Chem. So.c *106*, 1722 (1984).

[10] R. Bechtold, C. Kuehn, C. Lepre, S. Isied, submitted to Nature, 1985.

[11] K. Yocum, J.B. Shelton, W.A. Schroeder, G. Worosila, S. Isied, E. Bordignon, J.B. Gray, Proc. Natl. Acad. Sci. USA *19*, 7052-55 (1982).

[12a] C. Cruetz, N. Sutin, Proc. Natl. Acad. Sci. USA *70*, 1701-03 (1974).

[12b] C. Creutz, N. Sutin, J. Biol. Chem. *249*, 6788-95 (1974).

[13] G. Moore, Z. Huang, C. Eley, H. Barker, G. Williams, M. Robinson, R.J.P. Williams, Faraday Disc. Chem. Soc. *74*, 311-379 (1982).

[14a] I. Tabushi, K. Yamamura, T. Nishiya, J. Am. Chem. Soc. *101*, 2785-87 (1979).

[14b] I. Tabushi, K. Yamamura, T. Nishiya, Tet. Lett., 4921-4924 (1978).

Isied Discussion

Wolynes: You may have found another example of one of these conformationally gated reactions as Miller was talking about earlier. I believe this always has to be suspected when you have a nearly activationless process, probably also in the photosynthetic case. One of the results from the theory of the influence of the environment on reactions, is that if the thing really is nonadiabatic, then the dynamical change in the solvent shouldn't affect things. So, I wonder if people looked at, say the viscosity dependence of the cytochrome c reduction, if you add glycerol to the solution or something like that, or perhaps pressure changes which would presumably speed up electron transfer by moving things close together if it's not adiabatic, but slow it down if it was determined by a conformational change.

Warshel: As we always say, viscosity: no, electrostatic: yes! I believe (and I might be right or wrong) that proteins are simpler than small molecules in solutions. The only difference is that in solutions you could use the same dielectric for the donor and acceptor. In protein you already know the "solvent" structure (the x-ray data of the protein) and what probably happens is that the local polarity around the donor and acceptor is different. I believe that what happens is that when you put your acceptor or donor in a protein site you expect that it will have the free energy as it has in water. This is clearly an unsafe assumption. The free energy depends on the local dielectric. It's very, very simple. It's like putting the donor and acceptor in different solvents. You cannot do this without the protein by standard physical organic chemistry techniques, because you cannot play with two environments.

Isied: I don't think that I understand your comment. I have shown two systems that differ by one variable. I might accept the fact that there are differences in the motion of these two complexes, but the fact remains that our electrochemical experiment measures the free energy difference between the donor and acceptor sites in both complexes.

Warshel: No, I did not follow the details and I will have to think about it. But in addition to polarity, which is different, there is also the dielectric relaxation which is different in both of them. If we exclude miracles then the environment must determine the observed effects.

Wolynes: Let me say that I basically agree with Dr. Warshel's picture, although I disagree with his view about the inefficiency of looking at viscosity because, as we've seen, many internal processes in proteins are relatively sensitive to viscosity.

Jortner: I agree to a large extent, but not entirely, with Dr. Warshel because the dielectric relaxation will be involved as you have shown us this morning, Dr. Wolynes, provided that the nonadiabatic process will become adiabatic. Now, if you are in the range where very large viscosity effects prevail but the process is still nonadiabatic, in other words, the Landau-Zener factor is still smaller than unity, than local viscosity effects and dielectric relaxation effects will still not be critical. This is what I feel and I am curious to hear a counter opinion. Now I would really like to raise the issue of the structural effect which accompanied the energy transfer, or which promoted electron transfer. Because, I have adopted the point of view that the structural effects of protein are very important in promoting the process by changing the energetics, changing the energy gaps and so on. I would appreciate hearing your view.

Isied: I think what I have shown is that you can get the electron transfer in one direction, but not in the other direction if you change the free energy of the process. These two ruthenium complexes have very similar reorganization energies and the same charge. The only process that changed is the direction of electron transfer. There are many possible explanations that can account for this observation. We are not violating microscopic reversibility. For example, structural or conformational changes between the reduced and oxidized cytochrome c, including possibly the breaking of a hydrogen bond can account for this observation. It is known that the structure of reduced cytochrome c is different than that of the oxidized form.

This is not a unique observation; the change in the CD of cytochrome c(III) upon reduction also indicates conformational changes. In fact, the work of Tabushi et.al. used the words 'directionality in electron transfer' before we did, based on their CD studies. Their work showed that the rate of reduction of cyt c(III) proceeds with intermediates different than those observed for the rate of oxidation.

Miller: Well, I'd like to thank Dr. Isied for some really stimulating data. That's an awfully interesting reaction. There are some interesting interpretations, one could imagine. One thing that would be nice to find out is the following: If you can find both couples to be reversible when you increase the sweep speed in your voltameter, you could really find out that both groups are completely normal and reversible. The ideal way would be to get up near the speed of the reaction. This may not be easy, but another way to do that would be just to run some homogeneous oxidation reduction equilibrium of cytochrome c in solution which you can to operate at that kind of speed. Then you'll really be ready and posed to say: Hey, there's something extremely interesting and important going on here. Has any of that been done yet?

Isied: A lot of work you have described has been done. We have done redox titrations which are, I think, very important. We have used various oxidants and reductants; we started with the reduced form and went to the oxidized form and started with the oxidized form and went to the reduced form. We have done electrochemical experiments under a variety of conditions. But the electrochemistry of these proteins, on electrodes, as you know, is very tricky and only works under limited conditions. Studies on the kinetics and thermodynamics of the redox reactions of these sites are the way to answer your concern. In our experiments, the two sites behave normally towards external oxidants and reductants. Most of the concerns you expressed are concerns that I share and we have done a large number of experiments to answer these and more are in progress.

Naqui: What happens to the reaction of cytochrome c with cytochrome oxidase when you modify your cytochrome c? *Isied*: It turns out that it does not change very much. If you determine the turnover for a sample of the native cyt c and a sample of the modified proteins, you get about 80-90% of the turnover number for the modified as for the native. And we have done that with the cytochrome c peroxidase also which is another one of these systems that is even more well defined. Even then, most of the activity is maintained. The CD, the ε°, and the 695 band do not change. All these observations indicate that little change, if any, has taken place upon modification.

Simulating the Dynamics of Electron Transfer
Reactions in Cytochrome c

A. Warshel

Department of Chemistry

University of Southern California

Los Angeles, CA 90089-0482

A detailed understanding of the relation between the structure of proteins and their biological function requires realistic microscopic models. In the past years we have invested a significant effort in developing such models [1]. Here we would like to present some of the results from our microscopic modeling of biological Electron Transfer (ET) reactions. We will consider the two main directions of our study: static estimates of activation free energies and dynamical simulation of the actual reactions.

(a) A Direct Static Correlation

Our static modeling is based on the assumption that the activation barriers for chemical reactions can be correlated with the energetics of the reactants and products. These energies can be estimated directly using the x-ray structures of the protein [1]. Our static studies of ET reactions in proteins are presented elsewhere [2,3]. Here we only consider briefly our study of cytochrome c. In the case of cytochrome c we have the actual x-ray structure of the reduced and oxidized cytochrome [9]. Taking these structures we evaluated (Figure 1) the electrostatic energy of the reduced and oxidized heme along the reaction coordinate (obtained by connecting the coordinates of the reactants and products). The electrostatic calculations we performed using our microscopic dielectric model [1]. This model considers explicitly the electrostatic interaction between the heme charges and the protein permanent and induced dipoles, as well as the surrounding water molecules. Our calculated potential surfaces give the activation barrier for exchange reaction between two cytochrome c molecules held at a *fixed* distance. This activation barrier is obtained from the intersection of the product and reactant diabatic potential surfaces (Figure 1). The calculated barrier is consistent with the role of the protein as a catalyst of ET reactions. It is interesting to note that our potential surfaces are nearly quadratic. In the quadratic limit one can obtain the activation barrier, $\Delta g^{\ddagger}$, from the reorganization energy by Marcus' relation [4]

$$\Delta g^{\ddagger} = (\alpha + \Delta G_0)^2 / 4\alpha \tag{1}$$

where α and Δg_0 are the reorganization free energy and reaction free energy, respectively.

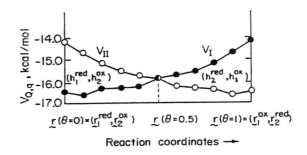

Figure 1. *A direct interpolation of the protein potential surface for electron exchange, using the observed x-ray structures of cytochrome c. The reaction coordinates are given by:* $\underline{r}_1(\theta) = \underline{r}_1^{red} + \theta(\underline{r}_1^{ox} - \underline{r}_1^{red}), \underline{r}_2(\theta) = \underline{r}_2^{ox} + \theta(\underline{r}_2^{red} - \underline{r}_2^{ox})$ *where subscripts 1 and 2 refer to molecules "1" and "2", θ is a number between 0 and 1, and r^{ox} and r^{red} are the corresponding observed x-ray coordinates. The potentials V_I and V_{II} are the electrostatic interaction, $V_{Q,q}$, between the heme partial atomic charges, Q and the protein dipoles, q: "I" labels the reactants (heme$_1^{red}$, heme$_2^{ox}$) and "II" labels the products (heme$_1^{ox}$, heme$_2^{red}$).*

(b) Dynamical Studies

Our static approach [2] provides a reasonable approximation for the energetics of the reaction. However, in order to *verify* the microscopic correlation between $\Delta g^{\ddagger}$ and ΔG_0 it is important to simulate the actual reaction [5].

Our dynamical simulation is based on a semiclassical trajectory approach that was introduced in our previous studies of radiationless [6,5] and radiative transitions [7] in many dimensional systems. In order to realize the main points of this approach it is useful to consider the solvent (protein) as a single fluctuating dipole that interact with the donor and acceptor (Figure 2). The fluctuations of the dipole changes the energy of the reactant and product states (V_I and V_{II}). The energies V_I and V_{II} fluctuate along the classical trajectories of the nuclei of the system (the dipole). The ET process occurs when the system passes through an intersection of V_I and V_{II}. The evaluation of the reaction rate is reduced to a calculation of the number of times the surface intersect and the probability to cross at each intersection. To calculate the surface crossing probability we write the time dependent wave function of the system as:

$$\psi(t) = \sum_k b_k(t)\phi_k(\underline{r}, \underline{R}) \exp\left\{-(i/\hbar)\int^t V_k[\underline{r}(t'), \underline{R}(t')]dt'\right\} \qquad (2)$$

where r are the coordinates of the solvent (protein) and R are the coordinates of the donor and acceptor. The ϕ_k are the diabatic electronic wavefunctions, satisfying $H(r, R)\phi_I = V_I(r, R)\phi_I + \sigma(R)\phi_{II}$ where $\sigma(R)$ is the coupling matrix element between state I and II. Substituting $\Psi(t)$ in the time dependent Schrodinger equation gives:

$$\dot{b}_{II}(t) = -(i/\hbar)\sigma(\underline{R}(t))b_I(t) \exp\left\{-(i/\hbar)\int^t \Delta V[\underline{r}(t'), \underline{R}(t')]dt'\right\} \qquad (3)$$

where $\dot{b}$ denotes db/dt and $\Delta V = V_{II} - V_I$. Given the initial conditions $|b_{II}(0)|^2 = 0$, $|b_I(0)|^2 = 1$, it is possible to determine the probability $|b_{II}|^2$ of being in state II for any path $(\underline{r}(t), \underline{R}(t))$ of the solute solvent system by integrating eq. (3). This is done by starting a trajectory near the minimum of V_I, adjusting the velocities until the system equilibrates with an average kinetic energy that corresponds to a given temperature, and then integrating eq. (3) along the classical trajectory $(\underline{r}(t), \underline{R}(t))$. Our semiclassical approach implies that the trajectories are split at points t_i where $|b_{II}|^2$ increases by a significant amount (these points occur when $\Delta V(t) = 0$). A fraction $|b_I|^2 - |\delta b_{II}|^2$ stays on V_I while a fraction $|\delta b_{II}|^2$ is trapped in state II (see figure 2). Dealing with a multidimensional system we neglect interference between trajectories that cross to state II at different times, because such trajectories are likely to be trapped in different vibrational states. Thus the overall probability that the system is in state II is calculated by propagating a long-time trajectory on state I and summing the crossing possibilities $|b_{II}|^2$ (evaluated by integrating eq. (3) around the individual crossing points, t_i, where $\Delta V = 0$).

The overall rate is given by:

$$k_{I,II} = \lim_{\Delta t \Rightarrow \tau} \left\{ \frac{\Delta}{\Delta t} |b_{II}(\Delta t)|^2 \right\} = \lim_{\Delta t \Rightarrow \tau} \left\{ \frac{\Delta}{\Delta t} \sum_i |\delta b_{II}|^2 \right\}$$

$$= \lim_{\Delta t \Rightarrow \tau} \frac{\Delta}{\Delta t} \sum_i \left| \int_{t_i^-}^{t_i^+} \left\{ (i\sigma/\hbar) \exp\left[-i/\hbar \right) \int_0^t \Delta V(t') dt' \right] \right\} dt \right|^2 \qquad (4)$$

where $t_i = (t_{i+1} - t_i)/2$ and is the trajectory time after which the computed k reaches a constant value. As shown in ref. [5], the analytical continuation of $|\delta b_{II}|^2$ can be approximately by the Landau-Zener transition probability; and k can be approximated by

$$k = (\pi/\hbar^2 k_b T \alpha)^{1/2} \sigma_{I,II}^2 \exp\left\{ -\Delta g^{\ddagger}/k_b T \right\}. \qquad (5)$$

The activation free energy $\Delta g^{\ddagger}$ is obtained in the dynamical approach by counting the number of times the two surfaces intersect. When the intersection of the two surfaces is very rare one has to use an umbrella sampling method to evaluate $\Delta g^{\ddagger}$ (see the Appendix of ref. [5]).

The semiclassical approach might seem an *ad hoc* approach. Nevertheless a recent systematic study [8] have indicated that this approach is *exact* in the test case of the high temperature harmonic limit. Furthermore, in anharmonic cases it gives activation free energy rather than energy. This is of a major importance since none of the available quantum mechanical methods can be used to obtain the actual activation free energy.

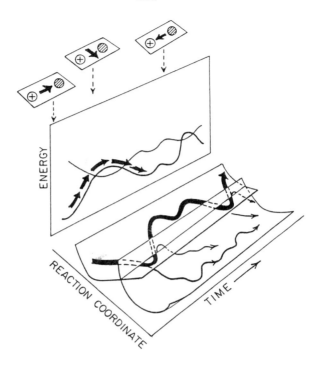

Figure 2. Sketch of solvent fluctuations that permit ET to occur. The solvent "moves" in the field of the reactants (hypothetically a neutral donor and a positively charged acceptor). The solvent trajectory is drawn as a thick black curve in the reaction coordinate vs. time plane, and as thick arrows in the energy vs. time plane. Three orientations of a hypothetical dipole are shown above the energy vs. time sketch: When the dipole points away from the + charge the potential energy is low; with the dipole perpendicular to the line joining the donor and acceptor the potential surfaces for the reactants and products intersect, and if the dipole points toward the charge, the reactant potential surface (the darker of the two curves in the energy vs. time plane) is at a maximum. At the intersections of the two potential surfaces there is a probability for ET (sketched as thin, wavy "trajectories" breaking away from the main trajectory).

Our preliminary simulation of an exchange ET reaction between reduced and oxidized cytochrome c at 300°K are presented in Figures 3 and 4. The quan-

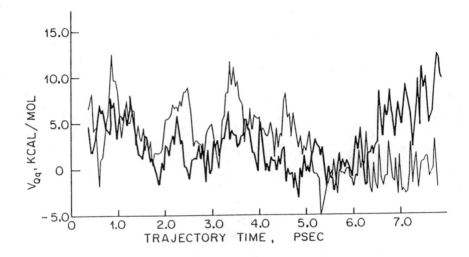

Figure 3. The fluctuations of the potentials V_I (dark tracing) and V_{II} (lighter tracing) as the protein atoms move in the field of charges $Q_I = (Q_1^{red}, Q_2^{ox})$. $V_I(t) = V_{I1}(t) + V_{I2}(t) + C; V_{II}(t) = V_{III}(t) + V_{II2}(t)$, where $V_{I1}(t) = V(Q_1^{red}, r_1^{red}(t))$, $V_{I2} = V(Q_2^{ox}, \underline{r}_2^{ox}(r)), V_{III}(t) = V(Q_1^{ox}, \underline{r}_1^{red}(t)), V_{II2}(t) = (Q_2^{red}, \underline{r}_2^{ox}(t)); r_1^{ox}(t)$ denotes "atoms of protein i moving in field of Q^{ox}". The energy required to switch the charges at time t is the energy gap, $\Delta V(t) = V_{II}(t) - V_I(t) - C = \Delta V_1(t) + \Delta V_2(t) - C$, where $\Delta V_1(t) = V_{III}(t) - V_{I1}(t), \Delta V_2(t) = V_{II2}(t) - V_{I2}(t)$. The constant C was defined by the following procedure: For the separate trajectories distribution functions $P_1(\Delta V_1)$ and $P_2(\Delta V_2)$ were calculated. $P_i(V_i) \cdot (\delta V_i)$ gives the probability that V_i is in the range $(V_i, \Delta V_i + \delta(\Delta V_i))$. The maxima of P_1 and P_2 were at $\Delta V_1^* = 4$ kcal/mol, $\Delta V_2^* = 11$ kcal/mol. For these simulations the most probable charge switching energy is $\Delta V_1^* + \Delta V_2^* = 15$ kcal/mol, in disagreement with the static model which gives 2.5. Assuming that the fluctuations of the system around the incorrect minimum are similar to those around the correct one, we can correct the energy gap by using $C = 12.5$ kcal/mol. The discrepancy that requires the introduction of the correction C is probably due to the fact that the water molecules surrounding the protein were not included in the present dynamical simulation. Proper simulation of the water molecules or introduction of an effective constraint is needed to prevent protein 2 from contracting around the partial charges of the oxidized heme. Note also that the water contribution should be included in the reorganization energy [2].

tities labelled V_{Qq} in the upper panel are the electrostatic potentials V_I (dark) and V_{II} (light). The two protein conformations fluctuate (independently) under the influence of bond stretching, angle bending, Van der Waals, and electrostatic forces. The trajectories propagate under the influence of charges of state I only. That is, V_{II} is the computed value of V_{Qq} if the heme charges were to be switched, but the trajectory is propagated with the reactant charges. The probability for ET is accumulated near the crossings of V_I and V_{II}, which correspond to $\Delta V(\underline{r}(t), \underline{R}(t)) = 0$. The number of crossings should be related to the reorganization energy (α in eq. (1)) which is given by the most probable value of ΔV. Using this principle, the curve $V_I(t)$ was shifted relative to $V_{II}(t)$ (using a calibration described in the caption of Figure 3). The numerical values of V_I and V_{II} shown in the figure were then used to compute the ET rate according to eq. (4), setting $\sigma^0_{I,II} = .01$ kcal/mol. The resulting transition probability, $|b_{II}|^2$ as a function of time is shown in the lower panel of Figure 4. Accordingly, the ET rate is $(\sigma/\sigma^0)^2 \times (3.5 \times 10^{-2}/8^{-12} sec) = (\sigma/\sigma^0)^2 \times 4.4 \times 10^9 sec^{-1}$, in excellent agreement with $(\sigma/\sigma^0)^2 \times 4.2 \times 10^9 sec^{-1}$ obtained from eq. (5) with the $\Delta g^{\ddagger}$ of the static modeling presented in Section.

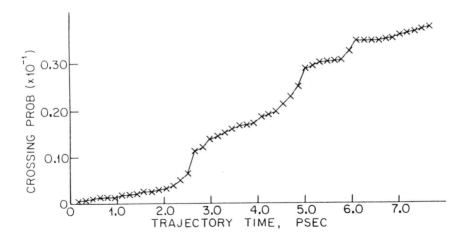

Figure 4. The simulated ET probability accumulated over the course of the cytochrome trajectory. ET probability accumulated (using eq. (4)) at the crossings of V_I and V_{II} in Figure 3.

The main point of the above calculation (and the more complete analysis presented in ref. [10]) is not so much in evaluating the actual rate of ET reaction in cytochrome c, but in demonstrating that such calculations are possible. The key for such calculations is the evaluation of the time dependent energy gap $\Delta V(t)$ which contains all the relevant information about the microscopic dynamics of the given ET reaction. Of course, if the distance between the donor and acceptor is allowed to vary then one has to also evaluate the time dependence of σ and to introduce it in eq. (3).

References

[1] A. Warshel and S.T. Russell, Quart. Rev. Biophys. *17*, 283 (1984).

[2] A.K. Churg, R.M. Weiss, A. Warshel and T. Takano, J. Phys. Chem. *87*, 1683 (1983).

[3] A. Warshel, Proc. Natl. Acad. Sci. USA *77*, 3105 (1980).

[4] (a) R.A. Marcus, J. Chem. Phys. *24*, 966 (1956); (b) *Ibid 24*, 979 (1956).

[5] A. Warshel, J. Phys. Chem. *86*, 2218 (1982).

[6] A. Warshel, Nature *260*, 679 (1976).

[7] (a) A. Warshel, P. Stern and S. Mukamel, in *Time Resolved Vibrational Spectroscopy*, G.H. Atkinson, ed., Acad. Press Inc., NY (1983) pp. 41. (b) A. Warshel and J.K. Hwang, J. Chem. Phys. *82*, 1756 (1985).

[8] A. Warshel and J.K. Hwang, J. Chem. Phys. (Submitted).

[9] T. Takano and R.E. Dickerson, J. Mol. Biol. *153*, 79 (1981).

[10] A.K. Churg and A. Warshel, in *Structure and Motion: Membranes, Nucleic Acids and Proteins*, eds. E. Clementi, G. Corongiu, M.H. Sarma and R.H. Sarma, Adenine Press, 361 (1985).

Excitonic Ion and Auger Photoemission in Organic Crystals

M. Pope

Chemistry Department

and

Radiation and Solid State Laboratory

New York University

When an insulator is subjected to ionizing radiation of high intensity, relatively large concentrations of mobile excited states such as excitons and electrons (holes) can coexist. The possibility of creating a bound state (excitonic ion) between the electron (or hole) and an exciton was first suggested by Lampert [1], and several theoretical papers have since appeared that have discussed the question of the stability of such a state consisting of a free electron (or hole) and an exciton of either the Frenkel type or Wannier type [2-7]. The Frenkel exciton is an electrically neutral, electronically excited mobile state of a crystal, and it is the collective counterpart of the isolated excited molecule. The Wannier exciton is an electrically neutral, mobile electronic state of a crystal, and it is based on the correlated hole-electron pair [eh]. The spatial extent of the hole-electron pair can vary from near-neighbors (charge-transfer or CT exciton) to those of large-radius [8]. The first experimental evidence for the existence of an excitonic ion in an organic crystal was presented by Arnold, Pope and Hsieh [9]. These authors were looking into the suggestion of Pope and Kallman [10] that earlier experiments of Pope et al [11] could represent the interaction of a free carrier and a free or trapped CT exciton.

In this article, we briefly review the experiments that have been carried out in this laboratory that relate to the excitonic ion. These [8] consisted of the measurement of the energetics and kinetics of the photoemission of electrons from a small crystals of anthracene or tetracene suspended in a N_2 atmosphere ($p = 1$ atm) in a modified Millikan chamber [12]. The crystal was subjected to light of high intensity ($\sim 10^{15} cm^{-3}s^{-1}$), and with a photon energy sufficient to produce an internal ionization, but not an external photoelectric effect. It had been shown by Pope, Kallmann and Giachino [13] that an external photoelectric effect could be produced in organic crystals by a two-step process in which two electronically excited states of relatively low energy interacted to produce another state with an internal energy large enough to allow photoemission to occur. This process was referred to as a double quantum external photoelectric effect (DQEPE).

The general scheme for these DQEPE processes is as follows:

(1) $h\nu + S_0 \rightarrow X_1+$ heat; one photon is absorbed by the ground state S_0, producing state X_1.

(2) $h\nu + S_0 \rightarrow X_2+$ heat; another photon is absorbed producing state X_1.

(3) $X_1 + X_2 \rightarrow S^*$; states X_1 and X_2 combine to produce excited state S^*.

(4) $S^* \rightarrow S_0 + e^*$; S^* ionizes producing a crystal ground state and an electron in a high energy plane- wave state.

The electron in equation (4) has a maximum kinetic energy (E_{max}) in the threshold region as given by the Einstein photoelectric equation:

(5) $E^* - I_c = E_{max}$ where E^* is the energy state S^* and I_c is the ionization energy of the crystal. States X_1 and X_2 can be a photon, electron (free or trapped) or exciton (free or trapped); X_1 and X_2 can be identical. The only requirements are that $E^* > I_c$ and that at least one of the X state be mobile so that a recombination process is possible. From equation (3) it will be evident that E^* is in general an unknown, and must be calculated from the measured values of E_{max} and I_c, as shown in equation (5). A knowledge of E^* is essential for the determination of X_1, X_2 and S^*.

The DQEPE was first discovered in anthracene [13], and was shown unequivocally to be due to the interaction of two singlet excitons at 3.15 eV to produce an electronically excited singlet state intermediate (S^*) at 6.3 eV that subsequently ionized. Since the ionization energy (I_c) of the anthracene crystal is 5.8 eV [8], the state S^* has sufficient energy to produce a photoelectron with a maximum kinetic energy (E_{max}) of about 0.5 eV, which was indeed observed. In addition, the photoemission quantum yield varied as I^2, as was to be expected for a bimolecular annihilation of two singlet excitons. The photoemission quantum efficiency at this energy was estimated to be $10^{-25} I_0$, where I_0 is the photon intensity; thus, 10^{-10} electrons were emitted per absorbed low energy photon when $I_0 = 10^{15} cm^{-3} s^{-1}$. This DQEPE in anthracene was observed for photon energies $3.15 < h\nu < 4$ eV, and E_{max} was constant, independent of incident photon energy in this energy range, an observation critical to the conclusion that the same metastable intermediate state was responsible for photoemission in all cases. However, for photon energies $4 < h\nu < 5.8$ eV. The mechanism proposed for this DQEPE involved the generation of CT excitons which in turn implied the generation of hole-electron pairs. From a study of the action spectrum of CT exciton generation, it was concluded that the band gap of anthracene was close to 4 eV, and that the interacting excited states were CT excitons of energy 3.45 eV and with the relatively long lifetime of about 10^{-7}s. All of these conclusions received independent support from other investigators. However, in view of the fact that the DQEPE in anthracene can be produced by Frenkel excitons as well as CT excitons, the interpretation of the kinetics of the photoemission process was not unequivocal.

This difficulty in interpretation was not present with tetracene, where the ionization energy (I_c) of the crystal is 5.3 eV and the energy of the singlet state (S_1) is 2.4 eV [8]; it is thus impossible for two singlet Frenkel excitons to produce a DQEPE. Nevertheless, when the incident photon energy was in the range $3.2 < h\nu < 5.3$ eV, there was a DQEPE. In the case of tetracene [9] it was shown that the quantum yield of photoemitted electrons varied as $I^{1.5}$, where I is the incident light intensity, and that the energy E^* of the excited state intermediate S^* (the excitonic ion in this case) was 6.1 ± 0.1 eV, independent of photon energy. Furthermore, the superposition of light of energy 2.5 eV, which is capable of producing singlet and triplet excitons, upon the photoionizing light in the energy range $3.2 < h\nu < 5.3$ eV, had no effect on the rate of the photoemission process.

The metastable complex S^* can ordinarily be produced by the following interactions:

a) Frenkel exciton-exciton

b) Frenkel exciton-free carrier

c) Frenkel exciton-trapped carrier

d) Frenkel exciton-photon

e) Frenkel exciton-CT exciton (free or trapped)

f) CT exciton-photon

g) CT exciton-CT exciton (free or trapped)

h) CT exciton-free carrier

i) photon-photon

j) photon-carrier.

All interactions involving photons are eliminated by the photon energy independence of E^*. Similarly, all interactions involving Frenkel excitons are eliminated by the null result on the photoemission quantum efficiency of the superposition of exciton-generating light of energy 2.4 eV on the free electron generating light of energy $h\nu > 3.4$ eV. This leaves mechanisms (g) and (h) above. Since the yield $\alpha I^{1.5}$, (g) is excluded; this leaves (h). In this process, a free carrier receives enough energy as a result of an Auger-like interaction with a trapped CT state $(eh)_t$ to leave the crystal. The reaction scheme would be as follows:

(6) $h\nu + S_0 \rightarrow e + h$; generation of free holes and electrons by light

(7) $e + h \rightarrow (eh)_t$; recombination of holes nd electrons to product CT states

(8) $(eh)_t \rightarrow S_0$; decay of CT state to ground state

(9) $e + h \rightarrow S_0$; recombination of holes and electrons to the ground state

(10) $e + (eh)_t \rightarrow (e..h..e)_t$; formation of an intermediate excitonic ion

(11) $(e..h..e)_t \rightarrow e^* + S_0$; Auger decay of excitonic ion producing an energetic electron

(12) $e^* \rightarrow e$; hot electron thermalizes in the conduction band.

Under steady state conditions, it can be shown [14] that the light intensity dependence of e^* varies as $I^{1.5}$. Since the concentration $[e]$ varies as $I^{0.5}$ under recombination- limited lifetime, and $[eh]$ varies as I, the concentration $[e..h..e]_t$ varies as $I^{1.5}$. Although the CT exciton could be free or trapped, this makes no difference in the determination of the light intensity dependence.

A schematic representation of the DQEPE, the excitonic ion, and the Auger emission process is shown in Figure 1. The E_{max} of the photoemitted electron was 0.8 eV, which together with a value of $I_c = 5.3$ eV implies according to equation (5), an effective energy of the $(e..h..e)_t$ states of 6.1 eV. From equation (10), it may be seen that the energy of $(e..h..e)_t$ is equal to the energy of formation of the free electron e and of $(eh)_t$, less the binding energy of $(e..h..e)_t$. Previous experiments on the determination of the band gap E_g for tetracene had yielded a value of about 3.2 eV [8], leaving an energy of about 2.9 eV for the CT exciton. This value agrees reasonably well with that of Sebastian et al [15], who arrived at a value of 2.7 eV for a CT exciton in tetracene, obtained from the measurement of the electroabsorption spectrum of tetracene films. The existence of the $(e..h..e)$ state was proposed in order to explain the close coupling that was needed among the three members of the complex in order insure an efficient Auger process of transfer of energy from the $e - h$ recombination to the excess electron member of the excitonic ion, causing its photoemission from the crystal.

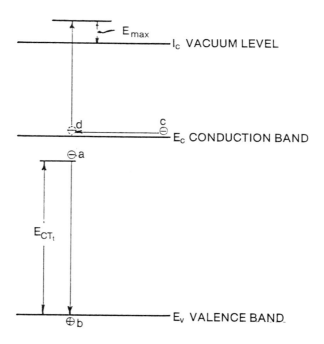

Figure 1. Schematic representation of an Auger emission of an electron. The trapped CT state is represented by a and b, c is a free electron in the conduction band that moves to position d, forming the transitory $(e..h..e)_t$ state. The recombination of a and b releases the energy E_{CT}, which is transferred to the electron c, ejecting it from the crystal with the maximum kinetic energy E_{max}.

The lifetime of the excitonic ion could not be calculated in these experiments but internal consistency requires that the lifetime of the CT state be relatively long - as much as 10^{-6}s. This long lifetime probably implies that the CT exciton lies in a crystal defect; long CT lifetimes have been observed by Popovic [16], Altwegg et al [17], and justified theoretically by Petelenz [18].

The cross-section for the annihilation of the trapped CT exciton by a thermalized electron in anthracene to produce a plane-wave electron state as shown in equations (10) and (11) has been estimated theoretically by Andreev and Pope [19]; a value of $10^{-9} - 10^{-10} cm^3 s^{-1}$ was arrived at. The quantum efficiency of photoemission was not measured but it probably is greater than $10^{-25} I_0$, where I_0 is the light intensity.

The binding energy of the excitonic ion in organic crystals has not yet been calculated or measured. Calculations have been made by Munschy and Stébé [20] for copper halide crystals; the value for CuCl was about 0.01 eV. In view of the much smaller dielectric constant of the organic crystals, the binding energy of the excitonic ion should be greater than the thermal energy at room temperature. This is all that would be required for the lifetime of the excitonic ion to exceed that of a typical vibrational period of 10^{-12}s.

The question arises as to the generality of this excitonic ion species; as an example it has not been observed in the merocyanine dye, 3-ethyl-5[2-(3-ethyl-2(3H)-benzothiazolydine)- ethylidene[-2-thioxo-4-thiazolydinone, also known as merocyanine A-10. In this material, a DQEPE is found [21] in which E_{max} is a linear function of the incident photon energy implying that the photon is directly involved as one of the metastable states, and the photoemission quantum yield varies as the light intensity $I^{1.5}$. In view of the recombination-limited lifetime of the electron, as illustrated in equation (7), which makes $[e] \alpha I^{0.5}$, and the linear light intensity dependence of the photon density, the observed $I^{1.5}$ dependence is easily justified. The DQEPE in merocyanine is shown systematically in Figure 2. As to the question of whether the electron is free or trapped, it was possible to measure E_g for the merocyanine; it was 2.5 eV based on the valence level as zero whereas the energy of the electron as calculated from the relationship

$$E_t + E_p = I_c + E_{max} \tag{6}$$

where 1.6 eV, where E_t is the energy of the trapped electron.

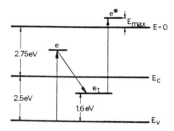

Figure 2. Schematic representation of the photoionization of a trapped electron from merocyanine dye A-10. The first photon excites an electron from the valence band to the level e. This electron thermalizes and is trapped at e_t. The second photon ionizes the trapped electron to e^*, which leaves the crystal with the energy E_{max}.

What then is the criterion for deciding whether one will observe the Auger decay of an excitonic ion rather than the photoionization of a trapped electron? The answer is not yet at hand, but a conjecture will be presented: In anthracene and tetracene, the CT exciton is more stable against dissociation into free carriers than in the more polar crystal of the merocyanine dye. Thus, in anthracene, $(E_g - E_{CT}) \simeq 0.6$ eV; in tetracene, $(E_g - E_{CT}) \simeq 0.4$ eV. However, in merocyanine, $(E_g - E_1) \sim 0.1$ eV, where E_1 is the energy of the lowest excited singlet state. Since $E_{CT} > E_1$, this implies that the binding energy of the CT exciton is $<$ 0.1 eV, making for a low concentration of this intermediate state. In addition, it is anticipated that surface traps in the merocyanine would lie deeper, and hence be more long-lived than in the non-polar crystals, permitting more efficient photoionization.

This work was supported by the Department of Energy. Acknowledgement is made of the assistance of Marek Zielinski in preparing this paper.

Bibliography

1. Lampert, M.A., Phys. Rev. Lett. *1*, 450 (1958).

2. Gerlach, B., Phys. Stat. Sol. (b) *63*, 459 (1974).

3. Stébé, B. and Munschy, G., Solid State Comm. *17*, 1051 (1975).

4. Thomas, G.A. and Rice, T.M., Solid State Comm. *23*, 359 (1977).

5. Agranovich, V.M. and Zakhidov, A.A., Chem. Phys. Lett. *68*, 86 (1979).

6. Singh, J., Phys. Stat. Sol. (b) *103*, 423 (1981).

7. Gumbs, G. and Mavroyannis, G., Solid State Comm. *41*, 237 (1982).

8. See Pope, M. and Swenberg, C.E., "Electronic Processes in Organic Crystals"), Oxford University Press, 1982.

9. Arnold, S., Pope, M. and Hsieh, T.K.T., Phys. Stat. Sol. *B94*, 263 (1979).

10. Pope, M. and Kallmann, H., Disc. Far. Soc. *51*, 7 (1971).

11. Pope, M., Burgos, J. and Giachino, J., J. Chem. Phys. *43*, 3367 (1965).

12. Altwegg, L., Pope, M., Arnold, S., Fowlkes, Wm.Y. and El Hamamsy, M.A., Rev. Sci. Instr. *53*, 332 (1982).

13. Pope, M., Kallmann, H. and Giachino, J., J. Chem. Phys. *42*, 2540 (1965).

14. Altwegg, J., Fowlkes, Wm. Y. and Pope, M., Chem. Phys. *86*, 471 (1984).

15. Sebastian, L., Weiser, G. and Bässler, H., Chem. Phys. *61*, 125 (1981).

16. Popovic, Z.D., Chem. Phys. Lett. *100*, 227 (1983).

17. Altwegg, L., Davidovich, M.A., Funfschilling, J. and Zschokke-Gränacher, I., Phys. Rev. *B18*, 444 (1978).

18. Petelenz, P., Chem. Phys. *94*, 407 (1985).

19. Andreev, V. and Pope, M., Phys. Stat. Sol. (b) *125*, 573 (1984).

20. Munschy, G. and Stébé, B., Phys. Stat. Sol. *64*, 213 (1974).

21. Jing, X.-F., Zielinski, M. and Pope, M., Chem. Phys. Lett. *119*, 173 (1985).

22. Eastman Kodak Co. private communication.

Panel Discussion: How Well Do We Understand Inter-Protein Electron Exchange?

Ephraim Buhks Session Leader

Introduction: Ephraim Buhks

Buhks: I would like to open the discussion by reviewing the progress in our understanding of quantum-mechanical aspects of electron transfer processes.

Let us discuss, for example, the redox reactions of inorganic systems, such as transition metal ions. There is a wealth of information related to their vibrational frequencies, structural and electronic changes due to charge exchange. Here are examples where the non-adiabatic electron transfer theory, which incorporates both electronic and nuclear reorganizations, explains experimental data.

Consider spin effects in electron exchange reaction between $Co(NH_3)_6^{3+}$ and $Co(NH_3)_6^{2+}$. The ground electronic states of these ions are singlet and quartet, respectively, and therefore, the electron exchange reaction between them would be forbidden, however due to spin-orbit coupling the true ground states of the cobalt ions would include excited electronic states and the corresponding transition probability is reduced by 10^{-4} due to this spin effect. The large intramolecular reorganization (0.18 Å change in a Co-N bond length) reduces the activated factor by 10^{-9}, while the tunneling effects of metal-ligand vibrational modes ($h\omega = 500cm^{-1} > kT$) slightly increase the Franck-Condon factors (10^{-8}). Altogether, the spin effect and intramolecular reorganization are responsible for reduction of 12 orders of magnitude in the rate of $Co(NH_3)_6^{3+/2+}$ exchange as compared to a typical $Ru(NH_3)_6^{3+/2+}$ exchange reaction (E. Buhks, M. Bixon, J. Jortner and G. Navon, Inorg. Chem. *18* 2014 (1979)).

Another example is related to kinetic isotope effects in electron transfer reactions. Quantum-mechanical calculations predict that electron exchange is slower between transition ions with deuterated ligands and the ratio k_H/k_D depends on the redox potential. This ratio is close to unity for barrierless reactions and increases with decrease in exothermicity, reaching its maximum for symmetric reactions (E. Buhks, M. Bixon and J. Jortner, J. Phys. Chem. *85*, 3763 (1981)). The decrease of electron exchange rate in a deuterated system is explained by a widened potential barrier due to a lower metal-ligand vibrational frequency (heavier ligand) and, therefore, decreased tunneling probability. The width of the barrier decreases with increase in exothermicity going from symmetric reactions to the barrierless reactions.

The theoretical predictions were confirmed in the study of kinetic isotope effects in the redox reactions of $Fe(H_2O)_6^{2+}$, $Fe(D_2O)_6^{2+}$ and $Fe(^{18}OH_3)_6^{2+}$ (T. Guarr, E. Buhks and G. McLendon, J. Am. Chem. Soc. *105*, 3763 (1983)). Experimental data could be explained if the high-frequency frozen 0-H vibrational modes (3600 cm^{-1}) dominate kinetic isotope effect upon deuteration. A change of 0.03 Å to 0-H bond distance due to charge transfer was estimated from the data. This small structural change determines the observed 20% decrease in the redox rate upon deuteration due to decrease in the tunneling probability through a widened barrier along the 0-H normal coordinates. It is interesting to note that the absolute contribution of the 0-H modes to the Franck-Condon factors almost does not affect the redox rate as compared to the contribution of the iron-ligand vibrational modes undergoing 0.14 Å change in the bond distance (0.25 vs 10^{-6}, respectively).

Metal-ligand vibrational modes (500 cm^{-1}) also were found to dominate the temperature dependence of electron transfer in biological systems, such as cytochrome $\underline{c}$ oxidation in bacteria. Below 100 K electron transfer proceeds entirely by tunneling along the metal-ligand normal coordinates and the rate is temperature independent. The rate increases with increase in temperature above 100 K due to better overlap of the wave functions of the higher vibrational states and their partial activation.

A similar quantum-mechanical multiphonon theory was applied in the study of temperature dependence of CO-Hemoglobin recombination. In this case the onset temperature of 10 K separating the temperature independent and activated regions of the rate is determined by the low-frequency heme-protein vibrational modes in the range of 50 cm^{-1}. On the other hand, higher-frequency vibrational modes, such as iron-histidine mode (200 cm^{-1}), C-0 stretch mode (2000 cm^{-1}) and Fe-C- 0 bending mode (600 cm^{-1}), would explain the quantum-mechanical nature of $^{12}CO/^{13}CO$ isotope effect and its temperature dependence (1.5 at 20 K; 1.2 at 60 K). E. Buhks and J. Jortner, J. Mol. Struct. *123*, 241 (1985); J. Chem. Phys. *83*, 4456 (1985). The non-adiabatic multiphonon group transfer theory fits the value of the CO-Hb recombination rate by incorporating spin-orbit coupling between initial high spin and final low spin electronic states of the five-coordinate heme iron.

Beratan: I'd like to make a few general points about the non-adiabatic electron transfer theory within the Golden Rule formalism. There are still several open questions which exist in the formalism as it is now applied. Analogous questions will undoubtedly crop up in the more modern theories as well. The first one concerns the fact that the tunneling matrix element is dependent on the geometry and the structure of the bridge between donor and acceptor. The tunneling matrix element decay with distance is not a universal constant. There are experimental examples for that. For example, in the work of Stein, Lewis, Seitz and Taube where they connected ruthenium pentaamines with saturated cyclobutane linkers, they found the decay parameter α to be about 0.4 Å if you write $T_{ba} = Ae^{-\alpha r}$. However, if you look at the work of Joran, Leland, Geller, Hopfield and Dervan, where you're tunneling through the bicyclo [2.2.2] octane, you'll find an α of approximately 0.9. Now since this appears in the exponential, you get in a very different dependence of the rate on the geometry of the bridging medium. How can we extrapolate from these model systems to the biological ones which include some through bond and perhaps some through space interactions; how can we put the whole picture together?

The second question which I think exists concerns the role of electron vs. hole tunneling in proteins. If you have a highest occupied - lowest unoccupied energy gap in proteins of 5-10 electron volts, the decay length of the tunneling matrix element depends on the energetic distance of the localized state from the states of the protein. Since you can put these trap levels anywhere between the highest occupied and lowest unoccupied levels of the protein, there remains the question of where these levels belong and whether it is indeed hole transfer or electron transfer which assists charge mediation in proteins.

My third point is that within the Franck-Condon formulation of electron transfer theory, you must choose an energy for your tunneling electron. As you change the driving force of a reaction, you should, in principle, also change the electronic energy of the tunneling electron. How can we include that in a calculation of biological electron transfer rates and how important is this effect?

We've discussed today the fact that we must calculate wave function tails at a large distance from the electron donor. The question comes up, when the electron is far from its otherwise localized site, do the nuclei begin to realize that they are part of a developing ion on the time scale of the transfer? If the nuclei react to the fact that the electron is leaving when it is far from the nuclei, there are possibilities that the Born- Oppenheimer approximation may break down. The question, is, in biological reactions, are we in the right portion of parameter space for this breakdown to be important or not?

Margoliash: I would like to present an observation of the rate of electron transfer within a preformed complex between two heme proteins whose physiological function is the oxidation of one by the other. The common reaction kinetics that one measures with mitochondrial heme proteins are limited by the formation and breakdown of the protein complexes, not by the much faster electron transfer within the appropriate protein complex once it is formed. In a program to examine structure-function relations in cytochrome *c* by directed mutagenesis and expression of mutant genes, which would allow to vary the primary structure of the protein at will, it becomes essential to distinguish direct effects of the machinery of electron transfer from effects on the kinetics of formation and breakdown of the complexes between electron exchange partners, hence the work I am describing.

That baker's yeast cytochrome *c* peroxidase (C*c*P) forms a complex with cytochrome *c* (C*c*) has been evident since the work of Mochan and Nicholls in the early 1970's (for example, *Biochem. J. 121*, 69-82 (1971), a conclusion since verified by at least three different experimental approaches. So, when Poulos and Kraut solved the spatial structure of cytochrome *c* peroxidase, they proceeded to examine the likely structure of a complex between C*c*P and C*c* by computer model fitting, employing the coordinates of tuna C*c*, the one for which the most precise data were available (*J. Biol. Chem. 255*, 10322-10330, 1980).

The structure they obtained is pleasing, because the surface of cytochrome *c* is found to fit against the peroxidase in the region that was earlier demonstrated to constitute the so-called enzymic interaction domain, on the "front" surface of the molecule, namely that containing the solvent accessible edge of the heme prosthetic group, and centered on the point at which the positive end of the dipole axis of cytochrome c crosses the front surface of the protein near the α-carbon of phenylalanine 82. Also many of the positively charged lysyl side-chains in that surface have found complementary negatively charged residues of the system on its kinetics of reaction (see brief review by Margoliash and Bosshard, *TIBS 93*, 316-320 (1983).

To study electron transfer within the $C c$P-$C c$ complex we made use of zinc-substituted $C c$P ($ZnC c$P) in which zinc protoporphyrin IX replaces the ordinary heme of the enzyme. Reversible electron transfer within the $ZnC c$P-$Fe^{III}C c$ complex is initiated by flash photoexcitation. This forms the zinc-protoporphyrin triplet state which can either decay back to the ground state (Eq. 1) following a rate constant k_D, or reduce the ferricytochrome c, following a rate constant k_t:

$$\begin{array}{ccc} \left[ZnC c P - Fe^{III}C c\right] & \xleftarrow{\;k_D\;} & \left[^3ZnC c P - Fe^{III}C c\right] \\ (A) & & (A*) \end{array} \tag{1}$$

$$\begin{array}{ccc} \left[^3ZnC c P - Fe^{III}C c\right] & \xrightarrow{\;k_t\;} & \left[^3ZnC c P + -Fe^{II}C c\right] \\ (A*) & & (B*) \end{array}. \tag{2}$$

The intermediate B formed in this reaction, namely the complex between the thermalized π cation radical of the $C c$P and ferrocytochrome c, is obviously unstable; the π cation radical being strongly oxidizing $(E'_o+0.7V)$, will regenerate the ground state, at a rate constant $k_b \gg k_t$, by oxidizing the reduced cytochrome c (Eq. 3):

$$\begin{array}{ccc} \left[ZnC c P + -Fe^{II}C c\right] & \xrightarrow{\;k_b\;} & \left[ZnC c P - Fe^{III}C c\right] \\ (B*) & & (A*) \end{array} \tag{3}$$

The interesting point is that this reaction is directly analogous to the physiological enzymic process, in which ferrocytochrome c transfers an electron to the higher oxidation state of the peroxidase obtained by oxidizing the ferric enzyme with hydrogen peroxide.

This kinetic scheme can be diagramed in the following way:

The decay of the triplet state $^3ZnC\underline{c}P$ within the complex is first order, and its rate constant, k_p, will be the sum of the intrinsic decay rate constant, k_D, and the rate constant for electron transfer to the ferricytochrome $\underline{c}$ in the complex, k_t:

$$k_p = k_D = k_t. \tag{4}$$

Here, k_p is the observed rate constant of the triplet decay, k_D can be determined employing the complex with ferrocytochrome $\underline{c}$, in which electron transfer to the cytochrome $\underline{c}$ is blocked because the heme is already reduced, and k_t is the difference. For $k_t \leq k_D$ when one titrates $\underline{ZnC - cP}$ with a ferricytochrome $\underline{c}$, the triplet decay rate remains first order and increases linearly with added cytochrome $\underline{c}$, until a 1:1 molar ratio of the two proteins is reached. Thereafter, the decay rate constant remains invariant at the value, k_p, characteristic of the complex.

Our first surprise occurred when surveying a number of different cytochromes $\underline{c}$. It was found that the baker's yeast iso-1 and iso-2 cytochromes $\underline{c}$, as well as the cytochrome $\underline{c}$ of another fungus, *Candida krusei* all showed relatively similar large values of k_t, such as $138 \pm 12s^{-1}$, at 20°C in 10 mM phosphate buffer pH 7.0, for the iso-2 protein. A second group of cytochromes $\underline{c}$ showed rates 1/8th to 1/10th those of the fungal proteins. Among these were horse cytochrome $\underline{c}$, and the tobacco hornworm moth protein. For example, under the conditions just given for the fungal proteins, k_t value for horse cytochrome $\underline{c}$ was $17 \pm 3s^{-1}$. Finally, under these same conditions, the rate with tuna cytochrome $\underline{c}$ was too slow to determine. When the ionic strength was lowered to that of a 1 mM phosphate buffer, k_t for tuna cytochrome $\underline{c}$ was measured at $25 \pm 9s^{-1}$, and $210 \pm 25s^{-1}$ for the yeast cytochrome $\underline{c}$.

Furthermore, it is also possible to solve equation 1 to 3 for the time course of the concentration of the electron transfer intermediate complex between the π cation radical of the ZnC$\underline{c}$P and the ferrocytochrome $\underline{c}$, denoted B, as follows:

$$B(t) = \frac{A^*(t = 0)\, k_t}{k_b - k_p} \left[e^{-\,k_p t} - e^{-\,k_b t} \right]. \tag{5}$$

This expression corresponds to an exponential rise and then a fall, with the maximal concentration of B reached being much smaller than that of $A^*(t = 0)$, the initial concentration of the triplet, ^{3}ZnC$\underline{c}$P. The formation and decay of this intermediate can actually be followed spectrophotometrically as a small but well defined transient absorbance at 444.5 nm, the isosbestic point of ^{3}ZnC$\underline{c}$P and ZnC$\underline{c}$P. The reference compounds, ZnC$\underline{c}$P, the complex ZnC$\underline{c}$P - Fe^{II}C$\underline{c}$, and Fe^{III}C$\underline{c}$ show no signal at that wavelength. The kinetic behavior of the transient absorbance behaves exactly as predicted by Eq. 5, except that for tuna cytochrome $\underline{c}$, in 1 mM phosphate pH 7 and 20°C, the value of k_b was found to be $12 \pm 4s^{-1}$, while for yeast iso-2 cytochrome $\underline{c}$ in 10 mM phosphate, it was $10^4 s^{-1}$. This compares to the values of k_t, already cited, of $25 \pm 9s^{-1}$ (1 mM phosphate) and $138 \pm 12s^{-1}$ (10 mM phosphate) for the tuna and yeast cytochrome $\underline{c}$, respectively. It should be noted that with yeast cytochrome $\underline{c}$ taken to lower ionic strength k_b increased even beyond the approximate $10^4 s^{-1}$ measured in 10 mM phosphate.

To summarize, from the biological point of view, these observations are indeed remarkable in that they emphasize the extreme degree to which individual physiological electron exchange partners of different species are adapted to each other, even in a set of such highly homologous proteins as the cytochromes $\underline{c}$. That reduced yeast cytochrome $\underline{c}$ would transfer an electron to a thermalized π cation radical form of yeast cytochrome $\underline{c}$, peroxidase fully 1000 times faster than tuna cytochrome $\underline{c}$, could not have been expected from their very high degree of similarity in structure and function (see Margoliash and Schejter, *Adv. Prot. Chem.* *21*, 113-286, 1966). Even more importantly, with the present capability for changing single amino acid residues in cytochrome $\underline{c}$ at will be recombinant DNA procedures, the present observations open the way for a detailed examination of the physical basis for this adaption.

From the biochemical point of view, the present results again emphasize that in electron transfer reactions in physiological systems, particularly with those consisting of proteins, steady- state and presteady-state kinetic parameters most often represent rates of formation and/or separation of functional protein complexes, not the very much faster rates of electron transfer within preformed complexes. Since both types are referred to as electron transfer rates, confusion easily ensues. In the present case, the reaction analogous to the physiological oxidation of reduced yeast cytochrome $\underline{c}$ by the higher oxidation state of yeast cytochrome $\underline{c}$ peroxidase, was measured directly for the first time, and found to be several orders of magnitude faster than other reactions measured in this and related systems.

From the physical point of view, it is obvious that yeast cytochrome c docks 'better' with cytochrome c peroxidase then does tuna or horse cytochrome c, as has been evident for some time from their directly measured binding affinities (Kang et.al, *J. Biol. Chem.* *252*, 919-926 1977). However, the observed differences in long-range electron transfer rates, for the various cytochromes c within the performed 1:1 complex with yeast cytochrome c peroxidase, clearly represent more than a change in the 17-18 Å edge-to-edge distance between the hemes of the two proteins in the complex, inferred for the peroxidase-tuna cytochrome c complex by computer modelling. Indeed, such a change could not explain why for tuna cytochrome c, the rate of electron transfer from the triplet state ^{3}ZnCcP to the ferricytochrome c is twice as fast as the electron transfer rate from the ferrocytochrome c to the cation radical form of the ZnCcP, k_b/k_t being approximately 0.5, while the difference is in the opposite direction when yeast cytochrome c is in the complex, k_b/k_t being about 500, even though in both directions, for both cases, the reactions are comparably exoergic. These differences, which appear to violate microscopic reversibility, are likely to result from a rapid conformational rearrangement in the complex following electron transfer to the ferricytochrome c (Eq. 2) and prior to electron transfer from the ferrocytochrome c (Eq. 3). They may represent nothing more than the change in spatial conformation known to attend the reduction of cytochrome c. Furthermore, just as one has to consider with some hesitation the significance of complexes between the enzyme and relatively unreactive cytochrome c, it would appear that any generalized description of a mechanism of electron transfer in this case, is not likely to be particularly illuminating, unless an analysis can be provided on the influence of the structure of the proteins on the parameters which define the postulated mechanism. The identification of which protein structures are significant in this respect may have to await the results of the recombinant DNA approach to the influence of structure on function, the solution of the crystal structure of yeast cytochrome c, a computer modelling of the complex of that cytochrome c with peroxidase, or, even better, a solution of the structure of a co-crystallized yeast cytochrome c and peroxidase. From a biologist's point of view, the type of situation described for yeast cytochrome c peroxidase is likely to be the common one among physiological electron transfer systems, not an unusual one, and one must be prepared to deal usefully with the intricacies of biological specificity.

This work stems from a collaboration with Brian M. Hoffman on the mechanisms of action of cytochrome $\underline{c}$ peroxidase. Our colleagues involved in it were P.S. Ho, C.H. Kang, N. Liang and C. Sutoris, and some of these results have been presented in preliminary form in two communications to the *J. Amer. Chem. Soc.* (*107*, 1070-1071, 1985 and *in press*).

Fischer: I'd just like to come back to the challenge because I feel it is very nice to propose some open questions rather than showing that almost everything is understood and so I'd like to comment on the four points that brought up here.

The first point I interpret in the way that actually we should not just always draw these nice reaction coordinates and by that way forget that the electron wave function depends formally upon the coordinate; that means it is very crucial, I think, to consider the so called non-Condon effects. Actually, I may tomorrow come a little back to that, but I am very glad to hear that there is experimental evidence that this is important. We should also remember that when we write down the formula for the nonadiabatic weight there is a Franck-Condon matrix. It seems to me a little bit inconsistent to say on one hand, that we separate the electronic matrix from the nuclear factor, and, on the other hand, we draw some potential curve and say it is the matrix of the transition state which may be something very different and much harder to calculate.

Now with regard to the second point considering the hole, as I just heard it, I think this really is important. In the simple description we never talk about the hole. We should distinguish between the electron transfer or, hole transfer where we actually separate charges, and then the coupling to the electron as well to the hole is very important. I think many processes, as you also heard from other examples, may actually follow the way that you fill up the hole while you just transfer an excited electron.

I think this again is an important point.

With regard to the third point, change in ΔG, how may this may effect the coupling? Well, I think in some way that goes back to the first problem, because once you simultaneously account for the electron wave function and the reorganization, that actually relates to the same question.

There was the fourth point, the Oppenheimer question, does it apply? I'd just like to remind you that in many of these electron transfer reactions, actually, one cannot start really with the full concept of the Oppenheimer approximation. If you, for instance, have the symmetrical system, the Born-Oppenheimer potential would be symmetric and we like to start out with the localized state. So, from the very beginning, one doesn't actually start out with the Born-Oppenheimer approximation, rather than with the localized state. One has to find means to define really the initial state. I may tomorrow have a chance to go a little bit into this question.

Friedman: I have another one of these open ended questions. Before I say anything about hemoglobin, we'll see that proteins are capable of quite a bit of motion and they're not rigid structures. The hemoglobin, once the ligand comes off, the structure about the heme starts to respond fairly quickly and in effect the protein responds to the de-liganded heme almost as self induced trapping. The barrier to the ligands that come back off builds up in time self trapping the deoxy heme. Now it seems for the electron transfer, one can envision a similar kind of thing, I guess, the kind of thing that Dr. Jortner showed earlier. Say with photo excitation, you get transfer to this potential circuit and there's some back reaction. The protein structure starts to respond once the electron comes over, it responds in a way to stabilize the transferred configuration. You in effect trap the electron over here and it seems quite plausible that some self induced trapping should occur in proteins. I'm wondering if any of the electron transfer pros have seen any evidence for such phenomena? Have they looked? Or, actually, I guess, there's some evidence in photosynthetic work that protein structure does respond. Again, an open question: is there any evidence for self induced trapping with regard to electron transfer processes?

Honig: I just want to point out that Dr. Margolish's remarkable observations serve to point out the difference between approach and inclination of physicists, chemist, and biologists. For biologist and biochemists to be interested in what's going on in this room, they'd want to hear an attempt to account for the difference in behavior between different types of cytochromes. Until problems of this type are addressed, physicists will be doing physics which is obviously perfectly legitimate, but what they do will be doing will be of absolutely no interest to biologists.

Wolynes: I think we see how important the question of the role of electronic motion vs. protein motion is in these problems and I think that there certainly is a limit in which the electronic motion is the most important aspect and I would add another open question to Dr. Beratan's list. Namely when we are going to stop using Hückel theory to calculate transfer matrix elements, when no one else uses Hückel theory. There clearly are correlation effects as occur in such systems like Mott insulator transitions.

The story on tuna cytochrome and also the earlier talk of Dr. Isied, I thought had a very exciting aspect to them that reminds me of the Sherlock Holmes story in which the important thing was that the dog didn't bark. Certainly the main reason why one wanted to have electronic control in these processes is to say that when two things are very far apart there is no electron transfer. Apparently, that does happen. If you get things very far apart, the electron transfer processes are very slow and perhaps in that limit, things are always nonadiabatic if they are infinitely far away. On the other hand, perhaps if the molecules are close enough to be on the electronic pathway the situation is a little more problematic. Whether it's a configurational change or a simple electronic structure change, I think, that in some sense nonadiabatic ideas have really helped us in one aspect of the control, but now there are new aspects of control within the complex.

Isied: A question which we raised earlier is the subject of this slide. This is a comparison between two series of compounds with the same bridging group and the same donor group, but with different acceptors. Although it is difficult to ascertain a trend from comparing only two series of compounds, nevertheless I would like to bring these results to this forum of theoreticians and experimentalists. The Co(III) acceptor (d^6) has a large reorganization energy and undergoes a spin change as it accepts the electron. For this series of reactions we obtain an $\alpha \sim 2\text{Å}^{-1}$, which agrees with earlier theory. When the same experiment is done with Ru(III) (d^5) as the acceptor, we find that the decrease in rate of electron transfer as proline residues are added is less than a factor of 50 (i.e. $\alpha < 0.5\text{Å}^{-1}$).

It seems from these experiments that wide variation of α can be obtained from the same process depending on the wave function of the electron acceptor. I would like to hear any comments on that.

Jortner: I would like to address the issue of the role of medium relaxation in electron transfer and let me make four points: (1) The role of medium dielectric relaxation time. One should consider the Landau-Zener model with the transition probability $[1 - \exp(-4\pi\gamma)]$, with the Landau-Zener parameter $\gamma = |V|^2/2\alpha\hbar^2$, where V is the matrix element and α is the measure of the rate of the change of the solvent fluctuations. α can be expressed by $\alpha = (2E_skT)^{\frac{1}{2}}/\hbar\tau$, where E_s is the solvent reorganization energy and $\tau = \tau_D(\varepsilon_\infty/\varepsilon_S)$, with τ_D being the constant field dielectric relaxation time, while ε_∞ and ε_S are the high frequency and the low frequency permittivity, respectively. Accordingly, $\gamma = \tau|V|^2/\hbar(2E_skT)^{\frac{1}{2}}$. The Landau-Zener factor is determined by the solvent relaxation time at constant charge. (2) As long as the electron transfer process is nonadiabatic, i.e., $\gamma \ll 1$, nothing interesting will happen and the solvent dielectric relaxation will not affect the rate constant. (3) Transition to the adiabatic limit when the solvent relaxation becomes very slow. For large values of τ, which can be realized in a highly viscous medium, $\gamma > 1$ and the pre-exponential factor in the adiabatic rate will be determined by the solvent relaxation rate $1/\tau$. Two years ago Peter Wolynes wrote a very nice paper on the application of the Kramer's model to this adiabatic limit of electron transfer. (4) The role of intramolecular modes. In a system where electron transfer dynamics involves coupling to both solvent modes and to high frequency intramolecular modes, the manifestation of slow medium relaxation on the "transition" from a nonadiabatic to an adiabatic reaction with increasing of the solvent relaxation time will become much more complex.

These are, in my view, the interesting questions that should be addressed in the area of electron transfer.

Wolynes: I agreed with your first point. I'm not sure I understood your second point, but I'll have to listen more carefully. And on the third and fourth points, I whole-heartedly agree. How much one has very slow modes contributing and how much one has quantum modes contributing is a very important problem and in any practical problem you really have to address that. I think one theme emerging a little bit from Dr. Warshel's talk is the role of simulation in understanding these processes and I think that it's quite clear that we're eventually going to have full quantum simulations on these systems in some sense. Fortunately, I think we have techniques for beginning to do that.

Jortner: Yes, that's interesting. Quantum molecular dynamics can be applied to the electron transfer problem. Recently, Landman, Scharf and myself studied a related problem of electron localization in clusters. The technique involves the Feynman path integral method, which boils down to a molecular dynamics calculation and results in detailed information about the quantum particle. Interesting information concerning electron trapping in bulk and surface states and configurational changes of clusters induced by electron localization emerged. It may be very interesting to extend this method for electron transfer in model biophysical systems. This can be done for an electron interacting with 50-100 classical particles, which may be sufficient.

Warshel: If you use a surface hopping approach rather than Landau-Zener approximation, and you do it correctly by keeping traces of trajectories any mistakes that you do are in the limits of dynamics of running reaction trajectories. This is my opinion, because I am trying to calculate trajectories. I think that there is extremely fundamental problem that we neglected to discuss. There is a Marcus relation which relates the rate to the free energy gap and Dr. Wolynes essentially wrote something nice about it. I think that the most fundamental question from the viewpoint of the experimentalist is what should one use there? Should we use free energy gap or what? I did one work, and Dr. Wolynes did another. I used umbrella sampling to check how the free energy of activation depends on the redox difference and the only conclusion was that it was in the error of simulation. Marcus' idea, that the solvent responses linearly is OK. But I think that the role of simulation in determining the role of free energy gap is extremely important.

Miller: I have to disagree a little bit, Dr. Jortner. I think because I believe that if you have a dielectric relaxation which precedes electron transfer, it can dramatically control the rate. This, as I was suggesting earlier and now I wish to make a little bit more provocative, could even be the basic reason for the break in the temperature dependence observed many years ago by Chance and De-Vault and others which was so nicely, initially attributed to some conformational changes. We have developed and learned more, and now it has been attributed to nuclear tunneling, and we conceivably find out in the future that it's really due to conformational changes. But this is still really shooting from the hip, and I admit that very much. Dr. Beratan, concerning your question about hole tunneling. There is I think, definite experimental information, where we can find systems where we get close to hole states. This number alpha that determines how fast fall off with distances begins to get dramatically smaller, just as we can find evidence when we get very close to the electron states, the excess electron states. However, in most systems, it's very difficult to know which is the more important.

There is just one last comment and that's to me, though, out of all that has been happening this afternoon, the most exciting and interesting things are what we heard from Dr. Margoliash and Dr. Isied. In proteins, there is something which may be simply changing the distance when things aren't coming together properly, although that wouldn't seem to explain DeVault's experiment. But right now, from what I've seen so far this afternoon, we have no suggestions that can accommodate these in our present understanding.

Jortner: Dr. Miller, I have an experimental question. You have a lot of data concerning the electronic interaction term which was alluded to. Can you say something on the basis of this information about what is the proper approach for relating the electron binding energy and the exponential parameters and what can we learn? Because I think it's very, very important.

Miller: The measurement of the quantity alpha, how rapidly the electron exchange decreases with distance is a very, very serious problem. When we found this one of the hardest things that we've tried to do because when we do it within a rigid media, we have a problem that the Frank-Condon factors change with distance and this can confuse the measurement of alpha. There are many wrong reports of measurements of this quantity in the literature. We've done recently something that we beat our heads against the wall until it was bloody. It took a few years and we feel we have some reliable measurements in the case of a radical anion transferring an electron to a neutral molecule. This is a case probably where, Dr. Beratan, the electron tunneling rather than the hole tunneling is more important. What we find is that the quantity alpha changes from about 1.1 $Å^{-1}$ to about 1.35 $Å^{-1}$, while the binding energy of electron to the donor changes from about 2 volts to about 4.5 volts. The simple expression which would give a relationship, the first one which I used some years ago (the interaction of particles and wells with nothing in between but a smooth potential) that fails completely to represent this kind of data. This is only a very small part of the story because probably this interaction going through the excess electron orbitals may behave differently from the way it goes through hole tunneling orbitals. It's something on which there isn't any experimental information on it, and it is an interesting but, probably very complicated problem.

Kleinfeld: I first want to address an old question, referring to Dr. Jortner's discussion on activation energy in electron transfer. Then I'd like to answer Dr. Freidman's question about self trapping.

There are many processes in photosynthetic reaction centers which appear to be independent of temperature. Most of these, for reasons I don't understand, involve porphyrins. The usual answer given to explain why these reactions are independent of temperature is that the differences in redox energies between the electron donors and acceptors matches the nuclear rearrangement energy. However, Marilyn Gunner, in Les Dutton's lab, addressed this problem by looking at the charge recombination rate between the primary electron donor and the primary quinone acceptor in photosynthetic reaction centers. What Marilyn did was to look at this transfer rate both as a function of temperature and as a function of which quinone was used as the acceptor. By varying the quinone, she varied the redox potential between the donor and acceptor by roughly 0.5 volt, which is huge on the energy levels relevant to biology. When she did this, except for some odd ball cases, she found that the transfer was activationless over the entire spread of energy. Somewhat related to this, we found from preliminary studies that the ratio of the transfer rate at low temperature to that at high temperature is about a factor of 4, independent of the change in redox energy.

The question is, how can this sort of finding, your varying the energy by such a great deal without changing the kinetics, by explained theoretically.

The second point changes topic completely and has to do with Dr. Friedman's question about self trapping; I think Dr. Feher may show the data tomorrow that deals with this topic. If I understand it right, Dr. Friedman's question is this: When you have an electron transfer, does the process of an electron going to some new site somehow cause a rearrangement of the nuclei or whatever, so that the electron makes itself a stable potential well to sit in and thus trap itself. There's some experimental data that suggests that this may happen in, again, photosynthetic reaction centers. You can look at the electron transfer between the primary quinone acceptor and the second of the quinone acceptors. At room temperature, this transition takes some 100 μs to go, but if you cool the reaction centers down to cryogenic temperature, say 80 K, and you try to study this reaction, you don't see any transfer occurring. We made some estimates and it takes longer than 10^3 seconds for the transfer to occur. Yet, if you were to take reaction centers and cool them, not in the neutral state, but under illumination so than an electron is on the secondary acceptor, then wait about a day or so for the reaction centers to return to the neutral state, you will find that the electron transfer goes in less than 10^3 seconds. So you've somehow changed the transfer dynamics by 8 orders of magnitude. The suggestion is that you really rearranged the protein in the region of where the electron sits on the secondary acceptor. By cooling under illumination, you've trapped this rearrangement and thus the reaction can occur.

Seigert Fischer: Reply to the first part of the question which also relates to the question of how we handle the slow motion of the medium and the fast motion of the internal degrees of freedom. The way I like to look at the weight is that one actually retreats to fast degrees of freedom quantum mechanically and then you actually can write up the weight as a sum to contributions where you go into the individual quantum states and then for every final state, you then treat the medium more or less classically if it's a slow motion. So then you end up with the sum of many such contributions, it very easily can happen that you get very little temperature dependence even though you may have a wide range of changes. The temperature dependence is always dictated just by the remaining energy difference, because most of the energy is picked up by the internal degrees of freedom with high frequencies. This way the temperature dependence really just mainly comes in from the medium. I also would like to say that if we discuss the temperature dependence, one really should think in terms of this weight, possibly in a different way because the medium can induce a transition. As well, the internal degrees of freedom can induce a transition and then, one has actually to write down the equations for this. But I think we have had formulations here which were very much simplified and I think it is very nice to hear that they cannot explain everything.

Jortner: Now I think that Dr. Fischer's conjecture is intriguing. I'm not sure whether it's correct, as the role of intramolecular quantum modes has also to be considered. But now let's put the problem into a broader context. I think it's amazing that weak temperature dependence is very rare in physical chemistry and in solution chemistry. We really have to work very hard to find a temperature independent electron transfer process.

On the other hand, the temperature independence is so ubiquitous in photosynthetic reactions centers, so that teaches us something. I would say that really the intramolecular engineering and the molecular and protein modes in the reaction center are optimized to ensure maximum efficiency of the electron transfer. Two aspects of the optimal nuclear rearrangement are apparent. First, an activationless process that we have already alluded to, and second, the very fast processes which beat vibrational relaxation and, therefore, again the rate will be temperature independent. The temperature dependence of two activationless processes was carefully explored; one is the old back reaction from the quinone-minus to the dimer that Dr. Feher and his people studied and the other one is the electron transfer from bacteriopheophytin-minus to the quinone recently studied by Parson and his group. Both reactions exhibit negative apparent activation energies, in accord with electron transfer theory.

Feher: Actually, Dr. Jortner, there is no temperature dependence between 80 K and 4 K, it is really 0 K, and the temperature dependence from 80 K to room temperature can easily be accounted for trivially by an expansion coefficient.

Jortner: What should happen is that there should be constant at temperatures where kT is lower than a quarter of the characteristic frequency and exhibit $1/\sqrt{T}$ dependence at higher temperatures. Obviously, as you were saying, thermal expansion or contraction effect played an important role. A few years ago, it was noted that the $1/\sqrt{T}$ dependence was too weak to explain the entire temperature dependence of $Q^- + P^+$ back recombination and thermal expansion effects have to be added. What I'm trying to emphasize is that the striking temperature independence (or weak temperature dependence) really originates in this case from an activationless process and that electron transfer theory works.

Feher: Well, I know Dr. Kleinfeld well enough to know he what he wants to ask. What he is asking is the following: In Hopfield's formalism there is an exponent $E_{donor} - E_{acceptor} - \Delta$, and what he is saying is that you accidentally can get an activationless process if $E_{donor} - E_{acceptor} = \Delta$ but since he is changing $E_{acceptor}$ by half an electron volt, the condition $E_{donor} - E_{acceptor} = \Delta$ cannot always hold. Yet, one always observes this activationless process.

DeVault: John Miller said something that gave me a thought. Two years ago A. Sarai (A. Sarai and D. DeVault: *Advances in Photosynthetic Research,* C. Sybesma, Editor, The Hague: Martinus Nijhoff/W. Junk, 1984, pp. I.5.653 - I.5.656) made more accurate measurement of the rates of cytochrome oxidation in Chromatium in my laboratory and worked both with the high potential and with the low potential cytochromes. We found in the intermediate temperature region between, I'd say, over 250° and 100°K, quite often evidences of biphasic timecourse curves which we attribute to a freezing in of conformation states. There is also a fall-off in amplitude which can be attributed to the same thing. The fall-off in amplitude is much more marked with the high potential cytochrome, but here is quite a bit of it even in the low potential cytochrome, which, in our previous measurements, we never paid any attention to and didn't report. But the amplitudes are smaller at the low temperatures and it's quite probable that there are conformational states which can be frozen in and which the electrons can no longer tunnel from cytochrome to the reaction center.

Hopfield: The only thing which I did want to remind people of is something that, as I turned around and talked with Dr. Margoliash about; namely cytochrome c has a good crystal structure studied in the oxidized and the reduced states, tuna is the questionable situation which has been done, and actually the internal structure of the molecule is different in those two states. Whether that is a consequence of a crystallography effect, in which case, of course, you may also have different structures in binding the different molecules, or whether actually the structure changes internally when you go between oxidized and reduced forms isn't clear. But that certainly is a kind of thing which you actually need to understand when you are talking about what is the difference between one particular cytochrome and one particular other cytochrome, or even, when you get down to it, when you try to understand what one particular cytochrome is doing.

Margoliash: There are now two cytochrome c for which reasonably good crystallographic data have been obtained, the tuna protein you mentioned, in both the oxidized and reduced states, and the cytochrome c of rice, as far as I am aware, only in the oxidized state (Ochi et.al, *J. Mol. Biol.* *166*, 407-418, 1983). Yeast cytochrome c is also being worked on. With regard to differences between ferric and ferrous forms of the protein, the tuna cytochrome c crystallography shows only very small changes. Of course, one must not overlook the fact that the crystals are examined in near-saturated ammonium sulfate $(\sim 4M)$ in which the structure of the protein may be significantly different from that in the dilute buffers, around 25 mM, commonly used for functional studies. For a small protein having a net charge of $+8$ and no covalent bonds stabilizing the tertiary structure, a major disruptive influence is electrostatic repulsion, which will be strongly affected by the ionic strength.

This effect is seen clearly when one studies the pK_a of the 695 nm band of the ferric form, itself a measure of the ease of opening of the methionine 80 sulfur to heme iron bond, as a function of the dielectric of the bulk medium and of the ionic strength. As expected, the two parameters counteract each other, the lower the dielectric the lower the pK_a, the higher the ionic strength, the higher the pK_a, within the appropriate limits. Whatever the crystallography shows, there is no question that there are significant differences in the structures of ferric and ferrous cytochrome $\underline{c}$. As early in 1966, we were able to list something like half a dozen lines of evidence showing this was the case (*Adv. Prot. Chem. 21,* 113-286, 1966). These included, for example, large differences in the rates of digestion of the two forms by proteolytic enzymes, such as trypsin and chymotrypsin, the kinetics of which we studied carefully, including the order of release of different fragments.

One would imagine that to identify these structural differences, procedures capable of determining structures in solution will have to be employed, such as NMR or others, rather than classical x-ray crystallography.

Roder: I would like to add to the last comment of Professor Margoliash some observations from our studies of the effect of the redox on hydrogen exchange in cytochrome c. In yesterday's talk Dr. Englander mainly focused on the method and on the nature of the structural transitions that we study with the exchange technique. Comparing reduced and oxidized cytochrome c we found a large number of differences in structural mobility throughout the structure. These differences are sizeable in terms of free energy; the largest effect corresponds to about 4 kcal lower local stability in the oxidized form. Those changes were not restricted to the axial ligands, as some models predict, they are distributed all around the heme. We found particularly large redox effects in the "bottom" part of the molecule for the NH protons involved in the hydrogen bonding network with the propionate groups. We are just beginning to put this all together, but I think we have here a technique with sufficiently high resolution to pinpoint the effect of the redox transition on structure and dynamics.

Another point I would like to address here is a question that has been brought up before in the history of cytochrome c, namely the possibility of a conformational change induced by the binding event with the redox partners. Since I am new in the field I would like to hear from the experts: Is there evidence in favor or against such an induced kind of allosteric effect. One might expect that the interaction between reduced cytochrome c and oxidase, for example, might force the reduced protein into something closer to the structure of oxidized cytochrome c, which then would promote electron transfer.

Margoliash: I certainly like your finding the major difference at the lower seam of the heme crevice, where the two heme propionyl side chains are held, because this appears to be the area where x-ray crystallography detects the clearest structural variation between ferric and ferrous cytochrome *c*. Hopefully, this is more than a coincidence.

The question of whether there is a change in the structure of cytochrome *c* upon binding with one of it physiological electron exchange partners, is, as far as I know, unresolved. There are very small spectral changes, such as a maximum of about 2% in absorbance in some region of the Soret band, upon reaction with the mitochondrial oxidase and yeast cytochrome *c*, peroxidase (see for example, Michel and Bosshard, *J. Biol. Chem.* *259*, 10085-10091, 1984), but I am not sure whether it is certain these represent changes in the enzyme or the cytochrome *c*. Please also note that such complexes cannot be the physiologically operative ones which would lead to an immediate electron transfer, as compared to the hours needed to record repeatedly such small differences. Here again, that a small globular protein with a net positive charge of +8 would show some change in structure upon binding to a negatively charged electron exchange partner, having in the process a good proportion of its positive charges neutralized, particularly on the "front" surface containing the enzymic interaction domain, is not unexpected. It would certainly be most encouraging if the H/D exchange procedure could be used to define this situation structurally, more precisely than in its present very unsatisfactory state.

Charge Separation in Model Compounds for Photosynthesis

T. Moore and D. Gust
Department of Chemistry
Arizona State University
Tempe, AZ 85287

The central event in photosynthesis is the conversion of excitation energy arising from the absorption of light into chemical potential energy in the form of charge separation. The entity responsible for this conversion, the reaction center, consists of small organic molecules such as chlorophylls, quinones and carotenoid polyenes embedded in a protein matrix. The protein is in turn incorporated into the photosynthetic membrane. Excitation of the reaction center pigments may either be direct, or via singlet energy transfer from antenna pigments. The excited reaction center pigment donates an electron to a nearby acceptor to generate a high energy charge separated state.

This state would be expected to quickly recombine with the resultant loss of the chemical potential energy stored therein. However, a cascade of electron transfer reactions intervenes to further separate the positive and negative charges and thereby prevent recombination. Ultimately, the charge separation spans the photosynthetic membrane. These multiple electron transfer steps result in some sacrifice of chemical potential, but lengthen the lifetime of the charge separated state to the point where it can be used for photosynthetic work. Thus, the net result of the early stages of photosynthesis is photodriven charge separation across a phospholipid bilayer membrane.

Most of the basic physics and chemistry of photosynthesis is carried out by the small organic molecules, rather than the protein. Chlorophyll molecules in their first excited singlet states are the primary electron donors. Tetrapyrrole pigments also act as electron acceptors and as light harvesting antennas. Quinones act as electron acceptors in the electron transport chain. Carotenoid polyenes are found in all green plants where they perform several functions. They act as photosynthetic antennas by absorbing light in regions of the solar spectrum where chlorophylls are not particularly efficient at transferring singlet excitation to chlorophyll where it can be utilized for photosynthetic work in the usual way. Carotenoid polyenes also provide photoprotection from singlet oxygen damage. This is necessary because under some circumstances, chlorophyll triplet states may be formed in reaction centers. These states are excellent sensitizers for the formation of singlet oxygen, which is a highly reactive species which can do great harm to a biological system. Plants which lack some form of photoprotection against singlet oxygen die when they attempt to carry out aerobic photosynthesis. Carotenoid polyenes provide photoprotection by quenching chlorophyll triplet states via triplet-triplet energy transfer before they are able to react with diffusing oxygen. Finally, carotenoid polyenes are known to be able to act as electron donors to the oxidized form of the reaction center pigment ($P680^+$) in photosystem II of green plants, under certain conditions. The exact role *in vivo* of this electron donation is unknown.

Given that many of the basic physical and chemical steps of photosynthesis involve small organic molecules, rather than proteins, it is interesting to speculate upon the role of the protein in this process. Is it really indispensable to the energy conversion process? One way to investigate this question is to attempt to build model systems which carry out some of the basic chemistry and physics of photosynthesis sans protein. The simplest approach to such models might be to dissolve one or more of the important small molecules in a solvent and to study the physics and chemistry of the resulting solution. Much has been learned from studies of this type, particularly about the basic photophysics and photochemistry of the individual small organic species. However, there are some important limitations to this approach. For example, both energy transfer and charge transfer are crucially dependent upon intermolecular distance and orientation, and these parameters are difficult to control in solutions. Secondly, it is difficult to study processes which occur more rapidly than diffusion. Thus, the initial charge separation steps of photosynthesis, which occur within picoseconds of excitation, cannot be mimicked in these solutions.

The same is true of carotenoid antenna function. Indeed, much of the solution photochemistry of photosynthetically important molecules occurs from the triplet state, which has a long lifetime, rather than form the photosynthetically relevant singlet state.

It is clear, then, that the protein does have a role in photosynthesis, and that a major facet of protein function is the imposition of order upon the system of small molecules. That is, the protein holds the small organic molecules in the correct orientation, at the proper separation, and in a suitable environment for the energy and electron transfer reaction in question. One can begin to model this role of the protein by designing covalent chemical linkages which force the pigments and other small molecules to assume certain relationships to one another. By studying ordered models of this type, one can both learn something about the dependence of energy and electron transfer upon distance, orientation, and energetics, and make predictions about the relative relationships of the small molecules within their natural protein environment. Relatively simple models of this type have proven quite successful. For example, by studying carotenoporphyrins, which consist of carotenoid polyenes covalently linked to porphyrins or chlorophyll derivatives, we have been able to mimic photosynthetic antenna function and photoprotection and have learned a great deal about the rather stringent distance and orientation requirements for singlet and triplet energy transfer between photosynthetic pigments [1-6]. With regard to electron transfer, both carotenoporphyrins and porphyrins covalently linked to quinones mimic the initial photodriven charge separation of photosynthesis. In the case of the porphyrin-quinone (P-Q) system, for example, the correct covalent linkage allows the porphyrin first excited singlet state to readily donate an electron to the quinone to form $P^{+\bullet} - Q^{-\bullet}$. Unfortunately, however, the charge recombination (back electron transfer) reaction in these two-part systems is extremely fast, and in solution at normal temperatures the charge separated state typically lives for at most a few hundred picoseconds. Thus, these models can mimic the initial charge separation of photosynthesis very well, but are not able to stabilize the charge separation in time so that it may be used for photosynthetic work.

The creation of a long-lived charge separated state evidently requires the imposition of still more order upon the model system. In particular, it may require mimicry of the multistep electron transfer of natural photosynthesis. In order to test this hypothesis, we recently synthesized the molecular triad **I** shown below [7-9]. The molecule consists of a synthetic porphyrin covalently linked to a quinone electron acceptor and a carotenoid donor.

As mentioned above, both electron and energy transfer rates depend strongly upon donor-acceptor separations and orientations. Thus, it was imperative to know something about the solution conformation of **I**. Fortunately, high resolution nuclear magnetic resonance (NMR) spectroscopy can provide detailed conformational information for molecules such as **I**. This is true because the porphyrin moiety gives rise to large aromatic ring currents when placed in the magnetic field of an NMR spectrometer. These currents in turn generate local magnetic fields which cause shifts in the NMR resonance frequencies of protons in the region of the porphyrin. Calculations based upon such chemical shift changes for the protons of the carotenoid and hydroquinone moieties of **III** reveal that the molecule assumes an extended conformation with the carotenoid and hydroquinone side chains directed out away from the porphyrin, rather than folded back over it. Figure 1 shows some approximate intramolecular distances based upon this conformation.

The absorption spectrum of **I** resembles a superposition of the absorption spectra of carotenoid, quinone and porphyrin model compounds. There is no strong perturbation of the bands, and the three chromophores are apparently independent of one another.

When a methylene chloride solution of **I** was excited with a 10-ns flash of 600 rm laser light, a transient absorption with $\lambda_{max} = 970nm$ and $t_{1/2} = 170ns$ was observed (Figure 2).

A similar experiment with the hydroquinone form of the triad (**II**) yielded no

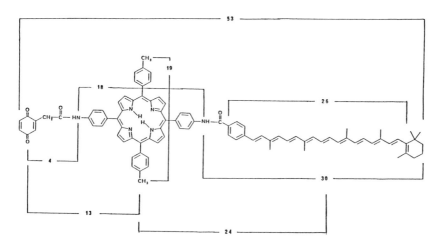

Figure 1. Approximate intramolecular distances in Angstroms for Triad I, based upon nuclear magnetic resonance measurements.

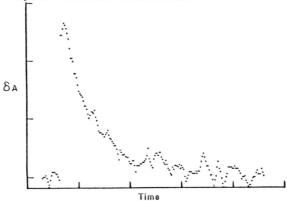

Figure 2. Transient absorption kinetics at 940 nm for I $(10^{-5}$ M, 1 cm path length) in methylene chloride (400 ns/div. 1.0 x $10^{-4}\delta A/div$). The sample was excited by a 10 ns, 600 nm laser pulse.

transient in this region. The spectrum of the transient revealed that it was due to the carotenoid radical cation $(C^{+\bullet})$. Thus, absorption of light by the porphyrin moiety produces the carotenoid radical cation, but only in the presence of the quinone moiety. A reaction scheme which accounts for these observations is

shown below. **Scheme I**

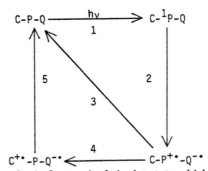

Irradiation of **I** produces the porphyrin first excited singlet state which donates an electron to the quinone to produce an initial charge separated state $C-P^{+\bullet}-Q^{-\bullet}$ (step 2). As one would expect by analogy with the porphyrinquinone systems mentioned above, the recombination reaction which regenerates the ground state and releases the stored chemical potential energy as heat (step 3) is extremely fast, with a rate constant of $> 10^{10}$ sec^{-1}. However, a second electron transfer from the carotenoid polyene to the porphyrin radical cation (step 4) competes with the back reaction and yields a second charge separated state $C^{+\bullet}-P-Q^{-\bullet}$ which decays slowly to the neutral starting material with a rate constant on the order of 10^6 sec^{-1}.

Picosecond absorption studies show that formation of the $C^{+\bullet}-P-Q^{-\bullet}$ state is complete within 100 ps of excitation. The quantum yield of $C^{+\bullet}-P-Q^{-\bullet}$ is about 0.04 in methylene chloride, but can be greater than 0.25 in some solvents. The lifetime of the final charge separated state (step 5) is also strongly solvent dependent. In butyronitrile, for example, it is ca. 2 microseconds.

The lifetime of the charge separation in **I** is many orders of magnitude longer than that of $P^{+\bullet}-Q^{-\bullet}$ charge separated states under similar conditions. A key reason for this long lifetime is the biomimetic multistep electron transfer which separates the positive and negative charges before they have time to recombine. Once the $C^{+\bullet}-P-Q^{-\bullet}$ state has formed, direct electron donation from the quinone radical anion to the carotenoid radical cation would be expected to be slow because of the relatively large separation between these moieties. Alternatively, charge recombination via the porphyrin might involve states such as $C^{+\bullet}-P^{-\bullet}-Q^{-\bullet}$. The formation of either of these states from $C^{+\bullet}-P-Q^{-\bullet}$, however, is thermodynamically unfavorable, and should also be slow.

It is interesting to note that although flash excitation of **I** in butyronitrile yields a $C^{+\bullet} - P - Q^{-\bullet}$ state with a 2 microsecond lifetime as mentioned above, a similar experiment at 77°K in a frozen butyronitrile glass yields no detectable charge separation. Fluorescence lifetime studies reveal that this is due to the failure of step 2 in Scheme 1 to occur. That is, the initial $C - P^{+\bullet} - Q^{-\bullet}$ state does not form.

A long lifetime for a photogenerated charge separated state is of interest only if that state preserves a significant fraction of the excitation energy as chemical potential energy. Electrochemical measurements of the redox potentials of **I** and related model compounds reveal that of the 1.9 eV inherent in the porphyrin first excited singlet state, ca. 1.4 V is preserved in the $C - P^{+\bullet} - Q^{-\bullet}$ state, and ca. 1.1 V remains in the $C^{+\bullet} - P - Q^{-\bullet}$ state. Thus, even the final $C^{+\bullet} - P - Q^{-\bullet}$ state is a highly energetic species with a strong thermodynamic driving force for recombination.

The microsecond lifetime of the $C^{+\bullet} - P - Q^{-\bullet}$ species suggests that it should be possible to harvest this chemical potential energy via reaction with extrinsic electron donors and acceptors. In fact, addition of 1,4-diazabicyclo[2.2.2]octane (Dabco) to a solution of **I** in butyronitrile quenched the carotenoid cation decay. A reasonable explanation for this shortening of the cation lifetime is the following reaction:

$$C^{+\bullet} - P - Q^{-\bullet} + \text{Dabco} \longrightarrow C - P - Q^{-\bullet} + \text{Dabco}^{+\bullet}$$

where in Dabco donates an electron to $C^{+\bullet} - P - Q^{-\bullet}$. Such experiments demonstrate the principle of harvesting photogenerated chemical potential energy and ultimately regenerating the photocatalyst. However, actual stabilization of chemical potential on the laboratory time scale is best achieved by separating the high energy chemical species by a phase boundary of some type. The biomimetic solution to this problem is to carry out photodriven charge separation across a phospholipid bilayer membrane.

Investigations of such charge separation have been begun using the apparatus diagrammed in Figure 3. Two electrochemical cells separated by a Teflon divider are electrically connected via Ag/AgCl electrodes and a current to voltage converter. Each cell contains buffers and electrolyte. In addition, the left hand cell contains ferricyanide ion, whereas the right hand cell contains ascorbic acid. The two cells are joined by a ca. 1mm hole drilled in the Teflon divider. Application of a phospholipid solution containing a triad molecule to this hole results in the formation of a phospholipid bilayer membrane (BLM) across the hole. Triad **I** is very insoluble in phospholipids. Therefore, a new, more amphipathic triad, **IV**, was prepared which was much more soluble. When black lipid membranes containing this triad were prepared in the apparatus described above, no current flow could be detected because such membranes are good insulators. However, upon illumination of the membrane with 600 nm light, a photocurrent was detected. The photocurrent appeared with the response time of the apparatus (milliseconds) and persisted until the light was turned off (typically many seconds).

IV

An explanation for this observation which is consistent with the experimental facts is as follows. The triad molecule will tend to reside in the membrane with the carotenoid polyene immersed in and nearly spanning the organic interior of the membrane. The amphipathic porphyrin, and therefore the attached quinone, will lie near the membrane surface.

Excitation of the porphyrin generates the first excited singlet state, which in turn produces the $C^{+\bullet} - P - Q^{-\bullet}$ charge separated state as in single phase solution. (Flash excitation of liposomes containing **IV** have revealed the transient absorption of the carotenoid radical cation with a lifetime on the ms time scale.) The

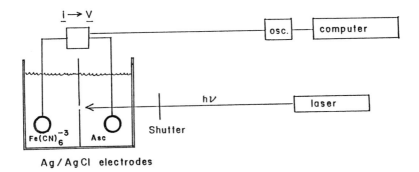

Figure 3. Photoelectrochemical apparatus for studying BLM photoconduction.

quinone radical anion of the left side of the membrane transfers an electron to the ferricyanide ion to regenerate the neutral quinone, and the carotenoid radical cation accepts an electron from the ascorbic acid to regenerate the neutral carotenoid. The result of these electron transfer steps is the net transfer of an electron from the right side of the membrane to the left side, and regeneration of the photocatalyst. An electron flows through the external circuit connecting the cells in order to maintain electrical neutrality, and is detected as a photocurrent. This experiment at least begins to mimic the transmembrane charge separation characteristic of natural photosynthesis. However, it is important to note that although the experiment demonstrates triad mediated photoconduction across a phospholipid bilayer, it does not represent photogeneration of chemical or electrical potential energy. The transfer of an electron form ascorbate to ferricyanide ion is an exergonic process.

In conclusion, studies of electron and energy transfer in photosynthetic model systems demonstrate that one can go rather far in the mimicry of natural photosynthesis without involving proteins at all. However, one would expect that the basic structural and environmental requirements for energy and electron transfer which are demonstrated by these crude models are satisfied with much more finesse by the native proteins.

This work was supported by the National Science Foundation under Grants

CHE-8209348 and INT-8212583, CNRS Grant 3064, and the North Atlantic Treaty Organization under Grant RG.083.82.

References

[1] Dirks, G., Moore, A.L., Moore, T.A. and Gust, D., *Photochem. Photobiol.* **1980**, *32*, 277-280.

[2] Moore, A.L., Dirks, G., Gust, D. and Moore, T.A., *Photochem. Photobiol.* **1980**, *32*, 691-695.

[3] Bensasson, R.V., Land, E.J., Moore, A.L., Crouch, R.L., Dirks, G., Moore, T.A. and Gust, D., *Nature* **1981**, *290*, 329-332.

[4] Gust, D., Nemeth, G.A., Lehman, W.R., Joy, A.M., Moore, A.L., Bensasson, R.V., Moore, T.A. and Gust, D., *Photochem. Photobiol.* **1982**, *216*, 982-984.

[5] Liddell, P.A., Nemeth, G.A., Lehman, W.R., Joy, A.M., Moore, A.L., Bensasson, R.V., Moore, T.A. and Gust, D., *Photochem. Photobiol.* **1982**, *36*, 641-645.

[6] Chachaty, C., Gust, D., Moore, T.A., Nemeth, G.A., Liddell, P.A. and Moore, A.L., *Organic Magn. Reson.* **1983**, *22*, 39-46.

[7] Moore, T.A., Gust, D., Mathis, P.,Mialocq, J.C., Bensasson, R.V., Land, E.J., Doizi, D., Liddelll, P.A., Lehman, W.R., Nemeth, G.A., and Moore, A.L., *Nature* **1984**, *307*, 630-632.

[8] Gust, D. and Moore, T.A., *Photochem.*, in press

[9] Moore, T.A., Mathis, P., Gust, D., Moore, A.L., Liddell, P.A., Nemeth, G.A., Lehman, W.R., Bensasson, R.V., Land, E.J., and Chachaty, C., in *Advances in Photosynthesis Research*, ed. Sybesma, C., Nijhoff/Junk, The Hague, 1984, pp. 729-732.

Electron Transfer Reactions in Bacterial Photosynthesis: Charge Recombination Kinetics as a Structure Probe

G. FEHER [a], M. OKAMURA [a], and D. KLEINFELD [b]

Introduction

During this conference the question of relevance to biology of the systems that were being investigated came up on several occasions. We are in the happy position of not having to defend our system on that score. Photosynthesis is essential to life; it is the source of energy of the entire living world.

Since this is the first talk on photosynthesis and reaction centers at this meeting, we shall start with a brief introduction to the subject. Photosynthesis deals with the conversion of light into chemical energy that is used by the organism to produce energy-rich compounds. The primary process of photosynthesis involves a charge separation, i.e., the formation of oxidized and reduced molecules. In photosynthetic bacteria this process occurs in a protein pigment complex called the reaction center (RC). The RC is composed of three polypeptide subunits called L, M, and H and a number of co-factors associated with the electron transfer chain. These are four bacteriochlorophylls (BChl), two bacteriopheophytines (i.e., a BChl without the central Mg), two ubiquinones (UQ-10) and one high-spin non-heme iron (Fe^{2+}) (for a review, see ref. 1).

Light induces a charge separation with an electron leaving the donor D, a specialized bacteriochlorophyll dimer, and passing via an intermediate acceptor, I, to the primary and secondary quinone acceptors, Q_A and Q_B, respectively (see Fig. 1); (for a review, see ref. 2). The remarkable thing about photosynthesis is that the quantum yield is close to unity. The high yield occurs because the forward reactions are 10^2–10^3 faster than the (energetically wasteful) charge recombinations reactions (see Fig. 1). We shall be hearing a great deal during this meeting about electron transfer reactions. We shall not discuss in detail the underlying theory here, but will mainly use one of the conclusions that seems to have been universally accepted by both theorists and experimentalists; namely, that *the kinetics of the electron transfer reactions are extremely sensitive to the spatial configuration of the charge separated species*. This enables us to use the kinetics as a

[a] Department of Physics B-019, University of California, San Diego, La Jolla, CA 92093.
[b] AT&T Bell Laboratories, Murray Hill, N.J. 07974.

structural probe. Although each of the transfer reactants can serve as such a probe we shall focus mainly on the recombination kinetics between D^+ and Q_A^- and D^+ and Q_B^-, characterized by τ_{AD} and τ_{BD}, respectively. We shall discuss the following topics:

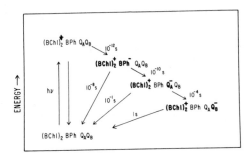

Figure 1 Schematic representation of the electron-transfer reactions in reaction centers from photosynthetic bacteria. After the absorption of a photon, the electron transfers through a series of reactants that are stabilized against charge recombination for progressively longer periods of time. Charged donor-acceptor species are in bold face. Transfer times are given for room temperature and are rounded to the nearest power of 10.

1. Do isolated RCs have the same structure as RCs *in vivo?*

2. The effect of removing the H-subunit on the charge recombination kinetics.

3. Conformational changes associated with the charge separation process.

4. The temperature dependence of the recombination kinetics.

5. The effect of electric fields on the recombination kinetics.

 a) Externally applied fields

 b) Fields due to intrinsic charges.

The first four topics deal with the relative spatial arrangement of the reactants; the fifth topic deals with the electronic level structure of the reactants.

1. Do isolated RCs have the same structure as RCs *in vivo* ?

Historically, the first complex that was called a reaction center was isolated by Reed and Clayton (3) and had a molecular weight of over one million. When we isolated a much smaller reaction center having a molecular weight of $\sim 10^5$, which resembled the "modern reaction center", we were astonished to find great opposition from several quarters when these findings were presented at the Gatlinburg Conference in 1970 (4). Some people just could not believe that such a small unit could perform the marvelous primary process of photosynthesis. They claimed that life had slipped through our fingers during the purification process and that these reaction centers probably bear little resemblance to what happens *in vivo*. Although this attitude has vanished by now, there still remain some lingering questions concerning the extent to which the isolated reaction centers have the same structures as RCs *in vivo*. We have, therefore, resurrected a table from an old piece of work by J. McElroy et al. (5) in which that question was addressed by measuring the charge recombination kinetics described by the scheme:

$$DQ_A \underset{\tau_{AD}}{\overset{h\nu}{\rightleftharpoons}} D^+Q_A^-$$ (1)

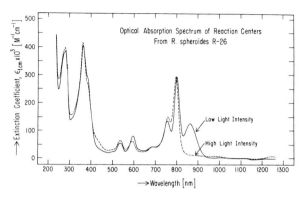

The state $D^+Q_A^-$ was formed by a short pulse of high intensity light. The charge recombination kinetics were monitored optically via changes in the absorption spectrum caused by the presence of D^+ (see Fig. 2). To prevent the electrons from leaving Q_A^-, these experiments were performed at 77 K. The results showed that whole cells, chromatophores and isolated reaction centers have, within experimental error, the same recombination time τ_{AD} * (see Table I). When reaction centers were

Figure 2 *Absorption spectrum of RCs from R. sphaeroides R-26 obtained under conditions of low light intensity (——) and with strong cross illumination (-----). The ordinate was normalized to the extinction coefficient at 802 nm, i.e., ($\epsilon^{802} = 2.88 \times 10^5 M^{-1}cm^{1-}$(6)). From Ref. 4.*

TABLE I. OPTICAL DECAY KINETIC OF WHOLE CELLS, CHROMATO-PHORES, AND REACTION CENTERS OF *R. SPHAEROIDES*, **R-26**

Whole cells and chromatophores were suspended in 50% glycerol, 0.05 M Tris, pH 8.0, to give A_{800nm} of approx. 0.1 (1 mm path). The presence of glycerol did not affect the kinetic behavior. It was used to insure the formation of a transparent glassy matrix of low temperatures. Reaction centers, unless otherwise specified, were suspended in the same buffer to give A_{800nm} of approx. 0.2 (1 mm path) at room temperature. Reaction centers which were exposed to 6 M urea for 4 h at 20°C in the dark had an absorbance A_{800nm} of approx. 1.0 (1 mm path). The decay of the optical change at 795 nm was monitored at a sample temperature of 80°K. The wavelength of the actinic illumination was 900 nm. The measuring beam intensity was approx. 5 $\mu W/cm^2$.

Preparation	1/e decay time	Treatment
Whole cells	30 ± 3 ms	No detergent
Chromatophores	32 ± 3 ms	No detergent
Reaction centers	29 ± 2ms	0.1% lauryl dimethyl-amineoxide
Reaction centers	52%, 28 ms, 48%, 93 ms	6 M urea, 0.1% lauryl dimethylamineoxide, $t = 4$ h
Reaction centers	60%, 260 ms, 40%, 1.8 s	0.1% sodium dodecyl-sulfate, 0.02% lauryl dimethylamineoxide

From: J. McElroy, D. Mauzerall, G. Feher (1974) Biochim. Biophys. Acta 333, 261-277.

treated with strong reagents (e.g., urea or sodium dodecyl sulfate) the decay kinetics changed significantly, indicating a structural change (see Table I). These experiments show that the structural integrity, at least with respect to the donor-acceptor complex, is preserved in isolated reaction centers.

2. The effect of removing the H-subunit

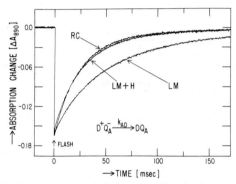

Figure 3 *Kinetics of charge recombination between* Q_A^- *and* D^+ (k_{AD}) *at cryogenic temperatures (77K). The formation and decay of the charge-separated state* $(D^+Q_A^-)$ *was monitored at 890 nm following an actinic flash. From Ref. 7.*

All the prosthetic groups associated with the electron transfer processes are bound to the L and M subunits. The question concerning the role of the H subunit, therefore, naturally arises. R. Debus in our laboratory was able to isolate the LM-pigment-complex and the H-subunit and to subsequently reconstitute LM and H to reform RCs (7). When the charge recombination time, τ_{AD}, was measured at low temperatures, as discussed before, the results shown in Fig. 3 were obtained. The recombination times in LM were about a factor of two slower than those found in RCs. The recombination time between Q_A^- and D^+ is believed to be critically sensitive to changes in the distance between D and Q_A (5,8,9), a point that we shall discuss in more detail in the next section. Since τ_{AD} changes only by a factor of ~ 2 (see Fig. 3), the relative configuration of D and Q_A is affected only to a relatively small extent by the removal of H. Thus, H does not play a major role in this charge recombination step. Incidently, note that the kinetics in the reconstituted LMH are practically identical to those in RCs. This shows that the change in the LM complex was not due to an irreversible denaturing effect accompanying the isolation procedure.

The charge recombination kinetics at room temperature in RCs containing *two quinones* exhibited a more dramatic change when H was removed. The kinetic properties of this system are described by:

$$DQ_AQ_B \underset{\tau_{AD}}{\overset{h\nu}{\rightleftharpoons}} D^+Q_A^-Q_B \underset{\tau_{BA}}{\overset{\tau_{AB}}{\rightleftharpoons}} D^+Q_AQ_B^- \tag{2}$$

$$\underset{\tau_{BD}}{\longleftarrow}$$

In RCs or reconstituted LMH complexes the recombination time τ_{BD} was $\sim 1s$, indicative of the characteristic time of recombination between Q_B^- and D^+ (see Fig. 4). In LM, on the other hand, the recombination time was ten times shorter, as expected from the recombination between Q_A^- and D^+ (see Fig. 4). This suggests that the electron transfer from Q_A^- to Q_B was impaired. Indeed, independent experiments have shown that the electron transfer time, τ_{AB}, is about three orders of magnitude longer in LM than in RCs (7). This, of course, is a large effect that would likely be detrimental to the physiological well-being of the bacterium. From a structural point of view, this means that the distances (or angles) between Q_A and Q_B have been significantly changed upon removal of H. Another effect shown in Fig. 4 is the lack of recovery of the absorbance change in LM. This presumably is due to a loss of an electron to exogenous acceptors and may again be a consequence of the opening up of the structure.

3. Conformational changes associated with the charge separation process

There exists some evidence for bulk structural changes in the charge separated state. It comes from the calorimetric study of Arata and Parson (10,11) who found that during charge separation the volume of the RC-solvent system decreased. In a different set of experiments, Noks et al. (12) found that incubation of chromatophores with the cross-linker glutaraldehyde affected the electron transfer kinetics only if incubation was performed in the presence of light. We addressed the question of a conformational change during charge separation by analyzing the charge recombination kinetics in samples prepared under different conditions (13).

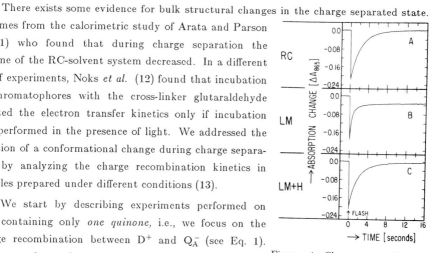

Figure 4 *Charge recombination between Q_A^- or Q_B^- and D^+ in RCs, LM, and reconstituted LMH at 4°C. The lack of complete recovery in the LM subunit is attributed to a loss of the electron from the quinone acceptors. From Ref. 7.*

We start by describing experiments performed on RCs containing only *one quinone*, i.e., we focus on the charge recombination between D^+ and Q_A^- (see Eq. 1). Two sets of samples were prepared. In one, RCs were cooled to cryogenic temperature under illumination, i.e., in the charge separated state. Thus any possible light-induced structural changes may be trapped when RC conformations are immobilized at low temperatures. The second sample was cooled to cryogenic temperature in the dark. The results of the kinetics of charge recombinations in the two samples is shown in Fig. 5a. There is a significant difference in the recombination time τ_{AD} between the two samples, i.e.,

$$\tau_{AD}^{light} = 120 \text{ ms}, \qquad \tau_{AD}^{dark} = 25 \text{ ms}.\ddagger$$

---

‡ The value of k_{AD} for UQ differs in recent RC preparations by $\sim$ 15% from that quoted earlier (see Table I). The origin of this discrepancy is not understood; it may be due to a changed binding site.

In Fig. 5b the change of absorbence is plotted logarithmically; a single exponential recombination process should give a straight line on this plot. We see that RCs cooled under illumination have not only a longer recombination time, but their kinetics are much more non-exponential than those for RCs cooled in the dark. For comparison we show also the recombination kinetics at room temperature; in this case a good exponential recovery is observed.

Qualitatively, we attribute the non-exponential behavior to a distribution of structural states. Evidence that such distributions may exist in proteins comes from the detailed work of Austin, Frauenfelder and collaborators (14,15), and Woodbury and Parson (16). Furthermore, we see that RCs cooled under illumination deviate much more from exponentiality, i.e. they will have a broader distribution of conformational states. Their recombination time is also longer, indicating that the average distance between D^+ and Q_A^- has increased during illumination.

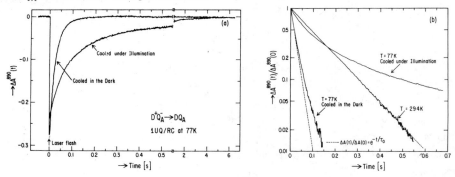

Figure 5 *Electron Donor Recovery Kinetics for 1UQ/RC following a laser flash. (a) Kinetics at 77K for RCs cooled in the dark and under illumination (b) semilog plot of the kinetics shown in part (a) together with kinetics obtained at room temperature. Dashed lines represent fits to an exponential function, with $\tau_0 = 22$ ms for RCs cooled to 77 K in the dark and $\tau_0 = 132$ ms for RCs at 294 K. Note the large deviation from an exponential of the kinetics in RCs cooled under illumination. From Ref. 13.*

To treat this problem quantitatively, we parameterize the recombination kinetics in terms of the D^+–Q_A^- electron transfer distance, r_{AD}. If all the donor acceptor pairs had identical separation distances, the observed absorption change would be given by a single exponential

$$\Delta A(t) \,/\, \Delta A(0) = e^{-t/\tau(r_{AD})} \tag{3}$$

where $\tau(r_{AD})$ is the characteristic recombination time, given by:

$$\tau(r_{AD}) = \tau_0 e^{-r_{AD}/r_0} \tag{4}$$

the value of $r_0 \simeq 1\text{Å}$ (17,18). If r_{AD} varies between different donor acceptor pairs, Eq. 3 is no longer valid and $\Delta A(t)$ is described by a normalized distribution function of distances $D(r)$, i.e.:

$$\Delta A(t) \, / \, \Delta A(0) = \int\limits_{0}^{\infty} D(r)e^{-t/\tau(r)}dr \qquad (5)$$

To solve for the distribution function $D(r)$, we fit the data with an analytic function given by:

$$\Delta A(t) \, / \, \Delta A(0) = [1 + t \, / \, (n\tau_0)]^{-n} \qquad (6)$$

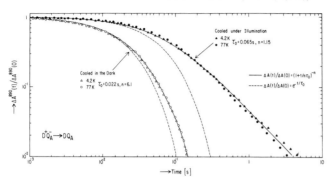

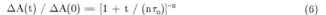

Figure 6 *Log-log plot of the donor recovery kinetics at 4.2 and 77 K in 1UQ/RC samples cooled in the dark and cooled under illumination. Dashed lines represent fits of the initial slopes of the data to an exponential; solid lines are fits to a power law (Eq. 6). The values of parameters τ_0 and n are given in the figure. Note that τ_0 is the same for both functions. From Ref. 13.*

This function is the same as used by Austin *et al* (14); it fits our data well as shown in Fig. 6. Equating this to the expression given by Eq. 5 one can solve for the distribution $D(r)$, i.e.,

$$r_0 D(r) = \frac{1}{\Gamma(n)} \left[n \frac{\tau_0}{\tau(r)} \right]^{n} e^{-n[\tau_0/\tau(r)]} \qquad (7)$$

where $\Gamma(n)$ is the gamma function.

The result of the calculation of the distribution function is shown in Fig. 7. It bears out our qualitative discussion given before, i.e.:

1.) the average electron transfer distance in RCs cooled under illumination is larger than the average distance in RCs cooled in the dark,

2.) the width of the distribution in RCs cooled under illumination is two and a half times larger than in RCs cooled in the dark.

The shift and width of the distribution is of the order of 1Å, which is similar to the root mean square displacements determined from crystallographic studies on other proteins (19,20) and from model calculations (see, e.g., Refs. 21 and 22). The recombination kinetics were essentially temperature independent between 4.2 and 77 K (see Fig. 6). This indicates that the distribution remained constant with temperature, on the time scale, τ_{AD}, of the measurement.

It is interesting to speculate whether these light-induced changes have a physiological function analogous to those produced by allosteric changes in other proteins (for a review, see Ref. 23). Perhaps the structural changes accompanying the charge separation process act to inhibit the wasteful recombination reactions by stabilizing the charge separated states. Such stabilization processes have also been discussed by Warshel (24), and Woodbury and Parson (16).

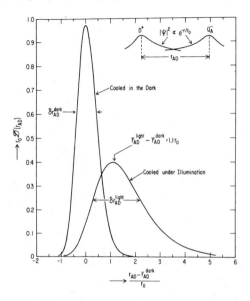

Figure 7 *Calculated distributions (see Eq. 7) of the electron transfer between* D^+ *and* Q_A^- *in 1UQ/RC samples cooled in the dark and under illumination. This distribution describes the nonexponential decay kinetics of* $D^+Q_A^-$, *shown in Figs. 5 and 6. The experimental parameters, n and τ_0, given in Fig. 6, were used together with Eqs. 4 and 7 to calculate the distributions. Insert shows the exponential decrease of the wave functions that leads to Eq. 4. Note that RCs cooled under illumination have a larger average electron-transfer distance as well as a larger spread in distances than RCs cooled in the dark. From Ref. 13.*

We now turn to the more complicated question of the recombination kinetics from the *secondary quinone*, Q_B, described by the scheme given by Eq. 2. The electron transfer time τ_{AB} is approximately 10^{-4} sec at room temperature but becomes unobservably long at 77 K. If we want to study, therefore, the recombination kinetics of $D^+Q_AQ_B^-$ at low temperature this state has to be trapped at 77 K by cooling RCs under illumination. This was done and the result of the recombination kinetics are shown Fig. 8. The solid line represents a theoretical fit to the same function as was used for the one quinone case (see Eq. 6). However, in this case one observes two features that are distinctly different from those observed in RCs containing one quinone. The recombination time is highly temperature dependent and the spread in characteristic times is very large. (Note the logarithmic scale of the abscissa). If we extrapolate the 18 K data to longer times the recombination is less than half complete even after 10^7 s (1 year!).

The observed temperature dependence of the two quinone systems can be explained by the model of Agmon and Hopfield (25). Due to the dynamics of protein motion the RC passes through a number of structural states. The most favorable states for rapid recombination are those for which the distances between Q_B^- and D^+ are small. As the temperature is raised the probability that transitions to these favorable states occur is increased, thereby reducing the recombination time τ_{BD}. For RCs with one quinone the recombination time τ_{AD} is many orders of magnitude shorter than τ_{BD}. Consequently there is no opportunity to sample the different conformational states within the time τ_{AD}. This gives rise to an effective static distribution of distances between Q_A^- and D^+, resulting in temperature independent kinetics as observed in RCs with one quinone.

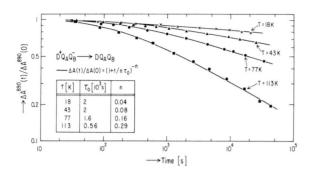

Figure 8 *Log-log plot of the donor recovery kinetics at different temperatures in 2UQ/RC samples cooled under illumination. Data were normalized to the maximum absorption change, $\Delta A^{890}(0)$, found by extrapolating the measured absorption changes back to zero time (more data were acquired at short times than are shown). The maximum absorption level [i.e., $A^{890}(\infty)$], which served as the base line for the absorption changes, was determined by warming and recooling the sample in the dark. Typically, $\Delta A^{890}(0)$ was 80% of $A^{890}(\infty)$. The parameters τ_o and n were found from fitting the data to Eq. 6 and are tabulated in the insert. From Ref. 13.*

The above model also explains the exponential behavior found for τ_{AD} at room temperature. If the transitions between structural states occur in a time that is much shorter than τ_{AD}, the individual states will not be expressed and a single, average, τ_{AD} will be observed. This situation is analogous to motional narrowing in magnetic resonance.

An interesting finding that we will not discuss in detail here is that the electron transfer time from Q_A^- to Q_B is at least 8 orders of magnitude shorter in RCs that have been illuminated while being cooled as compared to those cooled in the dark (13). It is difficult to see how such a large change can be produced by a light-induced conformational change. It is more likely that a proton that associates with Q_B^- at room temperature (26) remains trapped in the vicinity of Q_B^- after the RCs are cooled. This proton cannot associate with RCs at low temperatures. Thus, RCs cooled in the dark will remain unprotonated upon illumination.

4. The temperature dependence of the recombination kinetics

We next discuss the temperature dependence of the charge recombination rate $D^+Q_A^- \rightarrow DQ_A$ (see Eq.1). We shall inquire whether the experimental results can be explained by present theories of electron transfer or whether major contributions are due to temperature dependent structural changes, e.g., thermal expansion.

Measurements were made using RCs containing one quinone, i.e., 1UQ/RC. All samples were cooled to cryogenic temperature in the dark; the value of k_{AD} was stable with time at each temperature and was completely reversible as the temperature was cycled.[*] To insure the presence of an optically transparent sample at all temperatures, RCs were incorporated into a thin film of polyvinyl alcohol (PVA) [27].

The temperature dependence of k_{AD} is shown in Fig. 9a. The recombination rate was essentially temperature independent at low temperature, as discussed earlier (see Fig.

[*] For RCs cooled under illumination, a different behavior of the kinetics was observed. Above ~ 90 K, k_{AD}^{light} changed with time heading, toward the value of k_{AD}^{dark}. Apparently, the structural changes that had been trapped during illumination were annealing out at T > 90 K.

6). As the temperature was increased from $\sim$ 90 K to 300 K, k_{AD} decreased by a factor of $\sim$ 6 † (see also Ref. 28-31). The results are similar to the temperature dependence of k_{AD} observed by Loach *et al.* [30] and Mar *et al.* [31] with RCs from *R. rubrum*. These

temperature dependences are rather unusual; the rate k_{AD} *decreases* with *increasing* temperature. This is in contrast to the usual behaviour of thermally activated processes.

Can we understand the observed temperature dependence of k_{AD} in terms of the electron transfer theories of Hopfield [8] and Jortner [9,17]? In these theories, the electronic transition $D^+Q_A^- \rightarrow DQ_A$ is coupled to a vibrational mode(s) in the protein. The decrease in electronic energy during the charge recombination is compensated for by an equal increase in the energy of the vibrational mode coupled to the reaction. Thus, energy is conserved during the transition.

The theoretically predicted charge recombination rate can be expressed in a compact form under the approximation that both the electron donor and the acceptor are coupled to the *same*, single, vibrational mode. The recombination rate, valid for any temperature, T, is given by [9,17]:

Figure 9 *a.) Temperature dependence of k_{AD} for 1UQ/RCs embedded in a 0.1mm polyvinyl alcohol film (A^{800}=1.2). Dashed line represents Jortner's theory (Eq. 8) with ΔE_{redox}=E_{nuc} and T_o=500K. Solid line represents the best fit of the expansion model (Eq.13,14) plus Jortner's theory to the experimental data. b.) Temperature dependence of the distance between D^+ and Q_A^-, r_{AD}; computed from the thermal expansion model (Eq.14).*

$$k_{AD} = \frac{(2\pi)^2}{h}|M|^2 \frac{1}{k_B T_o} \left[\frac{\upsilon+1}{\upsilon}\right]^{p/2} e^{-s(2\upsilon+1)} I_p \left[2s\sqrt{\upsilon(\upsilon+1)}\right] \qquad (8a)$$

where

$$s = \frac{E_{nuc}}{k_B T_o} \qquad p = \frac{\Delta E_{redox}}{k_B T_o} \qquad \upsilon = \frac{1}{e^{T_o/T}-1} \qquad (8b)$$

The overlap integral M connects the electronic states of $D^+Q_A^-$ and DQ_A, T_o is the characteristic temperature of the vibrational mode (i.e., $k_B T_o = h\omega$), E_{nuc} is the energy required to rearrange the nuclear positions concomitant with the electron transfer, ΔE_{redox} is the

† The values for $\tilde{k}_{AD}$ obtained with RCs in PVA are $\sim$ 50% larger than those obtained with RCs in glycerol. This difference may be caused by the 1 Molar salt concentration in the dried PVA film.

difference in free energy between the D^+/D and Q_A/Q_A^- redox couples (i.e. the energy difference between $D^+Q_A^-$ and the ground state), $I_p(x)$ refers to the modified Bessel function of order p, k_B is Boltzman's constant and h is Plank's constant.

The temperature dependence of the predicted rate (Eqs. 8a and 8b) simplifies considerably in the limit of either high or low temperature. The recombinaton rate is expected to follow an activated temperature dependence when there is sufficient thermal energy available to excite the vibrational mode coupled to the reaction $D^+Q_A^- \rightarrow DQ_A$. In this limit, i.e., for $T >> T_o$, the rate is given by [8.9,17]:

$$\text{For } T>>T_o; \quad k_{AD} = \frac{(2\pi)^2}{h} |M|^2 \frac{1}{\sqrt{4\pi E_{nuc}k_BT}} e^{\frac{-(\Delta E_{redox}-E_{nuc})^2}{4E_{nuc}k_BT}} \tag{9}$$

At low temperature, i.e., for $T << T_o$, the vibrational mode coupled to the reaction remains in the ground state and thus the recombination rate is independent of temperature. In this limit, the rate is given by a Poisson distribution for the rearrangement energy, i.e [9,17]:

$$\text{For } T<<T_o; \quad k_{AD} = \frac{(2\pi)^2}{h} |M|^2 \frac{1}{k_BT_o} s^p \frac{e^{-s}}{p!} \tag{10}$$

Note that for large p, the term p! in Eq.10 makes the low temperature limit of k_{AD} very sensitive to changes in the redox energy difference, E_{redox}, between $D^+Q_A^-$ and DQ_A. This limit applies to RCs, where $E_{redox} \sim 500$ meV [2] and k_BT_o for proteins typically lies in the range 10 - 100 meV ($\sim$100 - 1000K).

The theoretical model we discussed predicts, in general, for temperatures near or above T_o an *increase* in the recombination rate with *increasing* temperature. This is in contradiction to the experimentally observed temperature dependence of k_{AD} (see Fig. 9a). A hypothesis often suggested (see, e.g., Refs.8,9,17) to circumvent this inconsistency between experiment and theory is that the redox energy difference between $D^+Q_A^-$ and DQ_A equals the nuclear rearrangement energy, i.e., $\Delta E_{redox} = E_{nuc}$. For this special condition, the theory predicts that k_{AD} is constant for $T<<T_o$ (see Eq. 10); and that k_{AD} *decreases* with *increasing* temperature for temperatures near of above T_o. For $T >> T_o$ and p! $>>1$, one obtains from Eqs.9 and 10 (with p! approximated by $\sqrt{2\pi p}\, p^p e^{-p}$) for the temperature dependence of k_{AD}:

$$\text{For } T>>T_o; \, p!>>1: \quad k_{AD} = k_{AD}(0)\sqrt{\frac{T_o}{2T}} \tag{11}$$

where $k_{AD}(0)$ is the low temperature limit of the recombination rate (see Eq. 10).

Since the observed recombination rate (see Fig. 9a) does not correspond to the high temperature limit we tried to fit the data with the general expression given by Eq. 8. We took the room temperature value of $\Delta E_{redox} = 500$ meV[2] and equated it to E_{nuc} (i.e., p

= s), with $k_{AD}(0) = 58$ s^{-1}. To estimate T_o, we plotted k_{AD} *versus* T/T_o and found that T_o corresponds to ~ 5 times the temperature at which k_{AD} changes from a temperature independent to a temperature dependent value. We see from Fig. 9a that this transition occurs at ~ 100 K, i.e., the characteristic temperature is

$$T_o \sim 500 \text{ K} \tag{12}$$

Using the above values of T_o, ΔE_{redox}, E_{nuc} and $k_{AD}(0)$, the predicted temperature dependence of k_{AD} (Eq. 8) disagrees with the observed behaviour (Fig. 9a). Changing the value of E_{nuc} away from the value of E_{redox} increases the disagreement even further. For redox energies outside the range of 400 meV $> E_{nuc} > 700$ meV (with $E_{redox} = 500$ meV), theory predicts a change in the *sign* of the temperature dependence, in accord with a thermally activated process. To reconcile the disagreement between experiment and theory, Sarai (32) and Kakitani *et al.* (33,34) modified the theories of Hopfield (8) and Jortner (9,17) by including a multiplicity of vibrational modes, as opposed to a single mode[‡]. An alternate mechanism to account for the observed temperature dependence of k_{AD} is the thermal expansion of the protein (29,35,36). Thermal expansion will cause the donor-acceptor distance, r_{AD}, to increase with increasing temperature. This in turn will decrease the value of the overlap integral M, thereby reducing k_{AD} (see Eq. 4). To estimate the magnitude of this effect we write

$$|M(T)|^2 = |M(0)|^2 \, e^{-|r_{AD}(T)-r_{AD}(0)|/r_o} \tag{13}$$

where r_o is the same scaling factor as used in Eq. 4. The change in lattice spacing for a simple solid with an anharmonic interatomic potential (see e.g. ref.37) is given by

$$\frac{r_{AD}(T)-r_{AD}(0)}{r_o} = \left[\frac{\gamma}{r_o}\right] \frac{T_o}{2} \left[\coth \frac{T_o}{2T} - 1\right] \tag{14}$$

where γ/r_o is an adjustable parameter related to the linear expansion coefficient, β, by $\beta = \gamma/r_{AD}(0)$. Assuming that the characteristic temperatures associated with the vibrational mode coupled to the thermal expansion is the same as that coupled to the electron transfer (i.e., $T_o = 500$ K) we fitted the thermal expansion model (Eqs. 13, 14) to the observed data with a value of

$$\gamma/r_o = 1.4 \times 10^{-2} \text{ K}^{-1} \tag{15}$$

This value corresponds to a thermal expansion coefficient that is an order of magnitude larger than that determined for a protein (38). It should be noted, however, that the relevant number is not the *average* expansion coefficient but the change in a *particular* distance, namely r_{AD} with temperature. This can be an order of magnitude larger that the change given by the average expansion coefficient (38).

[‡] Now that the three dimensional structure of RCs is being determined (Deisenhofer *et al.* (1984) *J. Mol. Biol.* **180**, 385 and Michel *et al.*, these proceedings), there is hope that one will be able to correlate the vibrational mode(s) with specific bonds in the vicinity of the primary reactants.

The temperature dependence of r_{AD} found from the expansion model is shown in Fig. 9b. The change in r_{AD} between low temperature and room temperature is $\Delta r_{AD} = 1.7 r_o \sim 1-2$ Å. This is the same as found for some atomic positions in a protein (38). Thus, the expansion mechanism seems to provide a possible mechanism for the temperature dependence of k_{AD}, although it certainly does not constitute a proof.

Before leaving this topic let us briefly discuss the assumption that $\Delta E_{redox} = E_{nuc}$. At first glance it seems odd that nature should have picked this equality since it maximizes the rate (k_{AD}) of a physiologically undesirable reaction. However, this constraint may be a consequence of maximizing the transfer rate for the forward reaction $D^+I^-Q_A \rightarrow D^+IQ_A^-$, where I is the intermediate acceptor (2). The redox energy difference between the I/I^- and Q_A/Q_A^- redox couples is approximately the same as that between the Q_A/Q_A^- and D^+/D couples (2).

The temperature dependence of k_{AD} in *R. rubrum* (30,31) was found to be similar to that observed in *R. sphaeroides* (Fig. 9a). However, Mar *et al.* (31) found a much weaker temperature dependence of k_{AD} in RCs from *Ectothiorhodospira* sp. The simplest explanation of this result is that in this bacterial species $\Delta E_{redox} \neq E_{nuc}$, although the alternate explanation that the temperature dependence of Δr_{AD} has been reduced cannot be excluded.

Can we test experimentally the applicability of the electron transfer theories to RCs, and in particular, whether our assumption $\Delta E_{redox} = E_{nuc}$ is justified? Eq. 8 predicts a parabolic-like dependence of k_{AD} verus ΔE_{redox} with k_{AD} peaking at $\Delta E_{redox} = E_{nuc}$. Thus the most direct test would be to vary ΔE_{redox} and to establish whether k_{AD} exhibits the expected parabolic dependence. ΔE_{redox} was changed by substituting quinones with different redox potentials for the native ubiquinone (39).

The low temperature (77K) values of k_{AD} are plotted together with the theortical curve (Eq.10) in Fig.10 with the assumption that $\Delta E_{redox}(UQ) = E_{nuc}$. The redox potentials of the quinones were taken from ref. (40); the accepted value ΔE_{redox} for UQ is 520 meV (11). The general parabolic feature of the theory are seen to be borne out by the experimental data. However, it should be kept in mind that the redox potentials used were obtained for quinones in dimethylformamide at *room* temperature and are likely to deviate from the values found in situ (41) at low temperatures. Furthermore, substitution of quinones may change other parameters (e.g. r_{AD}) besides ΔE_{redox} that affect k_{AD}. Consequently a quantitative agreement of the experimental results with theory cannot be expected at this point; the rather good agreement shown in Fig. 10 seems to us better than one has the right to expect.

An extensive and systematic set of substitution experiments have been performed by

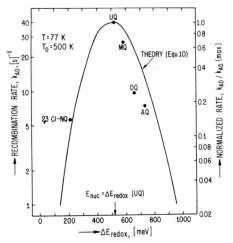

Figure 10 *The low temperature recombination rate*, k_{AD}, *as a function of* ΔE_{redox}. *Solid line represents Jortner's theory (Eq.10) with* $T_o = 500K$ *and* $E_{nuc} = \Delta E_{redox} = 520$ *meV. Dots represent experimental points obtained with RCs in which the native ubiquinone (UQ) was substituted with different quinones. The values of* k_{AD} *for menadione (MQ), duroquinone (DQ) and anthraquinone (AQ) were taken form ref. (39), the value for 2,3, dichloronaphtoquinone (2, 3Cl-NaQ) represents a new measurement. The values of the redox potentials were taken from ref. (40).*

Gunner *et al.* (42). They reported that at room temperature the value of k_{AD} was essentially constant over a large range ($\sim$ 0.7 eV) of redox potentials. These results seemed to be in disagreement with electron transfer theories as was pointed out during this conference. However, more recently, these authors concluded (43) that their room temperature measurements on the halogenated benzoquinones did not represent a direct recombination process but a transfer of electrons via excess quinones in solutions. Measurements of k_{AD} at low temperatures would eliminate this problem and would provide additional important data to compare with theory. An alternate way of changing ΔE_{redox} is to apply an external electric field across the reaction center as discussed next.

5. The effect of electric fields on the recombination kinetics

Before discussing the effect of an electric field on the direct recombination rate, k_{AD}, we shall consider the case in which the electric field changes the observed recombination via an indirect pathway. Thus, we shall discuss now how the kinetics can be used to probe *electronic energy levels* rather than *conformational changes*.

Fig. 11 shows the electron transfer reactions that we will be concerned with (44) The state $D^+IQ_A^-$ can decay via a direct or indirect pathway, as indicated. The observed decay rate of this state, k_{obs}, is in general a combination of the direct and indirect pathways. Assuming that the rates k_{AI} and k_{IA} are fast in comparison to k_{ID} and k_{AD}, the states $D^+I^-Q_A$ and $D^+IQ_A^-$ can be considered to be in equilibrium and the observed decay rate is given by:

$$k_{obs} = k_{INDIRECT} + k_{DIRECT} = k_{ID}\alpha + k_{AD}(1-\alpha) \qquad (16)$$

where α is the fraction of RCs in the thermally excited state $D^+I^-Q_A$ (see Fig.10). For $\alpha \ll 1$, the condition that prevails in RCs, Eq. 16 becomes

$$k_{obs} \simeq \alpha k_{ID} + k_{AD} = k_{ID}e^{-\Delta G^o/k_bT} + k_{AD} \qquad (17)$$

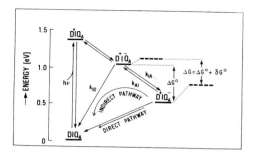

Figure 11 *Simplified energy level scheme showing electron transfer (arrows) in reaction centers of R. sphaeroides. The state $D^+IQ_A^-$ can decay either via the direct pathway (with rate k_{AD}) or via the intermediate state $D^+I^-Q_A$, depending on the value of the energy difference, ΔG^o (Eq. 17). An electric field changes ΔG^o by δG^o and affects, therefore, the recombination rate via the indirect pathway (Eq. 18). The change in energy levels, is illustrated for a direction of the electric field that reduces ΔG^o. When the field is reversed, the energies of the two states are lowered and ΔG^o is increased. From Ref. 44.*

Which of the two pathways predominates depends critically on the value of the energy difference ΔG^o. Substituting the measured values of k_{ID} and k_{AD} into Eq. 17 one can show that the two pathways will contribute equally, i.e., $k_{DIRECT} = k_{INDIRECT}$, for $\Delta G^o = 400$ meV. The energy gap, ΔG^o, in RCs containing the native ubiquinone (UQ) as the primary acceptor has been determined to be 500 - 600 meV (40,45) while for anthraquinone (AQ) $\Delta G^o = 340$ meV (40,44,46). Since one of these numbers is larger and the other smaller than the critical value of 400 meV, the direct pathway predominates for UQ whereas the indirect pathway predominates for AQ.

a.) Externally applied fields.

Effect on the indirect pathway: We shall first focus on RCs that have anthraquinone as the primary acceptor; in this case the observed (indirect) recombination rate will be given by the first term of Eq. 17, i.e.:

$$k_{obs} = k_{ID}e^{-\Delta G^o/k_bT} \tag{18}$$

Let us now consider the effect of an electric field on the energy levels. The two states $D^+I^-Q_A$ and $D^+IQ_A^-$ will be effected to a different extent since the magnitudes of their dipoles along the electric field are different. This is indicated by the dashed lines in Fig. 11. The energy difference between the two states has been changed by an amount δG^o producing a change in the recombination rate given by

$$k_{obs} = k_{obs}^o e^{-\delta G^o/k_bT} \tag{19}$$

where k_{obs}^o is the recombination rate in the absence of an electric field.

How is an electric field applied across the RCs? A. Gopher in our laboratory incorporated RCs into a lipid bilayer that separates two aqueous compartments (44,47). A voltage was applied across the bilayer and the current produced by the charge recombination following a pulse of light was measured. Fig. 12 shows the results of such an experiment performed on RCs containing AQ. The top panels show the current measured after the light is turned off, i.e. during the charge recombination process. The areas under the curves represent the total transferred charge and are, therefore, equal in all three panels. Consequently, the amplitude increases as the recombination time becomes shorter. In the lower panel the experimental data are plotted logarithmically. The data were fitted with a straight line given by

$$\tau = 1/k_{obs} = 8.5 \times 10^{-3} e^{-V/0.175} \text{ s} \tag{20}$$

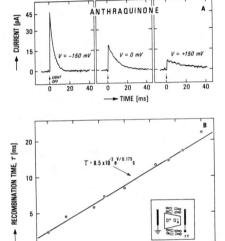

Figure 12 *The effect of an applied electric field on the kinetics of charge recombination in RCs with anthraquinone as the primary acceptor incorporated in a planar bilayer. A.) Time course of current after light pulse is turned off in the absence (V=0) and presence (V=±155mV) of an electric potential. B.) Dependence of the charge recombination rate on the applied voltage. Solid line represents least square fit to the data and obeys the relation $\tau = 8.5 \times 10^{-3} e^{V/0.175}$s. Inset shows the polarity of the voltage with respect to the functionally oriented population of RCs. The polarity of the output signal was inverted by an amplifier. From Ref. 44.*

We see that an e-fold change in the recombination time, τ, results when the applied voltage across the membrane is 175 mV. If I and Q_A were to span the entire membrane one would expect an e-fold change for 25 mV (i.e., $k_b T/q$), where q is the charge of the electron. The fact that we need a seven times larger voltage means that the component of the distance between I and Q_A along the normal of the membrane is only 1/7th of the width of the membrane.

It is interesting to speculate whether the effect of an electric field on the recombination kinetics has any physiological significance. We know that the outside of chromatophores is negative with respect to the inside and that the RCs are oriented in the membrane with the donors pointing towards the inside. Thus the membrane potential created during charge separation *in vivo* decreases ΔG°, thereby decreasing the quantum efficiency at high light intensity. Thus, nature may have build in a negative feedback to prevent detrimental effects at high light intensities.

Effect on the direct pathway: Let us now consider the case of RCs containing UQ. Since now ΔG° is larger than the critical value of 400 meV, the direct pathway predominates. Fig. 13 shows that the recombination kinetics remained uneffected within experimental error ($\pm 5\%$) over the range of applied voltages (± 150meV). How do we reconcile this result with electron transfer theories? If we assume again that $\Delta E_{redox} = E_{nuc}$, then k_{AD} is relatively insensitive to changes in ΔE_{redox} (i.e., $[dk_{AD}/d(\Delta E_{redox}-E_{nuc})]=0$). Under these conditions, Eq.8 predicts that for a 10% change in k_{AD} one needs a change in

$(\Delta E_{redox} - E_{nuc})$ of $\sim$ 200 meV (see Fig.10). Although we have applied 300 meV across

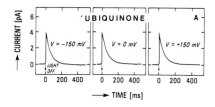

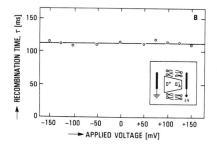

Figure 13 *The effect of an applied electric field on the kinetics of charge recombination in RCs with ubiquinone (UQ-10) as the primary acceptor. Compare the results with those shown in Fig. 12. (Note the difference in time scales and the logarithmic ordinate). From Ref. 44.*

the membrane, the effective voltage across the $D^+Q_A^-$ dipole is reduced by the ratio of the projection of the distance between D^+ and Q_A^- along the field to the thickness of the membrane. This will reduce the effective voltage below the required 200 meV. Thus, the experimental results are compatible with theory if $\Delta E_{redox} - E_{nuc}$ is close to zero.

We plan to repeat these experiments and determine k_{AD} with higher precision (48). We will also attempt to apply higher (pulsed) voltages to the membrane. If $\Delta E_{redox} - E_{nuc}=0$, the change in k_{AD} should be approximately independent of the direction of the electric field (note the near symetry of the theoretical curve in Fig.10). If $\Delta E_{redox} - E_{nuc} \neq 0$, k_{AD} should pass through a maximum for one field direction when the voltage across $D^+Q_A^-$ equals $(\Delta E_{redox} - E_{nuc})/q$.

Z. Popovic *et al.* reported at this conference experiments in which they applied considerably larger fields to RCs embedded in monolayers deposited on a substrate (see also ref. 49). Changes in k_{AD} have been observed although the analysis of the data is complicated by the fact that the RCs are randomly oriented and the decay has to be deconvoluted into a number of exponentials.

b) Fields due to intrinsic charges.

Instead of applying an electric field from an external source, we can also explore the effect of electric fields produced by charges associated with the protein. In particular, we can investigate the protonation of the quinones, a problem that has so far not been solved satisfactorily (see, e.g., Refs. 50-52).

The rationale of the experiment is as follows: The proton produces an electric field, thereby shifting the energy levels of $D^+I^-Q_A$ and $D^+IQ_A^-$ as described previously (see Fig. 11). This produces a change in k_{obs} given by the relation (in analogy to Eq. 19):

$$k_{obs}^{H^+} = k_{obs}^{\circ} e^{-\delta G^{\circ}/k_b T} \tag{21}$$

where $k_{obs}^{H^+}$ and k_{obs}° are the recombination rates in the presence and absence of a proton and δG° is the energy shift caused by the binding of the proton.

As the pH is varied, k_{obs} should change in accordance with the pK value for the protonation, i.e.:

$$k_{obs} = \frac{k_{obs}^{o} + 10^{(pK-pH)}k_{obs}^{H^+}}{1 + 10^{(pK-pH)}} \qquad (22)$$

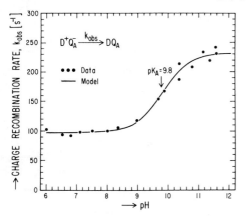

Figure 14 *The pH dependence of the charge recombination rate* k_{obs}*. The solid line (Model) was calculated using Eq. 22 with* $pK_A = 9.8$*,* $k_{obs}^{H^+} = 97s^{-1}$ *and* $k_{obs}^{o} = 230s^{-1}$*. From Ref. 53.*

Fig. 14 shows the pH dependence of k_{obs} in RCs containing AQ (53). The solid line represents a theoretical fit (Eq. 22) with $k_{obs}^{o} = 230$ s^{-1}, $k_{obs}^{H^+} = 97$ s^{-1} and pK $= 9.8$. The value of pK is in agreement with that found from redox titrations (54-56) and electron transfer measurements (26).

The interaction energy δG^{o} obtained from Eq. 21 is 22 meV. From this value one can make a rough estimate of the location of the proton binding site relative to Q_A^-. Assuming that the interaction of the proton with both Q_A^- and I$^-$ is electrostatic in origin, one calculates a distance that is *larger* than ~ 5 Å. Additional experiments are in progress to obtain this distance from ENDOR experiments on Q_A^- in RCs (57,58).

Summary:

We have shown how electron transfer reactions can be used to probe the spatial and electronic structure of photosynthetic reaction centers. Both "static" structural changes (e.g., produced by removal of the H subunit) and "dynamic" changes (e.g., produced by illumination) as well as the effect of an electric field on the energy levels were investigated. Several findings (e.g., the temperature dependence of the recombination kientics and the lack of dependence of an electric field on the recombination kinetics) can be reconciled with present theories of electron transfer reactions by assuming that the difference in redox energy, ΔE_{redox} is approximately equal to the reorganization energy, E_{nuc}. Additional experiments were suggested to investigate the validity of this assumption. The temperature dependence of the recombination kinetics was explained by a thermal expansion model. Although we have focused in this work only on a particular charge recombination reaction, the approach should be applicable to other electron transfer reaction as well.

Acknowledgement:

We gratefully acknowledge the contributions of the many students, post-docs and collaborators whose work was cited in this review. The work from our laboratory was supported by grants from the NIH (GM-13191) and NSF (DMB 82-02811).

NOTE ADDED IN PROOF

On the Determination of the Characteristic Temperature, T_o.

Bixon and Jortner (59) have tried to fit the temperature dependence of k_{AD} (see Fig. 9) with a characteristic frequency $h\omega = 100$ cm^{-1} (i.e. $T_o \simeq 140K$), which is considerably lower than the one we used (Eq. 12). Although their fit at temperatures below 200K is good, they fail to fit the temperature dependence between 200 and 300K. There is, of course, no justification (except simplicity) to fit the entire temperature dependence with one value of T_o, since the vibrations involved in the electron transfer are likely to be different from those associated with the expansion. In the absence of information about the characteristic temperatures of either set of vibrations, we had opted in Fig. 9 for the simple approach of fitting the entire temperature range with a single temperature, T_o. We have now been able to determine the characteristic temperature of one of the vibrations that we believe plays a role in the electron transfer and thus fit the observed temperature dependence of k_{AD} in a more logical way.

The vibrations in question are those of the hydrogens bonded to the two oxygens of the primary acceptor, Q_A (57,58) (see insert in Fig. 15). We have determined the temperature dependence of the O--H bond length by measuring the hyperfine interaction of the proton with the unpaired spin on Q_A^- (60). This interaction has been shown to be dipolar (58,61), i.e., it is proportional to r^{-3}, where r is the O-H bond length. Thus, for small changes, Δr, in the bond length, the change in the hyperfine coupling, ΔA, is given by

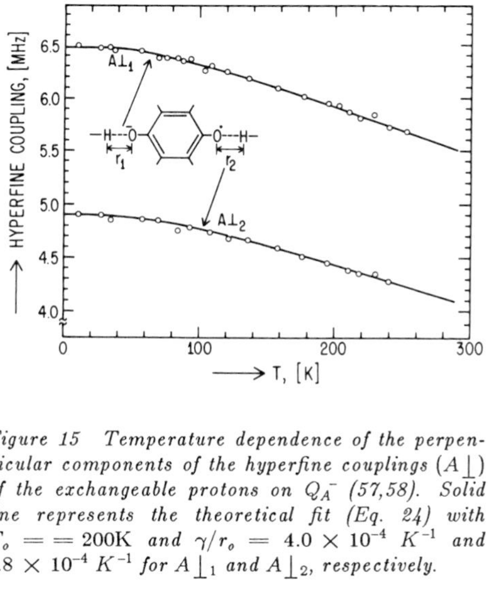

Figure 15 *Temperature dependence of the perpendicular components of the hyperfine couplings ($A\perp$) of the exchangeable protons on Q_A^- (57,58). Solid line represents the theoretical fit (Eq. 24) with $T_o == 200K$ and $\gamma/r_o = 4.0 \times 10^{-4}$ K^{-1} and 4.8×10^{-4} K^{-1} for $A\perp_1$ and $A\perp_2$, respectively.*

$$\frac{\Delta A}{A} \simeq -3\frac{\Delta r}{r} = -3\left(\frac{\Delta r}{r_o}\right)\left(\frac{r_o}{r}\right) \quad (23)$$

Substituting Eq. 14 for $\Delta r/r_o$ yields

$$\frac{\Delta A_{1,2}}{A_{1,2}} = -\frac{3}{2}\frac{\gamma_{1,2}}{r_{1,2}}T_o\left[\coth\frac{T_o}{2T} - 1\right] \quad (24)$$

where the subscripts 1 and 2 refer to the two protons. The temperature dependence of the perpendicular component of the hyperfine interaction, $A\perp$, of both protons is shown in Fig. 15. The solid line represents a fit of the data to Eq. 24 with $T_o = 200K$, $\gamma_1/r_o = 4.0 \times 10^{-4}$ K^{-1}, $r_1 = 1.55$Å, $\gamma_2/r_o = 4.8 \times 10^{-4}$ K^{-1}, $r_2 = 1.71$Å. Thus, the characteristic temperature is closer to the value favored by Bixon and Jortner (59). From the sensitivity of the fit to T_o we estimate a possible error in T_o of ± 50K.

We next have to show that the O-H vibration is involved in the electron transfer reaction under discussion (Eq. 1, Fig. 9). The evidence comes from the isotope effect, i.e., the observed change in k_{AD} when the protons were substituted with deuterons (62). The experimentally determined value of k_{AD} increased at 300K by 6% upon deuteration. A simple theoretical argument showed that in the low temperature limit a 20% effect is expected (62). At T = 300K and T_o = 200K Eq. 8 predicts an order of magnitude smaller isotope effect. Notwithstanding the lack of quantitative agreement between the observed and predicted isotope effect, we take the qualitative agreement as evidence that the hydrogen bonding protons associated with Q_A provide a vibrational mode that is important in the electron transfer reaction.

Having determined the characteristic temperature, T_o^{ET}, of the vibrations coupled to the electron transfer, we leave the other characteristic temperature, T_o^{exp}, associated with the expansion as well as the expansion coefficient γ/r_o as free parameters to fit the observed temperature dependence of k_{AD} with the expression (obtained from Eq. 8 and 14)

$$k_{AD}(T) = k_{AD}(0)\tanh\left(\frac{T_o^{ET}}{2T}\right)e^{-\frac{\gamma T_o^{exp}}{2r_o}\left(\coth\frac{T_o^{exp}}{2T}-1\right)} \qquad (25)$$

where $\tanh\left(\dfrac{T_o^{ET}}{2T}\right)$ is the strong cou-

pling limit (s >> 1) of Eq. 8a. It is plotted in Fig. 16 (dashed line) for $T_o^{ET} = 200K$. A fit of Eq. 25 to the experimental data (dots) with $T_o^{ET} = 200K$, $T_o^{exp} = 1000K$ and $\gamma/r_o = 0.036$ is shown by the solid line in Fig. 16. Although the fit is very good, the high value of γ/r_o is cause for concern. The expansion mechanism invoked probably represents an oversimplification of the situation; other mechanisms may contri-

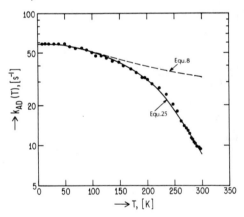

Figure 16 Temperature dependence of k_{AD}. Experimental data (dots) same as in Fig. 9. Solid line represents the best fit of eq. 25 with $T_o^{ET} = 200K$, $T_o^{exp} = 1000$, $\gamma/r_o = 0.036$. Dashed line represents Yortner's theory (eq. 8) with $\Delta E_{redox} = E_{max}$ and $T_o^{ET} = 200K$.

bute to the temperature dependence of k_{AD}. Clayton for instance, showed that k_{AD} in dehydrated RCs at T = 300K has a value that is similar to the one

observed at cryogenic temperatures (63). Similarly, we have found that k_{AD} of RCs in PVA films that have been thoroughly dehydrated by prolonged pumping fitted the dashed line of Fig. 16 (Eq. 8) rather than the solid line (Eq. 25) (64). Thus, the water of hydration must play an important role in the electron transfer. Clayton suggested that the observed temperature dependence is due to a phase transition of the bound water (63). The orientational polarizability of the water dipoles may represent another mechanism. It should also be noted that the distance between Q_A^- and D^+ is rather large ($\sim 20\text{Å}$) and consequently encompasses a large amount of protein structure. Any temperature dependent characteristic of the intervening space (e.g. a change in the conformation of the protein backbone) may contribute to the observed temperature dependence. X-ray structure analyses of RCs from *R. sphaeroides* performed at different temperatures should shed some light on this question (65,66).

References

1. Feher, G. and Okamura, M. Y.; in *The Photosynthetic Bacteria*, R. K. Clayton and W. R. Sistrom, eds. (Plenum Press, N.Y.), chapter 19, pp. 349-388 (1978).

2. Parson, W. W. amd Ke, B. (1982) In *Photosynthesis: Energy Conversion by Plants and Bacteria* (Govindjee, ed.), pp. 331-385, Academic Press, New York.

3. Reed D. W., and Clayton, R. K. (1968), *Biochim. Biophys. Res. Commun.* **30**, 471-475.

4. Feher, G. (1971) *Photochem. & Photobiol.* **14**, #3 (Supplement #1), 373-388.

5. McElroy, J. D., Mauzerall, D. C., and Feher, G. (1974) *Biochim. Biophys. Acta,* **333**, 261-277.

6. Straley, S. C., Parson, W. W., Mauzerall, D. C., and Clayton, R. K. (1973) *Biochim. Biophys. Acta* **305**, 597-609.

7. Debus, R. J., Feher, G. and Okamura, M. Y. (1985) *Biochemistry* **24**, 2488-2500.

8. Hopfield, J. J. (1974) *Proc. Natl. Acad. Sci. USA* **71**, 3640-3644.

9. Jortner, J. (1976) *J. Chem. Phys.* **64**, 4860-4867.

10. Arata, H. and Parson, W. W. (1981) *Biochim. Biophys. Acta* **636**, 70-81.

11. Arata, H. and Parson, W. W. (1981) *Biochim. Biophys. Acta* **638**, 201-209.

12. Noks, P. P., Lukashev, E. P., Kononenko, A. A., Venediktov, P. S., and Rubin, A. B. (1977) *Mol. Biol. (Moscow)* **11**, 1090-1099.

13. Kleinfeld, D., Okamura, M. Y., and Feher, G. (1984) *Biochemistry* **23(24)**, 5780-5786.

14. Austin, R. H., Beeson, K. W., Eisenstein, L., Frauenfelder, H., and Gunsalus, I. C. (1975) *Biochemistry* **14**, 5355-5373.

15. Frauenfelder, H. (1978) *Methods Enzymol.* **54**, 506-532.

16. Woodbury, N. W. T. and Parson, W. W. (1984) *Biochim. Biophys. Acta* (in press).

17. Jortner, J. (1980) *J. Am. Chem. Soc.* **102**, 6676-6686.

18. Redi, M., and Hopfield, J. J. (1980) *J. Chem. Phys.* **72**, 6651-6660.

19. Frauenfelder, H., Petsko, G. A., and Tsernoglou, D. (1979) *Nature (London)* **280**, 558-563.

20. Artymiuk, P. J., Blake, C. C. F., Grace, D. E. P., Patley, S. J., Phillips, D. C., and Sternberg, M. J. E. (1979) *Nature (London)* **280**, 563-568.

21. Karplus, M, and McCammon, J. A. (1981) *CRC Crit. Rev. Biochem.* **9**, 293-349.

22. Levitt, M., Sander, C., and Stern, P. S. (1985) *J. Mol. Biol.* **181**, 123-447.

23. Huber, R., and Bennett, W. S. (1983) *Biopolymers* **183**, 261-679.

24. Warshel, A. (1980) *Proc. Natl. Acad. Sci. USA* **77**, 3105-3109.

25. Agmon, N. and Hopfield, J. J. (1983) *J. Chem. Phys.* **78**, 6947-6959; **80**, 592 (erratum).

26. Kleinfeld, D., Okamura, M. Y., and Feher, G. (1984) *Biochim. Biophys. Acta* **766**, 126-140.

420

27. Eisenberger, P., Okamura, M. Y., and Feher, G. (1983) *Biophys. J.* **37**, 523-538.

28. Parson, W. W. (1967) *Biochim. Biophys. Acta* **153**, 248-259.

29. Hsi, E. S. P. and Bolton, J. R. (1974) *Biochim. Biophys. Acta* **347**, 126-133.

30. Loach, R. A., Kung, M., and Hales, B. J. (1975) *Ann. NY Acad. Sci,* **244**, 297-319.

31. Mar, T., Vadeboncoeur, C., and Gingras, G. (1983) *Biochim. Biophys. Acta* **724**, 317-322.

32. Sarai, A. (1980) *Biochim. Biophys. Acta* **589**, 71-83.

33. Kakitani, T. and Kakitani, H. (1981) *Biochim. Biophys. Acta* **635**, 498-514.

34. Kakitani, T. and Mataga, N. (1985) *J. Phys. Chem.* **89**, 8-10.

35. Hales, B. J. (1976) *Biophys. J.* **16**, 471-480.

36. Hopfield, J. J. (1976) *Biophys. J.* **16**, 1239-1240.

37. Feynman, R. P. (1972) *Statistical Mechanics: A set of Lectures,* pp. 53-55, W. A. Benjamin, Reading, PA.

38. Ringe, D., Kuriyan, J., Petsko, G. A., Karplus, M., Frauenfelder, H., Tilton, R. F., and Kuntz, I. D. (1984) *Am. Crystallographic Assoc. Transactions* **20**, 109-122.

39. Okamura, M. Y., Isaacson, R. A., and Feher, G. (1975) *Proc. Natl. Acad. Sci.* **72** #9, 3491-3495.

40. Prince, R. C., Gunner, M. R., Dutton, P. L. in *Function of Quinones in Energy Conserving Systems,* B. L. Trumpower, ed., Academic Press, Inc., New York., pp. 29-33 (1982).

41. Recently, N. W. Woodbury, W. W. Parson, M. R. Gunner, R. C. Prince, and P. L. Dutton (1986) *Biochim. Biophys. Acta* **851**, 6-22, have determined the redox potentials of different quinones *in situ* from the quantum yield of delayed fluorescence. They found significant deviations from the values determined in dimethylformamide. Unfortunately, their data were also obtained at room temperature and are, therefore, not strictly applicable to the data of Fig.10. We are indebted to N. W. Woodbury and W. W. Parson for making their data available to us prior to publication.

42. Gunner, M. R., Tiede, D. M., Prince, R. C., and Dutton, P. L. in *Function of Quinones in Energy Conserving Systems,* B. L. Trumpower, ed., Academic Press, Inc., New York., pp. 265-269 (1982).

43. Gunner, M. R. (private communication).

44. Gopher, A., Blatt, Y., Schönfeld, M., Okamura, M. Y., and Feher, G. (1985) *Biophys. J.* **48**, 311-320.

45. Arata, H., and Parson, W. W. (1981) *Biochim. Biophys. Acta* **638**, 201-209.

46. Gunner, M. R. Y., Liang, Y., Nagus, D. K., Hochstrasser, R. M., and Dutton, P. L. (1982) *Biophys. J. (Abstracts)* **37**, 226a.

47. Schönfeld, M., Montal, M., and Feher, G. (1979) *Proc. Natl. Acad. Sci. USA* **76**, 6351-6355.

48. We have recently been able to measure changes in k_{AD} with a precision of $\sim 0.5\%$ [M. Y. Okamura and G. Feher (1986) *Biophys. J. (Abstracts)* **49**, 587a].

49. Popovic, Z. D., Kovacs, G. J., Vincett, P. S., and Dutton, P. L. (1985) *Chem. Phys. Letters* **116**, 405-410.

50. Crofts, A. R. and Wraight, C. A. (1983) *Biochim. Biophys. Acta* **726**, 149-185.

51. Maroti, P. and Wraight, C. (1985) *Biophys. J. (Abstracts)* **47**, 5a.

52. Kleinfeld, D., Okamura, M. Y., and Feher, G. (1985) *Biochim. Biophys. Acta* **809**, 291-310.

53. Kleinfeld, D., Okamura, M. Y., and Feher, G. (1985) *Biophys. J.* **48**, 849-852.

54. Prince, R. C. and Dutton, P. L. (1976) *Arch. Biochem. Biophys.* **172**, 329-334.

55. Rutherford, A. W. and Evans, M. C. W. (1980) *FEBS Lett.* **110**, 257-261.

56. Wraight, C. A. (1981) *Isr. J. Chem.* **21**, 348-354.

57. Lubitz, W., Abresch, E. C., Debus, R. J., Isaacson, R. A., Okamura, M. Y., and Feher, G. (1985) *Biochim. Biophys. Acta* **808**, 464-469.

58. Feher, G., Isaacson, R. A., Okamura, M. Y., and Lubitz, W. (1985) *Antennas and Reaction Centers of Photosynthetic Bacteria: Structure, Interaction and Dynamics* (M. Michel-Beyerle, ed.) Springer-Verlag, Berlin, pp. 174-189.

59. Bixon, M., and Jortner, J. (1986) *J. Phys. Chem.* **90**, 3795-3800.

60. Feher, G., Isaacson, R. A., Okamura, M. Y. and Lubitz, W. (1986), unpublished results.

61. O'Malley, P. J., Chandreshekar, T. K. and Babcock, G. T., *Antennas and Reaction Centers of Photosynthetic Bacteria: Structure, Interaction and Dynamics* (M. Michel-Beyerle, ed.) Springer Verlag, pp. 339-344.

62. Okamura, M. Y. and Feher, G., *Proc. Natl. Acad. Sci. USA* (1986) **83**, 8152-8157.

63. R. K. Clayton (1978). *Biochim. Biophys. Acta* 504, 255-264.

64. Arno, T. R., McPherson, P. H., Feher, G. (1986), unpublished results.

65. Allen, J. P., Feher, G., Yeates, T. O. and Rees, D. C. (1986) presented at the VIIth International Congress on Photosynthesis, Brown University, *Proceedings*, Martinus Nijhoff/W. Junk (in press).

66. Allen, J. P., Feher, G., Yeates, T. O., Rees, D. C., Deisenhofer, J., Michel, H and Huber, R. *Proc. Natl. Acad. Sci.* (1986) **83**, 8589-8593.

Acknowledgements

Table 1 and Figure 2 (on page 401) are reprinted from Feher, G. (1971) *Photochem. & Photobiol.* 14(3, Suppl. 1), 373–388, with the permission of Elsevier/North Holland Biomedical Press, Amsterdam, and with the permission of Pergammon Press, Elmsford, New York, respectively.

Figures 5, 6, 7, and 8 (on pages 404, 405, and 407, respectively) are printed from Kleinfield, D., Okamura, M.Y., and Feher, G. (1984) *Biochemistry* 23(24), 5780–5786, with the permission of the American Chemical Society, Washington, D.C.

Figures 11, 12, and 13 (on pages 413, 414, and 415, respectively) are reprinted from Gopher, A., Blatt Y., Schönfeld, M., Okamura, M.Y., and Feher, G. (1985) *Biophys. J.* 48, 311–320, with the permission of The Rockefeller University Press, New York, New York

Figure 14 (on page 416) is reprinted from Kleinfeld, D., Okamura, M.Y., and Feher, G. (1985) *Biophys. J.* 48, 849–852, with the permission of The Rockefeller University Press, New York, New York.

Electric Field Modulation of Electron Transfer in Bacterial Photosynthetic Reaction Centers

Z. Popovic, G. Kovacs and P. Vincett
Xerox Research Centre of Canada
2660 Speakman Drive
Mississauga, Ontario
Canada L5K 2L1

and

G. Alegria and P.L. Dutton
Department of Biochemistry and Biophysics
University of Pennsylvania
Philadelphia, PA 19104

Abstract

Multilayer Langmuir-Blodgett (LB) films of reaction centers from the photosynthetic bacterium *Rhodopseudomonas sphaeroides* have been fabricated with partial net orientation. From measurements of the light induced electron transfer reactions in reaction center films, we have succeeded in quantitating the electric field dependence of (1) the quantum yield of charge separation and (2) the kinetics of charge recombination.

1. Introduction

A feature common to photosynthetic reaction centers is light- induced electron transfer [1]. In the reaction center of the photosynthetic bacterium *Rhodopseudomonas sphaeroides*, this process involves several redox components contained within the protein. Following light absorption a bacteriochlorophyll dimer, $(BChl)_2$, assumes an excited singlet state, $(BChl)_2{}^*$, and transfers an electron, possibly via a monomeric (BChl), to a bacteriopheophytin (BPh) to form $(BChl)_2^{+\cdot}$ $BPh^{-\cdot}$. Before useless recombination can occur, the $BPh^{-\cdot}$ reduces a ubiquinone-10 molecule, designated Q_A, to form $(BChl)_2^{+\cdot}$ $BPhQ_A^{-\cdot}$. A simplified representation of the rate constants, energy levels and distances between the different components in the reaction center are summarized in Figure 1. (The values for the forward rates are reported in [2-6]. The recombination rates are reported in [7-10]. The value for ΔG between $(BChl)_2{*}$ $BPhQ_A$ and $(BChl)_2^{+\cdot}$ $BPh^{-\cdot}Q_a$ is reported in [11,12]; the value for ΔG between $(BChl)_2^{+\cdot}$ $BPh^{-\cdot}Q_A$ and $(BChl)_2^{+\cdot}$ $BPhQ_A^{-\cdot}$ is given in [13-15]. The scaling distance of 4.4 nm between the non-heme iron and the cytochrome c iron was obtained by resonance x-ray diffraction [16].) In the native membrane the Q_A reduces a second ubiquinone (Q_B) in $100\mu s$ half-time and a cytochrome c reduces the $(BChl)_2^{+\cdot}$ in microseconds to stabilize the process further [1,17]. The reaction center is considered to span the cytoplasmic membrane [18-20] with cytochrome c and Q_B associated with the reaction center, but located on opposite sides of the membrane. Thus, the system is organized so that following light excitation electron transfer is coupled to the separation of charges within the protein directed across the membrane.

Expressions of the charge separation in photosynthetic bacteria have been seen *in vivo* as electrochromic responses of the carotenoid complement of the membrane [21,22], by enhanced fluorescence yield indicative of a reversal of the light reaction [23] and by shifts in redox equilibria between cytochrome c and $(BChl)_2$ [24]. Direct measurements of the charge separation have since been made *in vivo* with reaction centers incorporated into planar phospholipid bilayer membranes [25-30], as monolayers on solid supports [31] and on the interfacial region of immiscible liquids [32]. See also [33] for a review.

There is a consensus of agreement from the different approaches that the separation of charge across the membrane is effected by several distinct contributing electron transfer steps. Electron transfer from ferrocytochrome c to $(BChl)_2^{+\cdot}$ contributes in the region of 40-50% of the separation of charge across the membrane [22,24,27,29], while electron transfer from $Q_A^{-\cdot}$ to Q_B contributes little or nothing to the transmembrane charge separation [22,26,27,29]. Thus the charge

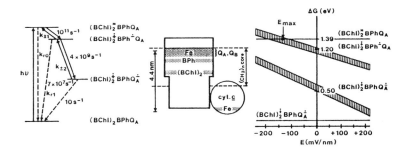

Figure 1a. Electron transfer rates in reaction centers of Rps. sphaeroides. b. Schematic representation of the "dielectric" distances between $(BChl)_2$, BPh, and Q_A, Q_B. The shaded areas indicate uncertainties in these distances. c. Electric field dependence of the energy levels of the states $(BChl)_2^{+\cdot}$ $BPh^{-\cdot}Q_A$ and $(BChl)_2^{+\cdot}$ $BPhQ_A^{-\cdot}$, according to Eqn. (4) with distances between chromophores (as given in Figure 1b) being 1.2 nm between $(BChl)_2$ and BPh and 0.6 nm between Q_A and BPh. For both states the shaded areas represent the range of field dependencies corresponding to the uncertainties in the zero field energy levels.

separation accounting for the remaining 50-60% of the membrane is provided by electron transfer from $(BChl)_2*$ to Q_A. Determinations of the electrical response from $(BChl)_2*$ to BPh have shown it to be two-fold greater than from $BPh^{-\cdot}$ to Q_A [32]; this indicates a 35-40% span contributed by electron transfer from $(BChl)_2*$ to BPh and a 15-20% contribution by the $BPh^{-\cdot}$ to Q_A reaction. Independent studies have suggested a similar (i.e. approximately 1/7 of the membrane thickness) value for the BPh to Q_A span [28-30].

The work reported here focuses on the light activated electrogenic reactions that occur between the $(BChl)_2$ and Q_A. The goal is to significantly alter the reactions by applying electric fields across planar arrays of reaction centers to obtain information of the energetics and factors that govern kinetics of electron transfer. Application of electric fields along the pathway of charge separation will change the relative energy gaps between each reaction step. However, since a) the reaction steps between $(BChl)_2$ and Q_A each contribute only small fractions of the total distance of charge separation within the protein, and b) substantial large free energy drops are apparent in each step [1,17] it was anticipated that substantial electric fields may be required to effect measurable changes in the electron transfer rates involved in charge separation and recombination.

An experimental opportunity to achieving our goal is provided by reaction centers as monolayer films deposited by Langmuir- Blodgett (LB) techniques on conductive substrates [31]. When coated with a blocking polymer layer, such films, in contrast to the systems based on planar bilayers separating aqueous phases, have been shown to withstand externally applied electric fields of up to 200 mV/nm [34]. With this range of electric fields at our disposal applied to either assist or impede charge separation and recombination, we have examined the quantum yield [35-37] of $(BChl)_2^{+\cdot}$ $BPhQ_A^-$ formation and the kinetics of charge recombination from this state.

2. Theory

2.1 Effect of an External Field on the Energy Levels

The attainment of the charge separated state $(BChl)_2^{+\cdot}$ $BPhQ_A^-$ involves the formation of at least two dipoles; d_1, associated with $(BChl)_2^{+\cdot}$ $BPh^{-\cdot}Q_A$ and d_2, associated with $(BChl)_2^{+\cdot}$ $BPhQ_A^-$. In the presence of an external field the energy of these dipoles will be given by:

$$U = -d_{1,2} \cdot E. \tag{1}$$

If one assumes that, a) the external field is homogeneous and equal to the induced local field and b) that the projected distances between the chromophores along the field axis are as given in Figure 1b, the positions of the energy levels relative to the ground and excited singlet states are presented in Figure 1c as a function of the field over the range used in our experiments.

2.2 Quantum Yield as a Function of External Fields

Let us assume that an LB film of reaction centers with an asymmetric up and down population is incorporated into a sandwich cell as seen in Figure 2. Let us call P_1 the up population and P_2 the down population. We can define the parameter, δ, describing the degree of asymmetry, as the fraction of unpaired reaction centers:

$$\delta = \frac{P_1}{P_1 + P_2} - \frac{P_2}{P_1 + P_2}.$$

(2)

The fraction of reaction centers with up and down orientations will then be given by $1/2(1 + \delta)$ and $1/2(1 - \delta)$, respectively.

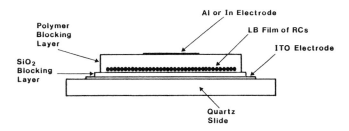

Figure 2. Schematic representation of the sample cell used in the measurements.

A light induced voltage is observed in the cell as a result of charge separation in the reaction centers. Its absolute value depends on the degree of asymmetry and its sign will depend on the direction in which the larger of the two populations is oriented.

Let us define the relative photoresponse:

$$R(E) = \Delta V(E)/I,$$

(3)

where I is the intensity of the pulsed illumination. For low light intensities the photo-induced voltage charge, ΔV, will be proportional to the light intensity, and the relative photoresponse, R, will be light intensity independent.

We now introduce the field dependent quantum yield for charge separation $\phi(E)$ which is a measure of the fraction of reaction centers that reach the state $(BChl)_2^+\cdot Q_A^-$ after excitation. The amount of charge generated at a given field by the population P_1 will be proportional to $\phi(E)P_1$, while that of population P_2 will be proportional to $\phi(-E)P_2$. For low light intensities the observed voltage drop will be given by:

$$\Delta V(E) = \frac{\phi(E)P_1 - \phi(-E)P_2}{c}I, \tag{4}$$

where c is a constant dependent on the cell geometry, capacitance and dielectric constants of the system. The relative photoresponses for positive and negative applied fields can then be written as:

$$R(E) = \alpha\left[(1+\delta)\phi(E) - (1-\delta)\phi(-E)\right], \tag{5}$$

$$R(-E) = \alpha\left[(1+\delta)\phi(-E) - (1-\delta)\phi(E)\right], \tag{6}$$

where $\alpha = (P_1 + P_2)/2c$.

In order to facilitate our subsequent analysis of the experimental data, we now introduce two new quantities $R_{av}(E)$ and $R_{diff}(E)$, defined by

$$R_{av}(E) = \frac{1}{2}\left[R(E) + R(-E)\right] = \delta\alpha\left[\phi(E) + \phi(-E)\right], \tag{7}$$

$$R_{diff}(E) = \frac{1}{2}\left[R(E) - R(-E)\right] = \alpha\left[\phi(E) - \phi(-E)\right], \tag{8}$$

which are respectively even and odd functions of the electric field. Eqns. (7) and (8) give the following dependencies of the yield on the electric field,

$$\phi(E) = \frac{1}{2\alpha}\left[\frac{R_{av}(E)}{\delta} + R_{diff}(E)\right], \tag{9}$$

$$\phi(-E) = \frac{1}{2\alpha}\left[\frac{R_{av}(E)}{\delta} - R_{diff}(E)\right]. \tag{10}$$

2.3. Kinetics of Charge Recombination as a Function of External Fields

The kinetic behavior of the reaction centers in an LB film shows a multiphasic character. We therefore consider a system of reaction centers whose time-resolved response to a pulse of bleaching light is characterized by rate constants $k_i(0)$ at zero applied electric field. For such a system the bleaching, ΔI, i.e., the change in the intensity of a probe beam, is described by the equation.

$$\Delta I(t, E = 0) = \Sigma_i A_i \cdot \exp[-k_i(0)t], \tag{11}$$

where the A_i are the light intensity amplitudes associated with the rate constants, $k_i(0)$, characteristic of the i'th "species" of reaction centers. We now assume that the reaction centers are oriented in either an up or down direction [31] relative to an applied constant electric field, E. Upon application of the electric field, the time decay will be given by

$$\Delta I(t, E) = \Sigma_i \left\{ A_i^+ \exp[-k_i(+E)t] + A_i^- \exp[-k_i(-E)t] \right\}, \tag{12}$$

where the $A_i^\pm$ are the amplitudes associated with the i'th "species" of reaction centers which are oriented in the up and down directions, and are related to the A_i by

$$A_i^+ + A_i^- = A_i. \tag{13}$$

In the case of a completely symmetric sample

$$A_i^+ = A_i^- = \frac{1}{2} A_i, \tag{14}$$

and the expression for the bleaching reduces to

$$\Delta I_{\text{sym. sample}}(t, E) = \Sigma_i \frac{1}{2} A_i \left\{ \exp[-k_i(+E)t] + \exp[-k_i(-E)t] \right\}. \tag{15}$$

The amplitudes, A_i, and the zero field decay rate constants, $k_i(0)$, are found by fitting eq. (11) to the experimental data obtained at zero field. The fitting of the experimental curves to obtain $k_i(\pm E)$ can then be performed in two steps instead of one, involving a reduced number of fitting parameters in each step. For any field, E, the steps are as follows:

(*i*) The bleaching after a saturating flash is measured while applying an oscillating bipolar square wave field of high frequency. Under these conditions each of the "species" of reaction centers will decay with an average rate constant

$$k_i^{\mathrm{av}}(E) = \frac{1}{2}[k_i(E) + k_i(-E)], \tag{16}$$

provided the oscillations of the field are much faster than any of the rates. (In eq. (16) E signifies the maximum amplitude of the bipolar square wave field.) Under the assumption that the pre-exponential factors A_i remain constant and are the same as the zero field ones, it can be shown that the bleaching is given by

$$\Delta I\,(t, E_{\mathrm{sq.wave}}) = \Sigma_i A_i \cdot \exp[-k_i^{\mathrm{av}}(E)t], \tag{17}$$

from which $k_i^{\mathrm{av}}(E)$ can be obtained.

(*ii*) The kinetics of the bleaching after a bleaching pulse are measured by biasing the sample with constant field $+E$ and these are then averaged with the decay under a $-E$ field. From Eqns. (12-15) it follows that

$$
\begin{aligned}
\Delta I_{\mathrm{av}}(t, E) &= \frac{1}{2}[\Delta I(t, E) + \Delta I(t, -E)] = \Delta I_{\mathrm{sym.\ sample}}(t, E) \\
&= \Sigma_i \frac{1}{2} A_i \{\exp[-k_i(+E)t] + \exp[-k_i(-E)t]\}\,.
\end{aligned} \tag{18}
$$

If one writes $k_i(\pm E)$ as

$$k_i(\pm E) = k_i^{\mathrm{av}}(E) \pm \Delta k_i(E), \tag{19}$$

where

$$\Delta k_i(E) = \frac{1}{2}[k_i(E) - k_i(-E)], \tag{20}$$

it follows that

$$\Delta I_{\mathrm{av}}(t, E) = \Sigma_i A_i \exp\left[-k_{\mathrm{av}}(E)t\right] \cosh\left[\Delta k_i(E)t\right]. \tag{21}$$

Under the assumption that the A_i retain their zero field values and by using $k_{\mathrm{av}}(E)$ deduced from oscillating field experiment one obtains $\Delta k_i(E)$ by fitting experimental data obtained under constant bias to Eq. (21). Equation (19) then provides the values for $k_i(\pm E)$.

3. Materials and Methods

Reaction Center Preparation: The reaction centers were isolated from the photosynthetic bacterium *Rhodopseudomonas sphaeroides* (R26) using the method developed by Clayton and Wang [38] and modified by Okamura et.al. [39]. The preparations used in this work contained variable percentages of secondary quinone, Q_B, ranging from 30% to 80% as estimated by the percent of slow phase charge recombination [40]. Nevertheless, it was observed that in the LB films used in these measurements all of them became equivalent with only a small percent of Q_B activity [41]. Furthermore, since Q_A to Q_B electron transfer does not contribute to the generation of electric potential across the membrane [27,29], the electrical measurements reported here are independent of the Q_B content of the preparation.

Electrical Cell Construction: The techniques used for the preparation of the quartz substrates, involving deposition of the conductive and blocking layers [42] in the sandwich cell (see Figure 2), were the same as described in [19]. Indium-Tin-Oxide (ITO) is sputtered onto a quartz slide followed by a sputtered SiO_2 blocking layer. The reaction centers are deposited as several monolayers on this surface from a Langmuir-Blodgett (LB) film balance. The cell is dip-coated with a polymer layer (hydrogenated polyolefin) out of hexane solution. After drying, an Indium layer is vacuum evaporated to complete the cell. The reaction centers were picked up from the trough as the coated quartz slide was lifted from the subphase through the surface and into the air. In this film the side of reaction center that was disposed toward the aqueous phase of the trough adhered to the SiO_2 blocking layer of the slide.

Electrical and Optical Measurements: Photo-induced electrical transients were measured using a modified RC circuit in which the noise associated with the power supply was eliminated by using a 1 μF capacitor and a set of relays [43]. In the signal averaging measurements possible charge accumulation in the sample was avoided by applying the following biasing sequence: positive voltage, zero voltage, negative voltage, zero voltage etc. For the charge recombination experiments the sequence was: positive field, high frequency square wave field, negative field, high frequency field, etc. It was possible to collect separate averaged data for either AC square wave, positive or negative bias as well as to average these last two together. As input amplifier, a Tektronix oscilloscope model 7633 was used equipped with the 7A22 differential amplifier plug in unit. Excitation light pulses were obtained from an electronic flash (Metz model Mecablitz 45 CT-1); the pulse duration used was 50 μs. The applied sample bias was turned on about 2 ms prior to the excitation light pulse and maintained for about 10 ms, the duration of the transient measurements. Light from the flash excited the sample at near normal incidence; the light intensity was adjusted to give 5% of the maximum observable voltage to insure a linear response.

When measuring the time resolved absorption recovery, the electric field was switched on within a few milliseconds after sample bleaching at zero field, and the subsequent decay in reflectance change was recorded. The total recording time was 1.0s and data were taken typically by averaging 40 measurement cycles.

Data were recorded using a Cromemco System Three microcomputer, which also controlled the timing of events in the measurement sequence. The electric field applied to the sample was determined from the known sample cell area, A, capacitance, C, applied voltage, V, and dielectric constant of the reaction centers proteins ($\varepsilon_r = 3$) [44,45] by using the equation

$$E = \frac{CV}{\varepsilon_0 \varepsilon_r A}. \tag{22}$$

A probe beam of 860 nm light was used in the second part of our experiment to determine the degree of sample bleaching after a saturating flash. The probe beam was incident at 45° onto the quartz slide side of the sample and was reflected back from the top metal electrode, thus making a double pass through the reaction center layer. In bleaching and action spectra measurements monochromatic illumination was obtained by using an Instruments S.A. Inc. H-20 monochromator with 1mm slits.

Results

It is important to show that the light induced electrical signal in the LB films is associated with the charge displacement in the reaction centers. This was done by obtaining the electrical action spectrum of the sample cell recorded without an applied electric field bias, and comparing it with the absorption spectrum of reaction centers in aqueous-detergent solution. The excellent correspondence between the two was obtained clearly indicating that the photovoltages are generated by light absorbed by the chromatophores of the reaction centers [46].

Figure 3 shows the field dependence of the relative photoresponse $R(\pm E)$ for two different reaction center LB films labelled A and B. The measured photovoltages were typically up to a few mV in the regime where photovoltage was proportional to the light intensity.

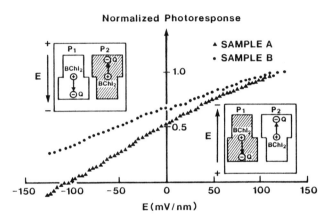

Figure 3. Normalized photoresponse as a function of the applied field, E; ($\triangle$)-sample A, ($\bullet$)-sample B. Populations P_1 (quinone side toward ITO) and P_2 have been schematically illustrated in the insets. The hatching indicates the population whose charge separation is assisted by the external field.

As observed previously [31,41], under zero field conditions, the photo-induced voltage yielded a negative ITO electrode relative to the Al electrode (see Figure 2). Since under zero field conditions the unpaired fraction of reaction centers, δ entirely determines the sign of the electrical response, from the sign of the observed response we conclude that the quinone- containing end of the reaction center protein is preferentially oriented toward the ITO electrode. The convention used in the present measurements is that δ is positive; therefore of the two vectorially up and down populations P_1 and P_2, $P_1 > P_2$, and P_1 is the population with the Q_A side disposed toward the water phase and hence the ITO electrode. The response under zero field conditions (i.e. ITO negative; A1, positive) corresponds by convention to a positive relative photoresponse ($R(0)$, Eqn. 3). Similarly, fields that assist the charge separation in the unpaired P_1 population are regarded as positive, i.e. the applied field makes the ITO electrode positive relative to the A1 electrode. In general, as shown in Figure 1c, fields that assist charge separation are considered to do so by increasing the free energy difference between the $(BChl)_2^*$ $BPhQ_A$ and $(BChl)_2^{+\cdot}$ $BPhQ_A^{-\cdot}$ levels while fields that hinder charge separation decrease this free energy gap.

Under the influence of positive fields the P_1 population is expected to saturate in photoresponse at high field values as the quantum yield approaches unity; at the same time the P_2 population will be subject to a hindering effect of the field and hence fewer reaction centers in this population will reach the charge separated $(BChl)_2^{+\cdot}$ $BPhQ_A^{+\cdot}$ state. As a result, the relative photoresponse $R(E)$ (Eqn. 6) increases as seen in Figure 4. Negative fields do the opposite: hinder P_1 and then increase the contribution of P_2 resulting in a smaller $R(-E)$.

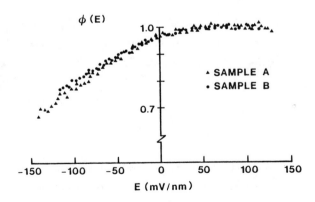

Figure 4. Dependence of the quantum yield of charge separation, ϕ, on the electric field, E, deduced from data presented in Figure 3.

In order to determine $\phi(E)$ we must know the different quantities appearing in Eqns. (10) and (11). $R_{av}(E)$ and $R_{diff}(E)$ are obtained directly from measurements of $R(E)$ and $R(-E)$, by using the same magnitude but opposite polarity of the bias applied to the sample. The quantum yield of charge separation in photosynthesis approaches unity [35-37] and saturation is expected for fields assisting the process. Therefore the unknown parameters α and δ in Eq. (12) and (13) were determined by stipulating the condition $\phi(E) = 1$ for 75 mV/nm $< E$ < 140 mV/nm and minimizing the function

$$F(\alpha, \delta) = \sum_i \left[\frac{1}{2\alpha\delta} R_{av}(E_i) + \frac{1}{2\alpha} R_{diff}(E_i) - 1 \right]^2, \qquad (23)$$

where the summation is over the experimentally employed field values which lie in the specified range. The determined values of δ for the samples were 0.134 and 0.170 which correspond to up and down populations of reaction centers of 57%-43% and 59%-41% for samples A and B respectively.

The foregoing considerations permit us to present what can be regarded as the field effect on the quantum yield (ϕ) in a reaction center population where all reaction centers are oriented in the same direction (Eqns. 9,10). This result is shown in Figure 4. At high positive fields ϕ is seen to saturate as expected. The saturation value in our monolayers has been assumed to be unity. With the introduction of this scaling factor we find that as the field is diminished to zero ϕ assumes the value of 0.96. This is in good agreement with previous measurements of absolute quantum yields in absence of external fields [35-37]. As the field becomes negative, hindering the charge separation process, the quantum yield decreases in a convex manner and assumes a value of ~ 0.75 at the field of -120 mV/nm.

The effect of the previously described electric field bias conditions on the kinetics of charge recombination is shown in Figures 5(a)-(d). Each figure shows the logarithm of variation in bleaching versus time for (i) the high frequency bipolar square wave field with amplitude, E, applied to the sample, (ii) for constant field, E, and (iii) for zero field. In each case, the lower curve corresponds to the bipolar square wave field bias, the middle curve to zero field and the upper curve to constant-field bias. (In Figure 5(a) the zero field and constant-field bias curves appear nearly coincident.) In this sample the constant-field curves showed little dependence on the sign of the field, indicating an approximately equal fraction of up and down RC populations; the constant-field curves in Figures 5(a)-(d) are an average over positive and negative bias conditions. It is clear that application of the electric field changes the recombination kinetics in the photobleached reaction centers. Three rate constants are necessary to fit the zero field decay curve in Figure 2. Their values are $20.8s^{-1}$, $6.8s^{-1}$ and $1.4s^{-1}$, with relative amplitudes 0.28(4), 0.55(3) and 0.16(3) respectively.

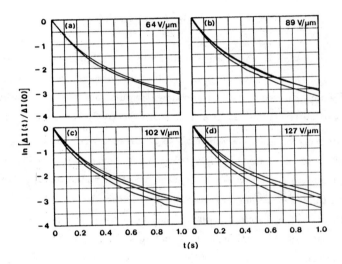

Figure 5. Absorption recovery of RCs as a function of time for various values of the electric field, applied under both constant bias and high frequency bipolar square wave conditions. The zero-field reference is given in each case. In part (a) the zero field reference and constant bias curves appear nearly coincident.

The values of the $k_i(E)$ derived from the data in Figure 5 using the procedure outlined in the theory section, are plotted as a function of E on a semi-log plot in Figure 6. The reproducibility of these measurements was tested using four different samples. All of them gave similar results for $k_i(E)$ [46]. All rate constants are approximately exponential in E over the range of data, with the fastest rate constant, $k_1(E)$, believed to describe the recombination from the first quinone to the special pair [27], showing the highest degree of linearity in Figure 6.

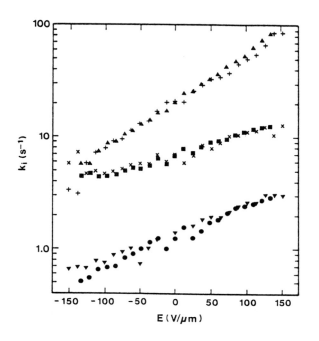

Figure 6. Electric field dependence of rate constants, $k_i(E)$ as determined from two different samples. The data bearing the labels $+, x$ and ∇ come from the same sample as the data in Figure 5. The data bearing the labels, $\triangle, \square$ and $\bullet$ come from another sample.

Most samples showed a small but variable degree of asymmetry in the relative populations of up and down oriented reaction centers. As was mentioned before, the photo-induced voltage invariably caused the top A1 electrode to become positive relative to the ITO electrode, indicating that the reaction centers are preferentially oriented with the quinone end toward the ITO. The absorption recovery kinetics of a particular sample, showing as large an asymmetry as that of any we tested, were determined for the high-frequency bipolar square wave field and for positive and negative constant fields separately. The results for a field amplitude of 155V/μm are shown in Figure 7. The application of a positive bias to the A1 top electrode has clearly increased the net recovery of absorption over the negative bias case. From Eqn. (12) one expects that the same rate constants are involved in the decay in the two cases, but that the corresponding amplitudes are interchanged. Therefore, more reaction centers decay at the field-hastened rates than at the field-hindered rates, with the A1 positive. For the unpaired fraction of reaction centers, which determine the electrical response, biasing the A1 top electrode positive increases the energy separation between the oxidized bacteriochlorophyll dimer, $(BChl)_2^{+\cdot}$, and the reduced quinone, $Q^{-\cdot}$, and assists electron transfer from $Q^{-\cdot}$ to $(BChl)_2^{+\cdot}$. One can therefore conclude that increasing the energy gap results in a faster decay. This result is consistent with the general trend found by varying the $(BChl)_2^{+\cdot}$ - $Q^{-\cdot}$ energy separation by varying the energy level of the quinone through chemical substitution of other quinone [47].

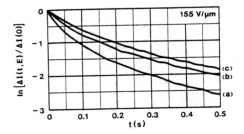

Figure 7. Absorption recovery kinetics for an asymmetric RC sample with the field applied under (a) high frequency bipolar square wave bias conditions, (b) constant bias conditions with the A1 top electrode positive and (c) constant bias conditions with the A1 top electrode negative.

By combining the information from the electrical and optical measurements described above, one can determine the projected electron displacement, Δx, normal to the plane of the LB film. A straightforward argument leads to the following formula

$$\Delta x = \frac{\varepsilon_0 \varepsilon_r \Delta V_{\max}(0)}{e \delta N_{RC}} \tag{24}$$

where ε_0 is the permittivity of vacuum, ε_r is the relative dielectric constant of the reaction center protein layer, e is the elemental charge, $\Delta V_{\max}(0)$ is the maximum photo-induced voltage at zero field, and N_{RC} is the number of active reaction centers per unit area of the film. This last quantity is determined from the maximum bleaching of the sample, the extinction coefficient of the RCs at the probe beam wavelength, and the geometry of the probe beam illumination [46]. Assuming $\varepsilon_r = 3$, one arrives at values of Δx of 1.9 nm and 2.1 nm for samples A and B respectively. These values are in good agreement with other measurements of the $(BChl)_2 - Q_A$ separation ($\Delta x > 1.5$ nm [20,48] and $\Delta x \approx 2.4$ nm [32]).

5. Discussion

According to notation used in Figure 1a, an expression for the quantum yield in terms of the rates associated with the process involved can be written as:

$$\phi = \frac{k_1 k_2}{(k_1 + k_{r0})(k_2 + k_{r1}) + k_{-1} k_{r0}}. \tag{25}$$

In deriving Eqn. (25) the rate k_{-2} has been neglected which will be justified later.

The field dependence of ϕ is governed by the field dependencies of the rates that determine it. However, from the experiments reported here it is not possible to uniquely determine which rates are responsible for the drop in yield. Information from Figures 1(b) and (c) allow us to predict expected free energy changes as a function of the applied electric field, from which only the ratio of the rates involved in each step can be determined.

It is clear from Figure 1(c) that the largest relative energy perturbation introduced by the external field is associated with first step, the $(BChl)_2^* \, BPhQ_A$

to $(BChl)_2^{+\cdot}$ $BPh^{-\cdot}Q_A$ transition; at a field of less than 150 mV/nm the energy gap is reduced from 180 mV to zero. Over the same range the energy between $(BChl)_2^{+\cdot}BPh^{-\cdot}Q_A$ and $(BChl)_2^{+\cdot}BPhQ_A^-$ has diminished by only about 15% of the zero field value, and the energy gap between $(BChl)_2^{+\cdot}BPh^{-\cdot}Q_A$ and the ground state has increased by only 15%. Finally the energy gap for recombination from $(BChl)_2^{+\cdot}BPhQ_A^-$ to the ground state increases by $\sim 50\%$.

Gunner et.al [47] have measured relative quantum yields for charge separation in reaction centers in which the native quinone was substituted by quinones with a wide range of midpoint potentials. Under this condition the first factor in Eqn. (25) remains unaltered. Assuming that k_{r1} is quinone independent, one sees that ϕ is a measure of k_2 as a function of the free energy between $(BChl)_2^{+\cdot}BPh^{-\cdot}Q_A$ and $(BChl)_2^{+\cdot}BPhQ_A^-$. Their results show that ϕ is independent of the energy gap for $\Delta G \gtrsim 430$ mV. It is clear from Figure 1(c) that, within the uncertainty limits in ΔG at zero field, the closest that $(BChl)_2^{+\cdot}$ $BPh^{-\cdot}Q_A$ and $(BChl)_2^{+\cdot}BPhQ_A^-$ ever get for the highest field is ~ 450 mV. Although the Q replaced experiments and the field work may not be entirely equivalent this result suggests that k_2 is not very sensitive to changes in ΔG, and also justifies neglecting the rate k_{-2} in deriving Eqn. (25) for $\phi(E)$.

From these considerations it would appear that the major source of quantum yield drop lies in the first step, $(BChl)_2^*BPhQ_A$ to $(BChl)_2^{+\cdot}BPh^{-\cdot}Q_A$. However the influence of the field dependencies of k_{r1} and k_2 on the yield can not be entirely ruled out. Experiments addressing the formation and time resolved decay of the $(BChl)_2^{+\cdot}BPh^{-\cdot}Q_A$ state as a function of the external electric field, which bear on this issue, are currently in progress.

We will now discuss our results on the field dependence of the recombination rates. The near-linearity observed in Figure 3 provides important information regarding the fundamental mechanism of charge transfer in photosynthesis. If one assumes the electron transfer distance (from $(BChl)_2^*$ to Q) projected along the electric field direction to be 2.0 nm, then on application of fields in excess of ± 150 V/μm one produces a shift in excess of ± 0.3 V in the energy level of the quinone acceptor relative to that of the ground state of $(BChl)_2$. Therefore, over the 0.6 V energy range covered, no evidence is observed for a maximum in the $k_i(E)$, as might be expected on the basis of electron transfer theories [49-53]. Important structure in the $k_i(E)$ may, however, be found in other energy ranges, accessible through the use of electric fields with alternate quinones substituted for the native ubiquinone. On the other hand it is possible that the predicted maximum in recombination rate cannot be observed at all in the RCs of photosynthetic bacteria at room temperature. This is because other mechanisms, such as indirect thermally activated recombination [54-55], may become dominant as the quinone energy level is shifted upward by application of the electric field. Low temperature work addressing these issues is currently in progress.

We can also ask the question whether field induced deformation of the RCs (electrostriction) may be responsible for the observed recombination rate changes. This possibility can be ruled out on the basis of measurements of the pressure dependence of the recombination rate constants [56-57]. For pressure of up to 5000 atm the recombination rates in solution were observed to be about $10 \pm 2s^{-1}$ [57]. We can estimate the field induced pressure in the RC interior as an electrostatic force acting on an elemental charge distributed over an area of about 1nm x 1nm. For an external field of 150 V/μm the estimated pressure is 240 atm, which is even much less than 5000 atm. Therefore the observed rate changes are not likely to be influenced to any appreciable extent by field induced polarization effects leading to changes in electron wave function overlap integrals.

Acknowledgements

We would like to thank Dr. D.K. Murti for supplying the sputter- coated substrates. This work has been funded in part by a grant to PLD from the Department of Energy (USA), grant number DE/AC02/80/ER 10590.

References

1. Clayton, R.K. and Sistrom, W.R. (1978) "The Photosynthetic Bacteria", Plenum Press, New York.

2. Kaufman, K.J., Dutton, P.L., Netzel, T.L. and Rentzepis, P.M. (1975) Science 188, 1301-1304.

3. Pockley, M.G., Windsor, M.W., Cogdell, R.J. and Parson, W.W. (1975) Proc. Natl. Acad. Sci. 72, 2251-2259.

4. Shuralow, V.A., Klenvanik, A.V., Sharkov, A.V., Matveetz, Y.A. and Krukov, P.G. (1978) FEBS Letts. 91, 135-139.

5. Holten, D., Windsor, M.W., Parson, W.W. and Thornber, J.P. (1978) Biochim. Biophys. Acta 501, 112-126.

6. Kaufman, K.J., Petty, K.M., Dutton, P.L. and Rentzepis, P.M. (1976) Biochim. Biophys. Res. Comm. 70, 839-845.

7. Cogdell, R.J., Monger, T.G. and Parson, W.W. (1975) Biochim. Biophys. Acta 408, 189-199.

8. Parson, W.W., Clayton, R.K. and Cogdell, R.J. (1975) Biochim. Biophys. Acta 387, 286-278.

9. Parson, W.W. and Monger, T.G. (1977) Brookhaven Symp. Biol. 28, 195-211.

10. Holten, D., Windsor, M.W., Parson, W.W. and Thornber, J.P. (1978b) Biochim. Biophys. Acta 501, 112-126.

11. Woodbury, N.T. and Parson, W.W. (1984) Biochim. Biophys. Acta 767, 345-361.

12. Schenk, C.C., Blankenship, R.E. and Parson, W.W. (1982) Biochim. Biophys. Acta 680, 44-59.

13. Prince, R.C. and Dutton, P.L. (1978) in "The Photosynthetic Bacteria" (Clayton, R.K. and Sistrom, W.R., eds.) pp. 440-453 Plenum Press, New York.

14. Prince, R.C. and Dutton, P.L. (1976) Arch. Biochem. Biophys. 172, 329-334.

15. Dutton, P.L. Leigh, J.S. (1973) Biochim. Biophys. Acta 292, 654-664.

16. Blasie, J.K., Pachene, J.M., Tavormina, A., Dutton, P.L., Stamatoff, J., Eisenberger, P. and Brown, G. (1983) Biochim. Biophys. Acta 723, 350-357.

17. Govindjee, (1982) "Photosynthesis: Energy Conversion by Plants and Bacteria", Academic Press New York.

443

18. Valkires, G.E. and Feher, G. (1976) J. Cell Biol. 95, 179-188.

19. Pachence, J.M., Dutton, P.L. and Blasie, J.K. (1979) Biochim. Biophys. Acta 548, 348-373.

20. Bachman, R.C., Gillies, K. and Takamoto, J.M. (1981) Biochemistry 20, 4590-4596.

21. Jackson, J.B. and Crofts, A.R. (1969) FEBS Letts. 4, 185-189.

22. Jackson, J.B. and Dutton, P.L. (1973) Biochim. Biophys. Acta 325, 102-113.

23. Evans, E.H. and Crofts, A.R. (1974) Biochim. Biophys. Acta 333, 44-51.

24. Takamiya, K. and Dutton, P.L. (1977) FEBS Letts. 80, 279-284.

25. Schoenfeld, M., Montal, M. and Feher, G. (1979) Proc. Natl. Acad. Sci. USA 76, 6351-6355.

26. Packham, N.K., Dutton, P.L. and Mueller, P. (1980) FEBS Letts. 110, 101-106.

27. Packham, N.K., Dutton, P.L. and Mueller, P. (1982) Biophys. J. 37, 465-473.

28. Gopher, A., Blatt, Y., Okamura, M.Y. and Feher, G. (1983) Biophys. J. 41, 212a.

29. Feher, G., Okamura, M.Y. (1984) in "Advances in Photosynthesis Research". Edited by C. Sybesma (Kliner Academic Publishers. The Hague) 155-164.

30. Gopher, A., Blatt, Y., Schoenfeld, M., Okamura, M.Y., Feher, G. and Montal, M., Biophys. J. in press.

31. Tiede, D.M., Mueller, P. and Dutton, P.L. (1982) Biochim. Biophys. Acta 681, 191-201.

32. Trissl, H.W. (1983) Proc. Natl. Acad. Sci. USA 80, 7173-7177.

33. Tiede, D.M. (1985) Biochim. Biophys. Acta Reviews, in press.

34. Popovic, Z.D., Kovacs, G.J., Vincett, P.S. and Dutton, P.L. (1985) Chem. Phys. Lett. 116, 405-410.

35. Loach, P.A. and Sekura, D.L. (1968) Biochemistry 7, 2642-2649.

36. Wraight, C.A. and Clayton, R.K. (1973) Biochim. Biophys. Acta 333, 246-260.

37. Cho, H.M., Mancino, L.J. and Blakenship, R.E. (1984) Biophys. J. 45, 455-461.

38. Clayton, R.K. and Wang, T.R., (1971) Methods Enzymol. 23A, 696-704.

39. Okamura, L., Sterner, A. and Feher, G. (1974) Biochemistry 13, 1394-1403.

40. Okamura, M.Y., Debris, R.J., Kleinfeld, D. and Feher, G. (1982) in "Function of Quinones in Energy Conserving Systems" (Trumpower, B.L., ed.) 299-317.

41. Popovic, Z.D., Kovacs, G.J., Vincett, P.S., Alegria, G., and Dutton, P.L., to be published.

42. Yang, C.C., Josefowicz, J.Y. and Alexandru, L. (1980) Thin Solid Films 74, 117-127.

43. Popovic, Z.D. (1983) J. Chem. Phys. 78, 1552-1558.

44. Momany, F.A., McGuire, R.F., Burgess, A.W. and Scheraga, H.A. (1975) J. Phys. Chem. 79, 2361-2381.

45. Van Krevelen, D.W. and Hoftyzer, P.J. (1976) "Properties of Polymers", 2nd Edition. Elsevier, Amsterdam, p. 231.

46. Popovic, Z.D., Kovacs, G.J., Vincett, P.S., Alegria, G., and Dutton, P.L., to be published.

47. Gunner, M.R., Woodbury, N., Parson, W.W., Dutton, P.L. Conference on Protein Structure posterbook (1985) p. 11.

48. Tiede, D.M. and Dutton, P.L. 91981) Biochim. Biophys. Acta 637, 278-290.

49. Redi, M. and Hopfield, J.J. (1980) J. Chem. Phys. 72, 6651.

50. Jortner, J. (1980) J. Am. Chem. Soc. 102, 6676-6686.

51. Rackovsky, S. and Scher, H. (1982) Biochim. Biophys. Acta 681, 152-160.

52. Marcus, R.A. (1964) Annu. Rev. Phys. Chem. 15, 155- 196.

53. Beitz, J.V. and Miller, J.R. (1979) J. Chem. Phys. 71, 4579-4595.

54. Gunner, M.R., Liang, Y., Nagus, D.K., Hochstrasser, R.M. and Dutton, P.L. (1982) Biophys. J. 37, 226a.

55. Gunner, M.R., Tiede, D.M., Prince, R.C. and Dutton, P.L. (1982) in "Functions of Quinones in Energy Conversion Systems", B.L. Trumpower, ed., Academic Press, New York and London, 271-276.

56. Clayton, R.K. and Devault, D. (1972) Photochem. Photobiol. 15, 165-175.

57. D. Kleinfeld (1984), Ph.D. thesis, University of California, San Diego, pp. 238-246.

Panel Discussion: How Well Do We Understand the Primary Event of Charge Separation

Les Dutton Session Leader

Introduction: Les Dutton

We are asking the question "How well do we understand the primary event of charge separation" and I think the answer is: at some levels, very well, but at the mechanistic level, not so well. I think that the talks this morning give a good example of how there have been some spectacular progress, not only in the way we look at the reaction center, but also in introducing new and very necessary techniques to look at these systems. What I just want to comment on right now is what we don't know, just to raise your consciousness about it. It's true to say that of the six chlorophylls that are present in the reaction center, we really can only be sure of three of them. So we can only say, we know what half the components do. The other monomeric chlorophyll and the other Fe function is completely unknown. It is known that energy transfer to the dimer is very effective through this system. So it may be that: a route in for photons. The monomeric chlorophyll, whether it's truly an intermediate that you can identify as a stable redox state, or just some kind of virtual state that the electron uses as a medium to get from chlorophyll to Fe, still remains to be seen. It would be very nice to know what the redox properties of these components are. In the equilibrium sense, we have no idea.

Everybody sat there looking at the brilliant work of Michel this morning, (Editors note: see Jortner's talk) and everybody sat there looking at the four hemes that where rolled out for you and I wonder how many people asked the question: What do they do? Well, the answer is, I have no idea and I don't think anybody else does. Even reaction centers that are equipped with this hydrophobic linkage which have these four hemes, at high and low potentials, that doesn't give them any benefit over the more advanced organisms like spheroides or captulata from which our mitochondria derive. They have this water soluble cytochrome c which makes the system work just fine. In fact, it turns out that even veridis with these extra four hemes still require diffusible cytochrome c.

These are some of the questions that are still very open to test as indeed are even the electron transfer pathways. We still have to do a lot more work to map them out. And then, one step beyond that, we are still asking the question, can we use this reaction center protein as a device to test tunneling theories? It's always been an eminent kind of experimental possibility. The eminency is always one step ahead of the experiment it seems to me.

I'd like to start this ball rolling by asking Jim Norris perhaps to explain himself with regard to his comment about P700 which is the green plant analogue of this system. It was, first of all, a dimer, in fact it was the first dimer proposed by Norris and Katz, then the bacteria caught up. In recent years, the P700, the archetype of this dimer, became a monomer. Today, we heard that it was back as the dimer. Would you like to explain yourself please.

Norris: Even when we said it was a monomer, we also said in the paper that one has to be very careful as to what one is talking about here, and let me tell you what I mean by that. As I said this morning, we've been trying to model these things for a long time. You try to model these compounds and you take models that are definitely dimers, I mean, they are chemically made as dimers. But you make a measurement on them, and they all appear to be a monomer. So, what's happened through the years is very simple. Some measurements that you make on this system show that it looks like it has the properties of a monomer. Other measurements you make on this system, show that it has properties associated with the dimer. Now, my theory is quite different from everybody else's theory. My theory is that once it became established that it was a dimer, the only choice that the rest of the world could make was to prove it was a monomer, if they wanted to publish in that area. So that's why we had all this conflict in the meantime as to what was going on. There was no percentage in proving again it was a dimer.

But now the specific example that I was talking about this morning. One way of probing it was with the magnetic nuclei in the system and with this asymmetry added into the system, it becomes very simple. What we had done in the past was that we had enriched the cholorphylls in vivo by growing them on 97%, 96% carbon 13. This put a very large magnetic moment in every site in the molecule essentially. It just turns out that if you have this model up here, you expect that the second moment will drop by a factor of 2, and if you do bacteria, it in fact drops by a factor of 2 and by bacteria I mean spheroides. If you do veridis, it doesn't drop by a factor 2 and once you take into account this asymmetry, this molecule dominates because it enters in as a square. So this molecule dominates. Instead of having the second moment drop by a factor of 2, say in P700, it drops only to 0.8. It doesn't drop by the 0.5 that we had expected it to. So on that basis, we could say it's definitely not this. I've always had in the back of mind that something like this can be going on; it's just that until the crystal was available, I had no direct way of trying to establish that there was asymmetry present. I think all that's happened is really rather simple. If you take into account that the system is somewhat more complicated, that we can have some asymmetry introduced into the problem, I think it goes a long way in explaining everything. I should just add, just like Herman said, you can actually take the linewidth of these systems and predict the chronological age of these organisms. That is, those organisms that have bacterial chlorophyll b in them, will have a larger linewidth, in my view, than those that have bacterial chlorophyll a in because they came later and this follows Herman's argument. The first ones that were made were symmetrical and as time went on, they became asymmetrical – in green plants all are asymmetrical. That's all.

But the confusion has arisen because you can get both monomer and dimer measurements out of the same system and it depends upon how you interpret your measurements.

Dutton: Are there any comments about this?

Q: Could it be that your measurements are already influenced by the ligand field and this would mean that the asymmetry that you propose here is not necessarily 2 for neutral action centers, or for action centers which are measured by ENDOR and takes a quicker time. The role of the negative charge on the quinone is heavily supported by the tremendous effect the charge has and that means a forward rate is changed in the presence of the negative charge on quinone. Thus the different rate change in spheroids and veridis.

Norris: Well, certainly, I think, that having the negative charge there must change some things. That, however, can be tested because we can actually remove the quinones from these systems and check to see if quinone removed is the same as with quinone. That's one possible way of going about checking it.

Q: Not in *veridis*.

Norris: No, but there are a variety of other ways that we can pursue this. For example, one of the things that I would predict that will happen is that if we can do something to the bridging molecules, which we know how to do, then we should change some of the properties of these systems. So we can do these things. But I would also agree with what George presented, that there is a lot of magnetic resonance experiments which we can do which suggest that what's happening in these systems is very close to what happens in the intact system. So, although I expect certain changes to take place because the charge is present, I don't think it's going to change my fundamental view of how the symmetry is broken on this C-2 structure even in the neutral species. I'd be very surprised to see that the symmetry is broken just because of the charge added by the quinone.

Kleinfeld: I'd just like to bring up a caveat in comparing the x-ray data on the donor with the ENDOR data. The x-ray data is presumably taken of the donor in the neutral form, and the ENDOR measurements are made on D+, it is the dimer of the donor: the donor when it shares an electron. So it maybe the geometry changes in going from the neutral to the plus form.

Norris: That's also possible, which is essentially the same thing I said, but the problem with that in my view is that the Mobious group has done the triplet state, in which you don't have to have a charged species present. They all seem to give approximately the same thing for spheroidies. So I would just believe that that's not such an important aspect of it myself. But you can be right.

Warshel: We tried to calculate the properties of the dimer for awhile and felt that when you do a supermolecular calculation, when you do the whole molecular orbital of the system, it's entirely unstable, you do the smallest change in the program and it becomes polarized. This is just for people who will try to reproduce this. However, more recently we repeated the same calculation using the observed dimer, talking a language of experimentalists which means using charge transfer states and the exciton states, in this language, you still find that the system is heavily polarized. I just trust the isolated molecule representation more than the supermolecule. You find that it is enough to have very small fields from the protein to polarize this system, which means that of course one has to wait until we have the x-ray structures, but small changes in the sequence, you don't even need a charge group, is expected to polarize the dimer and it has some functional advantages.

Dutton: Are there any comments to that?

If not, I think we should move to, not looking just at the dimer, but perhaps at the arrival of electron back to the dimer from wherever it can come from and Marilyn Gunner perhaps has some relevant data to this with regard to work done with reaction centers in which ubiquinone-10 has been replaced by a whole array of other quinones. This is following in the footpaths of George Feher's group in the '70's.

Gunner: This is an advantage of having your advisor be the chairman, so you get to talk.

Dutton: It's dangerous.

Gunner: Dangerous.

Hopefully you all remember what the reaction sequence is. What I'm going to talk about is the free energy effects on electron transfer for two different reactions in the reaction center. What I've done, and this is work that originated in George Feher's lab, but we've done a lot of work in the last five or six years on it, is to take the native ubiquinone-10 out and put other quinones in with different electrochemical midpoints. So that what happens, is you change the free energy of the P+Q- state which is basically here. Recently, we've gotten some pretty good measurements of what the midpoints are in the site rather than you have in the solution midpoints for the quinones, and that's actually given us a lot better data. It turns out that the midpoints in solution are only fair predictors of what the midpoints are in the site. The first process I want to talk about is the charge separation.

Kleinfeld: You measure the midpoint by redox titration?

Gunner: By delayed fluorescence and that is on the poster and I can talk about it later.

Kleinfeld: The midpoint at the 10 nsec point.

Gunner: That's right. What you do is you look at the electron transfer from the Fe minus to the quinone. What happens is that the dimer plus Fe minus charge recombines to the ground sate in 10 nsec. So the forward rate has to compete with that. If the forward rate is 1/(10 nsec), then you have 50% charge separation on a single turnover. If you're much faster, then you have a high quantum yield, if you're much slower, you have a low quantum yield. This is basically a quantum yield measurements; the open figures are picosec measurements where we actually measured the forward electron transfer from Fe to quinone with Robin Hochstrasser many years ago. This is as a function of the free energy of the semi-quinone in the site. This line here is what's predicted if the forward rate were to slow a factor of 10 for every 60 mV change in the free energy of the Q-a state. So as you raise the free energy of the semi-quinone, the difference in the equilibrium constant was affected by a slowing of the forward rate. This line up here would be if the forward rate were not affected at all, but only the back rate gave you the change in the equilibrium constant.

What you see is that for quite a wide range, perhaps 450 mV, there's almost no change in the forward rate. This is one instance where it looks as though, although you're changing the free energy by a fair amount, you're seeing very little change in the forward rate of electron transfer and then you begin to see some slowing of the rate.

The second process that I want to discuss is the charge recombination rate. This is where we have a lot more data and a lot better data. What is turning out is that this is completely midpoint dependent and completely temperature independent. Unfortunately the scale is different, but this is ubiquinone and in essence this is about 300 mV below ubiquinone and a little bit above ubiquinone. The open dots were done in George Feher's lab at 70 K, the closed figures were done by me at helium temperature at about 12 K and you can see that there's basically no temperature dependence and very minimal midpoint dependence over this entire range. So here are two electron transfer processes in the reaction center which seem to be not particularly dependent upon the free energy of the electron transfer reaction. The other things I would say is that the temperature dependence to the forward rate cannot be particularly great otherwise we couldn't make these measurements, that is, if the forward rate were slowing a lot with temperature, you wouldn't see any charge separation 300 mV different from ubiquinone.

Dutton: Are there any points to that? Yes.

Feher: The way to determine midpoint is by different acceptors in our case and electric field in their case with different ways to get the same situation, just to make that clear.

Dutton: The two sets of experiments claim to do the same thing with different agents because they may not be doing exactly the same thing. We hope to test that one day when the two come together.

Austin: I just want to know why the theorists are so quiet during this whole session because it seems to me that the theory is really breaking down as far as I can tell when you get away from the model systems. Is that a real problem or not?

Fischer: I think the first question the theorist always has to ask: Does the simple model apply, or in other words does a mechanism apply if one takes a certain model here. One simple way out, and this actually we discussed here too, is to argue that the weight determining step is not just the direct transfer from the quinone to the $P+$, but one could think possibly in terms of delocalization of the positive charge which extends to a small degree into the Fe and that is weight determining and then essentially just the process which is very fast from $Q-$ to the Fe but it's just the small delocalization of the positive hole and then you still can think of a simple process and bring in this idea. So, I only want to say, it's a very complex system and I think it is very difficult for the theory to say there's a conflict as long as one is not entirely sure about the mechanism.

Gunner: One thing I'd like to point out is that the mechanism of Fe transfer is basically up hill and I would think that would interfere with the temperature dependence (I'm talking from the naive experimentalist point of view).

Hopfield: I'm not sure it's such a big conflict with theory yet and I believe that the two of you really agree and are planning to take more data (laughter). I sometimes say it comes out differently than what I meant. Les or Marilyn will explain what that really means. But if you would take pairs of reactants like these quinones for example in rigid glasses and you'd have them near the maximum of the rate vs. free energy curve, there is for a reasonably large reorganization energy and especially if you have some changes of bond lengths and bond angles of these quinones involved, the rate doesn't depend very strongly on exothermicity over a region like 0.4 V, so a little wider range is needed and at present, I don't really think there's any conflict. Now there's even a weird cancellation that can occur that I won't mention because it's too strange and nobody will believe it.

Dutton: I agree with you with regard to extending the range here and it will be done. At least in this direction. That direction is getting tough. Joshua?

Jortner: Well you know at certain states there comes a moment of truth and every theory has to be modified. Now let me say what my feeling about it is in general. The striking result is not only the invariance but also the weak temperature dependence or rather the lack of temperature dependence. And, all the conventions are to give ad hoc explanation, I would be very careful, not that I am conservative, but you have really to account for all the wealth of information and not only theoreticians like us are working, but those experimentalists are very active and the magnetic interaction data in such a model of hole hoping, I am afraid, will not be reconciled easily with the data. So let's stick to a standard picture and here I must say, if you would see invariance about the onset of temperature dependence, then I would be very happy. Then there is no problem. But the beauty of this system is as we discussed at great length yesterday is the weak temperature dependence of the negative activation energy. Even adding some slow modes of the protein motion, will not be of great help because it will add quite a pronounced temperature dependence. So I think it's a beautiful experiment and for the time being we should say that there's usually some theory behind the experiment.

Honig: Theories that are applicable to model systems and not to proteins are good theories, its just we have bad model systems.

Wolynes: I guess I could wear two hats and answer that challenge. One is as a chemist, I think if an organic chemist saw a reaction that had a zero linear free energy relationship, you would say that the transition state didn't involve changing, breaking whatever bond it was that was affected by your substituent effect. So he would say that you're probably looking at a different process and maybe it is a conformational change. Now, I'm not sure he would be right. Joshua's point that he emphasized is about the activation behavior and is very important. I would also say though that if it is activation behavior, I'm a little bit picky about things, and I'm not really sure that all the simple nonadiabatic formulations really work in that limit, even Joshua mentioned that the other day when the vibration relaxation rates are comparable to these other rates. I think you really have to re-examine things.

Jortner: These are millisec.

Wolynes: Yes, sorry, exactly. But what I mean is I think still in the activationless regimen, I don't have confidence in the theory.

Dutton: The theory exists though and we are trying to test it.

Hopfield: There are a bunch of things that make one worry.

It is true that you can push the theory around just enough to accommodate the experiments and of course you can put slopes through this. At the same time, just look at the fact that these are different quinones, you also know the vibronic coupling really should have changed appreciably between these two and what these data are saying is really that the system is amazingly insensitive to exactly what that quinone is. The extreme flatness of the temperature dependence at low temperature (low is 100) is also something which you can fit with theory but, gee, you know, it's not very convincing. Now there are situations in classical physical where you know that simply there are rates which can't go any faster. Diffusion limited, in spite of the fact that there are still kinds of details going on, they don't show up in forward rate constants. They only show up in reverse rate constants. There are certain corners not normally looked at in quantum mechanics where you can get such things also, but it's very awkward to see why those corners should actually refer to the system. But for the same kind of reasons that are present in the classical systems, though, you can also get them in the quantum mechanical one.

Connection Between Biological Systems and Organic Conductors

E. Buhks

The B F Goodrich R&D Center

Corporate Research Department

Brecksville, OH 44141

Electron transfer processes in biological systems, conductive polymers and organic monolayers are described in terms of phonon-assisted charge carrier hopping between localized sites. The hopping distances ($\sim 10 \mathring{A}$) and electron-exchange interactions ($\sim 10^{-2} eV$) are similar in biological systems and conductive polymers. This theory accounts for the magnitude of conductivity in polymer, its temperature and dopant concentration dependence. The near-IR broad bands observed in conductive polymers are associated with light induced charge carrier hopping process. Tunneling effects in a transport across monolayer films at low temperature are discussed.

I. INTRODUCTION

A number of publications in recent years have demonstrated an active interest in the mechanism of electron transfer processes in biological systems [1-12]. This interest was stimulated by the extensive experimental information regarding temperature dependence of electron transfer rates between various donor-acceptor centers [13-18]. Tunneling effects were discovered in cytochrome-C oxidation process where the electron transfer rate was found to be temperature independent below 100K [13-15]. The rate increases according to Arrhenius law by 3 orders of magnitude from 10^{-6}s to 10^{-3}s with increasing temperature from 100K to 300K).

This temperature dependence was attributed to a strong electron-phonon coupling of high-frequency metal-ligand vibrational modes ($\sim 500 cm^-$) which determined the 100K separation temperature [7,8]. From the fit of multiphonon transition probability to the experimental data electron-exchange energy inter-action between cytochrome-C and reaction center of $\sim 10^{-2}$ eV was evaluated [8].

Weak temperature dependence of a fast electron transfer $(10^{-10}s)$ between pheophytin and quinone [15] and a slow back process from quinone to chlorophyll [16,17] $(10^{-2}s)$ were explained in terms of activationless process characterized by a redox potential being matched by the reorganization energy of normal vibration modes involved in the electron transfer process [5]. The rate of these reactions is determined by electron-exchange interactions, 10^{-3} eV and 10^{-7} eV, estimated for pheophytin-quinone $(R \sim 10\text{Å})$ and quinone-chlorophyll $(R \sim 20\text{Å})$ donor-acceptor centers, respectively [5].

Multiphonon theory of charge transfer between localized sites is successfully applied to biological systems. It correctly predicts the value of the charge trans-fer rate determined by electron-exchange interaction dependent on spacial sep-aration between two localized sites and Franck-Condon factors which include contributions from the normal vibrational modes. Franck-Condon factors give the correct account of the rate temperature dependence and low temperature quantum-mechanical tunneling effects observed in biological systems.

In view of the success of this theory applied to charge transfer in biopoly-mers one can consider the extension of these concepts to the field of conduc-tive polymers. Extensive experimental data has been collected in recent years on the properties of conjugated doped conductive polymers, such as polyacety-lene, polyphenylene, polythiophene, polypyrrole etc. (see reviews in ref. 19-22). Electrical conductivity in these materials changes by several orders of magnitude, depending on dopant concentrations, and reaches its maximum of $10^2 - 10^3 \Omega^{-1} cm^{-1}$. Conductivity usually increases with temperature, although the functional dependence deviates from a simple Arrhenius form. Broad optical bands in the near-IR spectrum $(\sim 1$ eV) appear under doping of polymers. The intensity and the peak position of these bands depend on the dopant concentra-tion.

The theoretical effort in the field of conductive polymers was concerned mostly with a one-dimensional conduction model, where various chain defects, such as solitons (cations), polarons (radical cations), bipolarons (dications) were assumed to participate in the intrachain transport [20]. A theory based on valence Hamiltonian technique, for example, provides a qualitative picture of the charge carrier and its energy position with respect to the band structure [20]. The transport properties of conductive polymers have been also analyzed from the viewpoint of dispersive transport in semiconductors [21]. In all these theoretical approaches a carrier is assumed delocalized and electron-phonon interaction is disregarded as well as a role of a dopant ion in the transport.

In the following it will be demonstrated a consistent quantitative interpretation of transport and optical data in conductive polymers by a multiphonon theory of charge carrier hopping between localized centers associated with dopant ions.

Multiphonon Theory of Charge Hopping Between Localized Sites

The transition probability W for charge carrier hopping between donor-acceptor ions (or localized sites) can be expressed in terms of the first order perturbation theory (for non-adiabatic process, i.e. weak electronic interaction) as a product of the square of the charge-exchange integral, V, and thermally averaged Franck-Condon factors $G(\{\omega\}, \{S\}, T, \Delta E)$

$$W = \frac{2\pi}{\hbar}|V|^2 G. \tag{1}$$

The charge exchange interaction V decreases exponentially with the distance R between the localized centers as

$$V = V_o e^{-\alpha R}. \tag{2}$$

Where α is typically $1 \pm 0.3 \text{Å}^{-1}$ and V_o is of the order of 10 eV [11]. From analysis of nearest neighbor interactions between aromatic molecules such as naphthalene and anthracene it was found $\alpha = 1 \text{Å}^{-1}$ and $V_o = 12.5 eV$ [7].

The Franck-Condon factors depend on the electronic energy gap between donor and acceptor centers, ΔE, the normal vibrational frequencies $\{\omega\}$ and their electron-phonon coupling parameters $\{S\}$ according to [2,7]

$$G(\Delta E) = (2\pi\hbar)^{-1}e^{-\phi(0)} \int_{-\infty}^{\infty} e^{i\Delta Et/\hbar}e^{\phi(t)}dt \tag{3}$$

where

$$\phi(t) = \int \rho(\omega)S(\omega)[n(\omega)+1)e^{i\omega t} + n(\omega)e^{-u\omega t}]d\omega \tag{4a}$$

$$n(\omega) = \left[\exp\left(\frac{\hbar\omega}{kT}\right) - 1\right]^{-1} \tag{4b}$$

$n(\omega)$ given by Eq. 5 is the equilibrium phonon occupation number, $\rho(\omega)$ is the phonon density of states.

Temperature dependence of cytochrome-C oxidation rate, displayed in Fig. 1, was fitted by Eq. (3) with the phonon density of states $\rho(\omega) = \rho(\omega_m) + \delta(\omega - \omega_c)$ corresponding to the medium (ice) optical vibrational modes 100 cm^{-1} [7]. In general $G(\Delta E)$ is calculated numerically using saddle point approximation [7]. A few useful relations which can be derived analytically [2,6-8,22] are given in the following text.

In low temperature approximation $kT \ll \hbar\bar{\omega}$ ($\bar{\omega}$ is a mean vibration frequency) nuclear tunneling take place and for small $|\Delta E|$, of the order of phonon energy, charge transfer occurs by one-phonon emission process described by

$$G(\Delta E) = e^{-\int d\omega\rho(\omega)S(\omega)}\hbar^{-1}\rho(-\Delta E/\hbar)S(-\Delta E/\hbar) \tag{5}$$

where $\Delta E < 0$ for exothermic processes. Transition probability assumes the lineshape of phonon density of vibronic states involved in charge transfer. One can vary ΔE by electrical field, for example, in order to get lineshape of ρS from the low temperature current/voltage experiment.

For larger values of $|\Delta E|$ 2-, 3-, etc., $-m$ multiphonon processes take place, as evident from Fig. 2 which displays numerical model calculations [22] of $G(\Delta E)$, Eq. (3), at $T = 0$ for rectangular phonon density of states in the range $\bar{\omega} - \frac{1}{2}\Delta\omega \le \omega \le \bar{\omega} + \frac{1}{2}\Delta\omega, \bar{\omega} = 100cm^{-1}$. Peaks in electronic transition probability

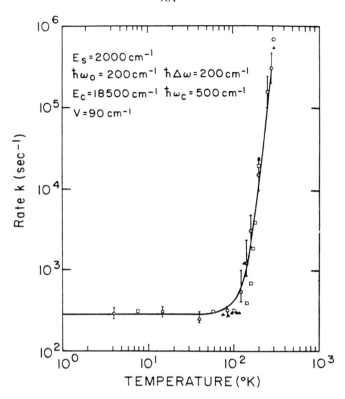

Figure 1. Theoretical fit of the temperature dependence of the rate of cytochrome oxidation in Chromatium [7]. Experimental data are taken from Refs. 13-15.

corresponding to $0 \to m$ phonon tunneling transitions remain even for a broad phonon density of states. In the single-mode approximation, $\rho(\omega) = \delta(\omega - \overline{\omega})$, Franck-Condon factors get the following analytical form:

$$G = (\hbar \Delta \omega)^{-1} \exp(-S \coth x - mx) I_m(S/\sinh x) \qquad (6)$$

where $x = \hbar\overline{\omega}/2kT, m = |\Delta E|/\hbar\overline{\omega}$, $\Delta\omega$ is a width of the phonon spectrum, and I_m is the modified Bessel function of the order m. At low temperature ($\hbar\overline{\omega} > kT$) Eq. (7) results in the Poisson distribution:

$$G = (\hbar \Delta \omega)^{-1} e^{-S} S^m/m! \qquad (7a)$$

which can be recast in the form of the energy gap law

$$G = (\hbar\Delta\omega)^{-1}(2\pi m)^{-\frac{1}{2}}\exp(-S - \gamma m) \quad \text{for } m \gg 1 \tag{7b}$$

where $\gamma = 1n(m/S) - 1$.

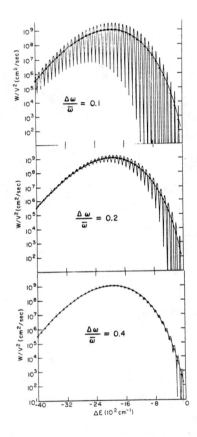

Figure 2. Model calculation, Eq. (3), of the electron transfer probability vs. electronic energy gap for rectangular phonon density of states $\rho(\omega) = $ const. for $\overline{\omega} - \frac{1}{2}\Delta\omega \leq \omega \leq \overline{\omega} + \frac{1}{2}\Delta\omega, \overline{\omega} = 100 cm^{-1}, E_r = 0.25 eV$. The envelope of the distribution is calculated using the saddle point approximation.

Saddle point approximation yields an analytical solution of Eq. (3) for isoenergetic processes characterized by $\Delta E = 0$

$$G(0) = \left[2\pi \int d\omega \rho(\omega) S(\omega)(\hbar\omega)^2 \operatorname{cosech} \frac{\hbar\omega}{2kT}\right]^{-\frac{1}{2}} x \exp\left[-\int d\omega \rho(\omega) S(\omega) \tanh \frac{\hbar\omega}{4kT}\right]$$

(8).

In the high temperature limit, $kT \gg \hbar\bar{\omega}$, Franck-Condon factors assume a well-known activated form followed from Eq. (3).

$$G(\Delta E) = (4\pi E_r kT)^{-\frac{1}{2}} \exp(-E_a/kT) \tag{9}$$

where

$$E_a = (\Delta E + E_r)^2/4E_r \tag{10}$$

$E_r = \int \rho(\omega) S(\omega) \hbar\omega d\omega$ is a reorganization energy, which in the continuum approximation yields

$$E_r = e^2(1/\varepsilon_{op} - 1/\varepsilon_{st})(1/a - 1/R). \tag{11a}$$

The appearance of optical (ε_{op}) and static (ε_{st}) dielectric constants of the medium in expression for reorganization energy reflects the polaron nature of charge transfer event. R is the distance between the redox centers, a is the radius of a redox center. For $R = 2a$, Eq. (11a) becomes

$$E_r = e^2(1/\varepsilon_{op} - 1/\varepsilon_{st})1/R. \tag{11b}$$

A transition temperature T_o, between the Arrhenius and low temperature rate forms is on the order [6] of $\hbar\omega/4k$ for the strong coupling case $S \gg 1$. For the activationless processes [5], characterized by a redox potential matched by the reorganization energy, $\Delta E = -E_r$ and $E_a = 0$, the transition temperature increases to $\hbar\omega/2k$, as seen in Fig. 3. The rate temperature dependence for activationless processes at high temperatures, proportional to $T^{-\frac{1}{2}}$, is characterized by a small negative apparent activation energy, which was observed in electron transfer between pheophytin and quinone [15], and quinone and chlorophyll [16,17].

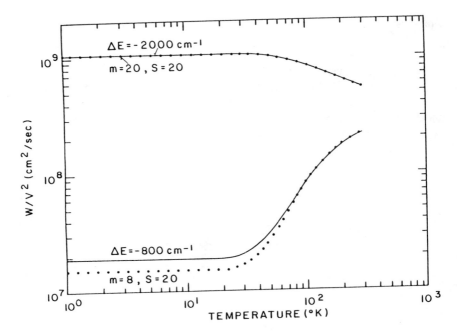

Figure 3. Theoretical prediction for the temperature dependence of the electron transfer rate for activated and activationless process [5]. Solid lines are calculated for a continuum of vibrational spectrum $\Delta\omega/\overline{\omega} = 0.4$ using Eq. (3), while the dotted lines represent the single-mode approximation, Eq. (7), for $m = |\Delta E|/\hbar\overline{\omega} = 8$ and 20, $\overline{\omega} = 100 cm^{-1}$ and $E_r = 0.25eV$.

The electrical conductivity σ of a p-type conductive polymer is given in its simplest form by

$$\sigma = ep\mu. \tag{12}$$

Here p is the hole concentration (equal to the anion dopant concentration) and μ is the mobility, which can be expressed through the drift velocity along and against the field F associated with the charge hopping process between isoenergetic sites, as following

$$\mu = R[W(-eFR) - W(eFR)]/F \tag{13}$$

In the case of weak fields, $eFR \ll kT$, use of Eq. (9) derived for the high temperature limit results in the Nernst-Einstein relation $\mu = eR^2W(0)/kT$ and conductivity temperature dependence is given by

$$\sigma(T) = AT^{-3/2}e^{-Ea/kT}$$

where

$$E_a = E_r/4 \text{ and } A = e^2(2\pi/\hbar)pR^2(4\pi E_r k)^{-\frac{1}{2}}V_o^2 e^{-2\alpha R}. \qquad (14)$$

The optical transition probability $I(\Omega)$ for photoinduced by a photon $\hbar\Omega$ charge hopping between isoenergetic localized states in the case of weak electron-exchange interaction [9,22], is expressed as

$$I(\Omega) = e^2 R^2 V^2 (\hbar\Omega)^{-2}(4\pi E_r kT)^{-\frac{1}{2}}x\exp\left[-(\hbar\Omega - E_r)^2/4E_r kT\right]. \qquad (15)$$

Absorption coefficient $\sigma(\Omega)$ includes $I(\Omega)$, light frequency Ω and index of refraction $n(\Omega)$ according to

$$\alpha(\Omega) = N(8\pi^3/3h)(\hbar\Omega/nc)I(\Omega) \qquad (16)$$

where N is the concentration of absorbing centers (equal to hole concentration p in conducting polymer if associated with anion).

Dielectric loss $\varepsilon''(\Omega)$ is directly proportional to $I(\Omega)$

$$\varepsilon''(\Omega) = N(8\pi^3/3)I(\Omega) \qquad (17)$$

and its peak position is approximately equal to the reorganization energy E_r. $\varepsilon''(\Omega)$ and $\alpha(\Omega)$, in general, have different shapes, especially in the near IR spectrum which is indicative of conduction mechanism in conductive polymers.

III. Polypyrrole. Results and Discussions

Recent data obtained in our lab on polypyrrole [23] will be analyzed from the viewpoint of charge carrier hopping between localized sites.

Conductivity temperature dependence of the electrochemically oxidized (30%) polypyrrole films doped with bisulfate and tetrafluoroborate anions is displayed in Fig. 4. The data is fitted very well by Eq. (14) yielding the value of activation energy of 0.06 eV. Low temperature measurements [24] (down to 4K) demonstrate 6 orders of magnitude drop in conductivity. No T-independent region, indicating nuclear tunneling, is observed due to the low frequency torsional/backbone modes involved in charge transport in polymers.

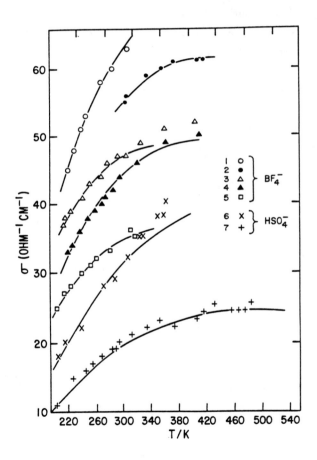

Figure 4. Temperature dependence of electrical conductivity of the oxidized polypyrrole doped with BF_4^- and HSO_4^-. Solid curves fitting the data were calculated from Eq. (14) with $E_a = 0.06eV$.

Using the experimental value of the conductivity pre-exponential factor A, reorganization energy $E_r = 4E_a = 0.24$ eV and carrier concentration p $= 3\mathrm{x}10^{21}$ cm^{-3} (anion concentration) one can plot out curves relating the elementary jump distance R and the distance decay parameter α of the electronic wave function tails, as presented in Fig. 5. From these empirical α vs R curves one gets $R = 6 - 8\text{Å}$ for $\alpha \simeq 1\text{Å}^{-1}$. Similar values of R are also calculated from the reorganization energy value using Eq. (11b) with $n = 1.5$ [27] and estimated permittivity $\varepsilon \simeq 2.8 - 3.7$. The estimated hopping distance is close to the average anion separation $p^{-1/3} = 7\text{Å}$ for oxidized polypyrrole (30%), which suggest that hopping occurs between localized sites associated with the dopant anions.

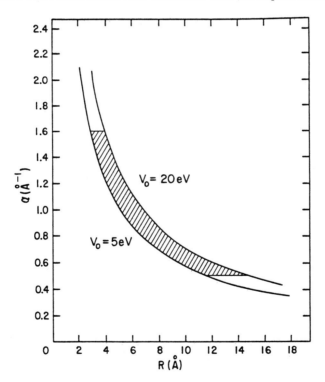

Figure 5. *Empirical relation between the distance decay parameter of the charge transfer rate α and elementary jump distance R in polypyrrole, as extracted from Figure 4 using Eq. (14).*

Then the functional dependence of conductivity on carrier concentration can be presented in the following form of Eq. (14).

$$\sigma \sim p^{1/6} \exp(-2\alpha p^{-1/3}). \tag{18}$$

A rapid increase in conductivity with p at low degrees of oxidation with a levelling off at the highest degrees, according to Eq. (18) is displayed in Fig. 6. Similar behavior was observed in O_2 doped polypyrrole [27]. Assuming $V_o = 10$ eV and $\alpha = 1\text{Å}^{-1}$ to be typical values for conductive polymers maximum conductivity of $10^2 - 10^3 \Omega^{-1}$ cm^{-1} is predicted for these systems.

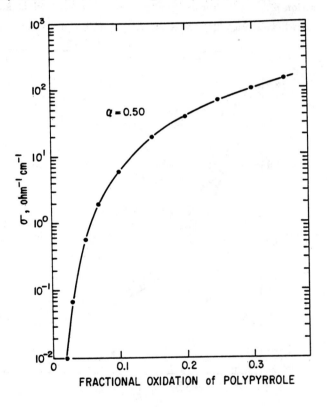

Figure 6. *Electrical conductivity in conductive polymers as a function of degree of oxidation, as predicted by Eq. (18) with $\alpha = 0.5\text{Å}^{-1}$.*

Optical spectrum of the oxidized polypyrrole (30%) presented in Fig. 7 exhibits two broad bands at 1 eV and 2.7 eV. The first band is associated with photoinduced carrier hopping since the corresponding dielectric loss peak [28], as predicted by Eq. (17), occurs at 0.25 eV which is close to the reorganization energy value estimated from the conductivity data.

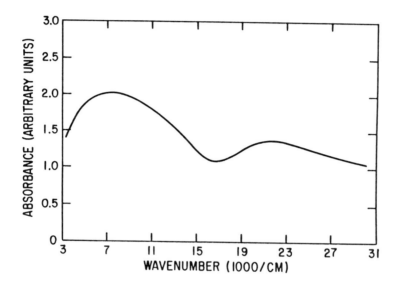

Figure 7. Optical absorption spectrum of oxidized polypyrrole (30%) doped with BF_4^-. NIR band corresponds to light induced charge carrier hopping between localized sites.

Photoinduced carrier hopping transition peak, as predicted by Eqs. (11b), (15)-(17), exponentially decreases in intensity and red shifts with the increase in the hopping distance. If the hopping occurs between defect sites associated with anions a change in polypyrrole degree of oxidation from 0.3 to 0.1 would result in the shift of the near IR peak from 1 eV to 0.7 with a corresponding reduction in intensity by $\sim 10^{-2}$ for $\alpha = 1 \mathring{A}^{-1}$. This is consistent with the published spectra of polypyrrole electrochemically prepared at various potentials [28].

From the experimental value of the conductivity pre-exponential factor A and the estimated hopping distance $R \sim 7\text{Å}$, the charge-exchange interaction V is estimated to be about 10^{-2} eV. This value is typical of similarly spaced donor-acceptor centers in biological systems, such as pheophytin-quinone [5]. The corresponding calculated charge transfer rate in polypyrrole is very fast $\sim 10^{-11}$s due to the low value of the reorganization energy in the isoenergetic hopping process in the polymer. In pheophytin-quinone electron exchange reaction similarly fast rate was attributed to an activationless process where $\Delta E = -E_r$ [5].

Charge Transfer in Organic Monolayers

Monolayers represent a model system for studies of the charge transfer process across thin films because of their high order and "molecular" thickness (10-30 Å). In such systems charge transfer probability across the film depends on the electronic energy gap (voltage), normal vibrational modes, and their electron-phonon coupling temperature. Electron-exchange interaction in charge transfer across an "insulating" organic film decreases exponentially with increase in the film thickness.

The electron tunneling effect was confirmed in photoinduced electron transfer studies between monolayers of donor (cyanine dye) and acceptor (viologen) separated by fatty acid interlayer [29]. Steady state fluorescence intensity from the cyanine dye monolayer was shown in increase exponentially with the donor-acceptor separation, in accord with Eq. (1), characterized by $\alpha = 1.1 \text{Å}^{-1}$.

The effect of distance and orientation between donor-acceptor centers in photosynthetic systems [30] was studied in monolayers by field modulation technique [31].

Nuclear tunneling effects have been recently observed in current-voltage characteristic measured across monolayer of polyvinyl stearate at low temperature (15K) [32]. As predicted by Eq. (5) and schematically reproduced in Fig. 8, current density peaks were found to be related to the IR vibrational frequencies of the polymer.

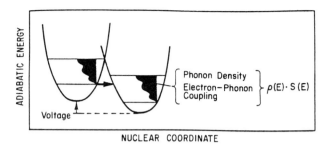

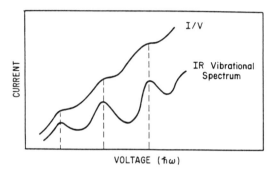

Figure 8. Phonon-assistant tunneling effects in charge transfer across polymer monolayer film. Peaks in current-voltage characteristic at low temperature correspond to polymer vibrational modes associated with charge transfer process.

Conclusion

In this paper charge transfer data in biological systems, conductive polymers and organic monolayer films were discussed in terms of nonadiabatic multiphonon charge transfer process between localized states. In the framework of this theory the Franck-Condon factors of the normal vibrational modes account for the temperature dependence of the charge transfer rate. The absolute magnitude of the rate is also determined by the electron-exchange interaction which exponentially decreases with donor-acceptor separation.

The charge transfer theory of the carrier hopping between the sites associated with dopant ions accounts for the magnitude of conductivity, its temperature and dopant concentration dependence in conductive polymers. Near IR broad bands in the optical spectrum of conductive polymers are generated by light induced charge transfer between the sites.

Electron tunneling and nuclear tunneling effects, observed in organic monolayers, were explained in the framework of the theory.

References

1. Hopfield, J.J. (1974) Proc. Nat. Acad. Sci. *71*, 3640.

2. Jortner, J. (1976) J. Chem. Phys. *64*, 4860.

3. Kuznetsov, A.M., Sondergard, N.A., Ulstrup, J. (1978) Chem. Phys. *29*, 383.

4. Sarai, A. (1979) J. Electroanal. Chem. *100*, 513.

5. Buhks, E., Jortner, J. (1980) FEBS Lett. *109*, 117.

6. Buhks, E., Jortner, J. (1980) J. Phys. Chem. *84*, 3370.

7. Buhks, E., Bixon, M., Jortner, J. (1981) Chem. Phys. *55*, 41.

8. Jortner, J. (1980) J. Am. Chem. Soc. *102*, 6676.

9. Redi, M., Hopfield, J. (1980) J. Chem. Phys. *72*, 6651.

10. Sarai, A. (1980) Biochim. Biophys. Acta *589*, 71.

11. Beratan, D.V., Hopfield, J.J. (1984) J. Am. Chem. Soc. *106*, 1584.

12 . DeVault, D., Chance, B. (1966) Biophys. J. *6*, 825.

13. Dutton, P.L., Kihara, T., McGray, J.A., Thornber, J.P. (1976) Biochim. Biophys. Acta *226*, 81.

14. Hales, B.J. (1976) Biophys. J. *16*, 471.

15. Peters, K., Avouris, P., Rentzepis, P.M. (1978) Biophys. J. *23*, 207.

16. Parson, W.W. (1974) Ann. Rev. Microbiol. *28*, 41.

17. Hoff, A.J. (1975) Phys. Reports *54*, 75.

18. Baughman, R.H., Bredas, J.L., Chance, R.R., Eisenbaumer, R.L. Shacklette, L.W. (1982) Chem. Rev. *82*, 209.

19. Greene, R.L., Street, G.B. (1984) Science *227*, 651.

20. Chance, R.R., Boudreaux, D.S., Eckhardt, H., Eisenbaumer, R.L., Frommer, J.F., Bredas, J.L., Silbey, R. (1984) in *Quantum Chemistry of Polymers - Solid States Aspects*, Ladik J. et. al. (eds.), Reidel, p. 221.

21. Epstein, A.J. in *Handbook on Conjugated Electrically Conducting Polymers*, Skotheim, T. (ed.), Dekker, N.Y. to be published.

22. Buhks, E. (1980) Electron Transfer Processes, PhD Thesis, Tel-Aviv University.

23. Buhks, E., Hodge, I.M. (1985) J. Chem. Phys., submitted.

24. Watanabe, A., Tanaka, M., Tanaka, J. (1981) Bull. Chem. Soc., Jpn., *54*, 2278.

25. Duke, C.B., Meyer, R.J. (1981) Phys. Rev. B *23*, 2111.

26. Arwin, H., Aspens, D.E., Bjorkland, R., Lundstrom, I. (1983) Synth. Met. *6*, 309.

27. Scott. J.C., Pfulger, P., Krounbi, M.T., Street, G.B. (1983) Phys. Rev. B *28*, 2140.

28. Yakushi, J., Lauchlan, L.J., Clarke, T.C., Street, G.B. (1983) J. Chem. Phys. *79*, 4774.

29. Möbius, D. (1981) Acc. Chem. Res. *14*, 63.

30. Rackovsky, S., Scher, H. (1982) Biochim. Biophys. Acta *681*, 152.

31. Popovich, Z., this book.

32. Burkhardt, C.W., Larkins, G.L., Lando, J.B. (1985) Bull. Am. Phys. Soc. *30*, 339, 402.

Why the Hell Are We Here?

B. Chance

Introduction by Hans Frauenfelder:

Hans Frauenfelder: Dr. Britton Chance is known to all of you and I've had the pleasure of having known him for many years. You all know that he is not just an outstanding biophysicist, but that he excels in many ways. But to me what is always impressive is to look back and to see whenever I or somebody else thinks we have found something new, Brit Chance has been there many years before, and in addition to that, if anyone has had the pleasure, or maybe displeasure, of working directly with Brit, he or she realized how hard it is because no matter how hard we work, Brit works two or three times as hard. Maybe one should introduce a logarithmic scale for this particular problem, which we should call pC for log Chance. But we'll leave that to another speech. So it is my pleasure to introduce Dr. Britton Chance.

Editor's Note: Professor Chance's slides are not reproduced in the text due to problems of reproduction. However, we feel that the absence of the slides will not detract markedly from the message of the talk.

Britton Chance: I am speechless after that but I thought I might have a joke for you. But Paul Mueller is the King of Jokes and I though it would becloud his performance tomorrow, so if you don't get his best joke, you should get your money back. No jokes.

Second, skits; well the skit we wanted to put on was a fight ring with Parak in one corner and in the other corner Anderson and this corner Hopfield and that corner Frauenfelder. Well, we didn't have the canvas quite clear for that one. So I finally came to the decision as to whether to tell you something about the history of biochemistry, but I saved that for the next meeting (grateful applause).

So there was then a question raised earlier by Hopfield and Spiro who are both bold and fearless and that's "Why the hell are we here ?" Well, I thought I would try to answer that and maybe talk a little bit about problems we're trying to solve, and this is something that is not part of the program but maybe it will indicate to you some interesting things.

Now you're looking at your own cardiac tissue, this is particularly for Vitalli. Those things in there are cytochromes – round raisins – 40% of your cardiac tissue is that of the topic of this symposium, you can see the mitochondria are just about as much of the myocardium as is the myofibrils. The astounding fact is that the heart spends just as much time on transporting oxygen to mitochondria and making ATP from the mitochondria as indeed it in contractions. Indeed, the myoglobin is in very high concentration and of course the heart is full of blood as well.

What can we learn from this? Well, we can look at its biochemistry and find out where we as biophysicists fit in: of course the heart breaks down ATP, it has a creatine kinase which restores the ATP level, it also makes ADP and phosphate which drives the oxidative machinery of the heart. Oxygen is delivered in the capillaries by hemoglobin and there is facilitated transport of the oxygen by myoglobin, and finally use of oxygen to oxidize NADH and make ATP.

So this is the machine: what can we contribute to it? We contribute most when oxygen is rate limiting because then all of the cytochrome change is interposed between NADH and oxygen, becomes reduced and the rate limiting step is the reaction of oxygen with ferro- cytochrome c, cytochrome oxidase on the one hand with deoxy hemoglobin in the lung, and deoxy myoglobin in the tissue. There is special interest in the deoxy states. In fact, we can measure the oxygen affinity of cytochrome a, a_3 or hemoglobin, myoglobin, by simply adding small amounts of oxygen and determining the disappearance of absorption by the reduced band or the appearance of the oxy bands from myoglobin in the visible. If we plot the oxygenation of myoglobin against the oxidation of cytochrome oxidase, you would expect that the bottom man on the ladder would use the oxygen better than myoglobin. And indeed, that's true, in other words, there is a sequential unloading from hemoglobin, which is about say 30 uM to myoglobin which is about 1 uM to cytochrome oxidase which has an affinity of 0.10 uM or less. So that's fine, God made everything right; from hemoglobin to myoglobin to cytochrome.

But when one gets into the actual tissue, and repeats this titration (these were suspensions of mitochondria with myoglobin added around them and progressively titrated from the deoxy in the reduced state to the oxy state) the system is very different. Here is the deoxygenation of myoglobin and its reoxygenation in a perfused heart. One sees that if the oxygen is turned off (that is, the rat is given nitrogen), the system reduces itself and cytochrome oxidase uses up all the oxygen and takes all the oxygen from myoglobin. If we add aliquots oxygen and plot oxygenation of myoglobin against the oxidation of cytochrome their profiles are identical. Well, two interpretations, one is that the tissue system is different from the in vitro system, or there are steep oxygen gradients. We choose the latter hypothesis because we can separate these profiles into a hyperbolic one if we decrease the respiration rate. So in other words, in spite of the fact that these pigments have intrinsically different oxygen affinities, they behaved in the tissue as if they had the same.

Well, that suggests that there are very steep oxygen gradients in the cardiac tissue and of course it is a very big user of oxygen. As a red blood cell loaded with oxygen traverses the arterial to the venous end of the capillary blood vessel, it unloads oxygen to oxygenate myoglobin, with 0.6 torr to transfer oxygen by facilitated diffusion to the mitochondria, which are shown here in these characteristic cells, and oxidize the cytochromes.

Well, what our result tells us is that as we decrease the oxygen concentration, it's almost a perfect sink and there's very little coming out of the venous end, let's say 3 torr out of 16 torr in, are typical examples for a hypoxic strained heart, and the fact that MbO_2 titrates the same as cytochrome oxidase even though that there is a 10-fold difference between their affinities for oxygen, suggest that the region which has oxygen concentrations between that of myoglobin and cytochrome is vanishingly small. In other words, between the fraction of the tissue volume which we see optically, which has tensions or concentrations between 0.6 torr and about 0.06 torr is vanishingly small. So this system now has been transformed into an all or nothing system. Above this point, we have MbO_2 and oxidized cytochrome oxidase making phosphocreatine, while right at this border zone we have myoglobin, high ADP, low phosphocreatine and essentially an anoxic region. So the effect, the stress, on the pigment we love, myoglobin, is the fact that it is deoxy down here where this system is making lactic acid, is giving you a pain if you have cardiac distress, and as the oxygen concentration gets lower, this border zone will move up to the arterial end and if it stays that way, you'll grow a cardiac infarct - hurray, hurray. That's a bad thing and this is the surprising fact, the system uses its high affinity to create a healthy tissue or an unhealthy tissue without a significant border zone between them. And of course, the border zone is interesting because there you have life or death, in other words, as the border zone is narrow, you'll be all right. If it's wide, you are threatened and if it engulfs the particular part of the tissue that is critical to the heart beat, you will get a heart attack.

We can visualize this by these macroscopic zones, this is a microscopic zone on a very small radius, by looking at NADH which is fluorescence. In fact, we can map the tissue which doesn't have oxygen because if there is no oxygen, the NADH will be reduced and it is fluorescent in the reduced state, so mapping of the zones on a macroscopic scale is done here. Here you can see a rat heart which has been perfused with fluorosine so you can see where the flow was. This little string tied off the coronary artery which simulates cardiovascular disease and this very blue part is blue with the fluorescence of NADH and that is the part that threatened. While it has a myriad of microscopic intracapillary zones in here, it has generally a sharp edge, which defined by analytical biochemistry, about 100 microns thick, about 1 intercapillary radius which is remarkable for confining this dead region, or threatened region, from spreading over the rest of the myocardium. So border zones are very sharp.

If one uses a technique which we've devised for looking at this, freeze trapped hearts, one can do a coronary tie off in three dimensions, and this, if you have stereovision without polaroids, is a stereo-pair and the apex of the heart is here and this is the infarct as visualized by redox scanning. So that's the tissue volume which was threatened by the coronary artery tie off. Thus, we can visualize these volumes in animal models, but of course you don't want to have your heart freeze trapped, cold. There is a way of determining the volume of threatened tissue or the state of the heart which is threatening the tissue from oxygen lack, and when we do that we are able to look at the parameters of the system. Phosphorus NMR will let you look at phosphocreatine, it will let you look at inorganic phosphate, and you can easily calculate that these two are complementary.

Let me tell you how you can observe the cytochrome function in vivo by looking at the NMR signals of the phosphate compounds. The simple way in which you can calculate the well-being of a tissue due to oxidative metabolism is by the ratio of ATP, which is at high potential, to ADP times P_i, which are low potential components. If one substitutes the elements of the creatine kinase equilibrium, one can substitute for ATP/ADP, phosphocreatine/creatine (PCr/Cr). So now we come to a new expression for the thermodynamic capability of the mitochondria, in terms of the ratio of phosphocreatine/phosphate which is determined directly, divide by the concentration of creatine, which is large compared to phosphate and can be taken as constant or calculated from other methods. So this gives an idea of how well the electron transfer and the oxidative metabolism is occurring.

Well, how do you explore this relationship? One of them is to load the system with work and cause, by cranking on an ergometer with arm, and observing with NMR, external to the tissue, the internal phosphocreatine level as set by oxygen delivery, myoglobin function and oxidative metabolism. This is a way in which we can assay the relationship between the physiological work output on the Cybex ergometer and the cost of such work in the arm muscle, of if this guy puts his leg the magnet to look at the solius muscle. In this case, the foot rests on a treadle, the coil here picks up the magnetic properties of the nuclei and is resting right against the solius muscle. The result that we obtain is shown here where we have a spectrum of phosphocreatine, which is very high because the system has been pumped up at rest to a high potential, a very small inorganic phosphate and a relatively constant ATP. These peaks would be of equal amplitude, but NADPH is under this peak as well as some other quantities, but generally its a very simple spectrum consisting of the high energy form and the low energy form. This is a peak which appears in mature individuals like myself, which is probably due to used up membranes. So if any of you think you have a lot of used membranes, well, try NMR to quantify it.

We can follow this as an example of what Hans Frauenfelder was calling sick. These are SICK blood vessels and these series of spectra show at rest a high phosphocreatine peak, and then during exercise the phosphocreatine peak is low while the inorganic phosphate is way up. This was a tremendous transformation in just 1 minute of exercise, the two peaks are equal here in recovery, and the phosphocreatine comes back. The balance of phosphocreatine and inorganic phosphate is constant and so forth.

You then come back to recovery which is about the same as the initial resting state. Well, this leg was diseased (it wasn't mine, could have been), because the arteries were clogged, so the hemoglobin was deficient in delivering oxygen to the leg and just a little over one minute of exercise caused pain and dropping of the phosphocreatine level. We can plot the recovery time post- exercise and we see that two things happen; it takes about 7 minutes to recover and to be able to walk comfortably again.

If you are interested in lactic acid, the pH dropped abruptly to 6.4 which is an ouch that would have hurt because the lactic acid causes discomfort, and then recovers in this way as the phosphocreatine reaches the base line. So here have a sick leg, sick blood vessels, but our surgeons are clever and they can operate on the leg, replace the blood vessels of the leg, the heart or any place that you want. And you can see now, the same person during exercise has a much better phosphocreatine phosphate ratio during exercise and is in the second is recovered. The kinetics profile for recovery now is much better and the oxygen delivery to the leg is improved, the grafted vessels are clear now and the leg improves and the acidity is much less, and he can go comfortably – what is it – climbing mountains or whatever you like to do for your favorite disaster sport.

So that's one kind of thing, and now the other thing is "how do you improve yourself?" Well, let us consider now a steady state system where you are going to a graded exercise at increasing levels of work, perhaps up to some comfortable level and we measure the efficiency in the graded exercise. This is all done in an ergometer and it was done with a series of people who might have been trying out for the Olympics, as this young lady was last summer. So she was super-trained and super eager. As she exercises more, she increases her phosphate and decreases her phosphocreatine. In other words, we plot the reciprocal of the phosphocreatine over phosphate and in this linear region, she is aerobic where she uses just the right amount of oxygen to match the work rate so that each one of these is a work load and the intersection with this profile is an operating point where she worked for five minutes here, five minutes here and so forth.

This is an endurance situation where she is able to go on fairly indefinitely in an equilibrium or near equilibrium state. But as the phosphocreatine falls to this point, we reach the maximum velocity permitted by oxygen delivery, myoglobin transport or cytochrome oxidation and then we start with glycolysis, because the level of the control system now turns on the auxiliary motor but of course the price of that is lactic acid. This is glycolysis. It costs you a bigger drop of phosphocreatine to run the glycolytic system. That is what Jane Kent could at the time and we quantified her performance in the slope of this line which I'll talk about in a little bit (50 units (watts) against cost) so this is a transfer function for the efficiency of the system.

Even more interesting was the Iron Man event which is I think bicycling, swimming and running in Hawaii for which five of the contestants trained with us and their slopes in this test were running from 265 down to 130, almost chronologically with their physiological performance. What was more important was their performance on the actual triathalon event, I think it was one from New York to Philadelphia. Since we already knew their capabilities we could have said that they really didn't need to do the triathalon, they could have just stayed home and read a book or just practiced in our system. But of course, they didn't agree, there was no prize here and the only sport here was a fellow who was good at rowing, so he hadn't developed his tissue for the endurance performance. These are endurance performers, because rowing is part sprint and depending upon whether it's sprint or endurance, you get different kinds of muscles, but I wanted to show to you the dynamic range of this system we are working on from a sedentary guy like you and me. I suppose Hans is up here, but to Olympic training people to these "elite" athletes is a factor of 10. Well, where does that 10 come from? That 10 comes from the fact that you can increase your oxidative capacity, your oxygen transport, over that range to give you a much more higher phosphocreatine/phosphate ratio, and you'll avoid lactic acidosis and you can keep doing this in the Iron Man or whatever your favorite event is.

Let us look for a minute at some equations (sorry about the ink dot, that should have been inorganic phosphate). One can show that the work rate in relation to the maximum feasible is expressed quantitatively as rectangular hyperbola. In other words, that's predicted by the simple Michaelis-Menton equation for an enzyme kinetic and one can expect this hyperbolic relationship and indeed in the case of Jane Kent, you can superpose an approximate hyperbola but you see, it shouldn't be an exact fit because glycolysis supplemented the activity which we didn't put into account. We have therefore a simple way of quantifying or calculating what fraction of maximal activity you are using, because this being a hyperbola, you can just take a work load and if the P_i/PCr has a certain value, you are certain of the way up to you're maximal capacity of the oxidative and the transport mechanism.

That's very useful and let's see how this might apply to another disease. These are sick mitochondria in a baby's heart. The so-called "ragged red" stain is a Gamori stain which simply is reduced when the mitochondria membranes have been broken in this actual autopsy tissue. It turns out that there are three babies of a couple in Boston, two of whom we saw. The first one died undiagnosed of a heart attack and the tissue showed that there was a mitochondrial disease, possibly in the respiratory chain itself. We saw this baby at about eight months with a very much enlarged heart and the surface coil that we used projected right over the sternum and got signals from the heart of baby, and we studied the NMR signals to find out whether that heart was overloaded because of a low energy state, a low functionality of the mitochondria in the heart. Indeed, we would expect that this heart would be overloaded, because it did hypertrophy and the baby was the sister of an older sibling who had died of an heart attack, and these mitochondrial diseases are usually transmitted by the mother because she has all the mitochondria and we provide only a few sperm tails or maybe only one, so that this is a female transmitted thing. Sure enough, we found that the baby (KP had)in her cardiac muscle hypertrophy at eight months where the inorganic phosphate was a very significant high peak and the ratio of phosphocreatine to inorganic phosphate was less than 1, which if we had been running, would have been a lactic acidotic, dangerous situation. Well, this baby died a month later of a heart attack. The couple persisted, and presented us nine months later with a new baby (OP) and this baby was symptomless and had a phosphocreatine/phosphate ratio between 1.7 and 2.2 in a month's interval, and is doing very well, which of course the parents were very eager to learn that this baby might not develop this mitochondrial disease, and of course we could tell that right away.

Well now did our analysis tell us anything about these two cases?

You can see the phosphocreatine/phosphate level for KP, who died, was low and therefore near the maximum velocity capable of the mitochondria myoglobin system, while OP, with a higher phosphocreatine/phosphate ratio plotted linearly in the reciprocal of 2, was significantly displaced from the maximum velocity. So this was what you might call a safe operation point. Now, why was this unstable, why did KP pass away? The point was, any further progress of the disease at this load line would have brought you closer and closer to the asymptote of the hyperbola and the minute the hyperbola is below the operating point, your heart has to stop or pump less blood and when it's in that state, this is what we could call a highly unstable region. So we would have logically expected that when the biochemical apparatus for oxidative metabolism fails to produce a phosphate potential consistent with the load imposed upon it, there will be a crash. You will either stop exercising if it's a muscle, or you'll stop beating if it's a heart, and you'll stop thinking if it's a brain. Going back to this initial model, we see that it all really does stem in a very interesting way from the properties of the mitochondria which can be overworked, or they can increase, you can get an adaptation to overwork, the hypertrophic heart, the hypertrophic athlete as indeed the iron men are, or you can just be content with things as you are and have a mild lactic acidosis and be careful at what you do.

In summary, there is a reason for what we're doing because we're trying to find the molecular nature of these reactions, but I would like to emphasize what is most interesting is the stressed condition and oxygen stress is a very important condition where the ferrous forms are reacting with oxygen so that anything that facilitates the entry of the ligand, any conformation that is more active, anything that helps myoglobin to do its this is worthwhile. So, let's have tomorrow's session instead of quitting now.

Discussion

George Lang: What are the physiological changes that cause an increase work output with exercise?

BC: Well, George, that's stress as far as we know. In the limbs where we know most about it, it's apparent that damage helps. It's terrible isn't it? That is, in other words, if you stress your arm or leg to the point where there is actual cell rupture, release of creatine, myoglobin into the blood stream, that then the body will make more. Fatigue itself doesn't seem to do it, but it is the stress beyond the normal point. We have a symposium next Monday at the FASEB meeting on just this point, fatigue and the relationship to muscle adaptation. I can give you an example; in a rabbit, if you stimulate a muscle at 10 Hz for three weeks and that's a lot, that muscle will completely change its characteristic from white to red, will be hyper-adapted in cytochrome, and myoglobin and oxidative capacity.

Audience Questions: Does the mass of the muscle increase?

BC: There will be mass increase, yes.

Peter Wolynes: In trying to settle one of the issues in this conference, what are the possibilities of getting Hans and John Hopfield to be visit on your machine?

BC: There are two machines in the hospital right across the street, and we'll settle the argument that way. Is that right? Agreed? Schedule it during the lunch hour.

Mark: Were there any changes after a heart attack in a patient? Has he exercised himself in a sense?

BC: Well, cardiac hypertrophy is a very common response to stress and in addition to having an infarct there will probably be partially occluded blood vessels which are causing just the kind of stress that George mentioned which will lead to cardiac hypertrophy. Whether or not the skeletal muscle analogy applies completely, we certainly don't know because, let me say that the heart is harder to measure than the limb because an NMR is something that you can stick any object into, and your heart doesn't project from the body – at least not normally. So to do that on adult humans, we need a big bore machine but most of our data is on babies or dogs, and I would say, the returns aren't in, but certainly the hypothesis is that will behave just like the skeletal tissue.

Thank you very much, have a good evening.

Submitted Abstracts

Evidence for Electrogenicity of the Q_A to Q_B Electron Transfer in Photosynthetic Reaction Centers

G. Alegria, P.L. Dutton

University of Pennsylvania

Philadelphia, PA 19104

It has recently been shown (1) that it is possible to transfer Langmuir-Blodgett (LB) films of reaction centers from *Rps. Sphaeroides* onto solid surfaces in arrays that permit the application of external electric fields. The effect that these fields have on the kinetics of charge separation and recombination processes by shifting the relative energy levels of the electrogenic states involved is being studied. Previous work has shown that the electron transfer reaction between the primary quinone (Q_A) and the secondary one (Q_B) was not functional, an effect that could be due to either yielding inactive configurations or alterations in the role of the proton thought to be involved in the reaction. Modifications in the film making technique have resulted in what appears to be a partial reconstitution of Q_A to Q_B electron transfer as judged from measurements of light induced electric currents. The following results have been obtained:

1) Treating the film with ubiquinone 10 to increase the $Q_A Q_B$ population yields a half time of $\sim$ 280 msec. Treatment of the film with orthophenanthroline, a well known inhibitor of the Q_A to Q_B electron transfer decreases the half time to $\sim$ 90 msec.

2) The charge recombination from Q_B appears to be very sensitive to external fields. Relatively small fields produce significant changes in the kinetics. Addition of orthophenanthroline made the recombination reaction field independent, observations that clearly show that in this system Q_A to Q_B e transfer has an electrogenic component. However, before a quantitative analysis of the field effect on this reaction can be started, structural work on the arrangement of the RCs in the solid film will have to be done.

The differences in the half times with respect to the native system values (1 sec for Q_B recombination) may be related to structural changes evident in the film although it is likely that the absence of protons is a major source of the difference.

In conclusion, the important Q_A to Q_B electron transfer reaction has been successfully reconstituted in reaction centers in the form of solid films. The reaction kinetics have been shown to be sensitive to applied electric fields.

(1) G. Alegria, P.L. Dutton, Z. Popovic, G. Kovacs, Biophys. J. (Abstr.) 47 (1985) 4a.

Directional Electron Transfer in Ruthenium Modified Cytochrome c

R. Bechtold, C. Kuehn, C. Lepre, H. Schwarz, S.S. Isied

Department of Chemistry

Rutgers, The State University of New Jersey

New Brunswick, NJ 08903

Horseheart cytochrome c has been modified at His-33 with several different ruthenium redox reagents. The resulting modified proteins undergo intramolecular electron transfer where an electron is transferred between the ruthenium label and the heme c site. Depending on the redox potential of the ruthenium to the heme or vice versa. Using the two ruthenium complexes, $[Ru(NH_3)_5(OH_2)]^{2+}$ and $[Ru(NH_3)_4(N \ O \ -C-NH_2)(OH_2)]^{2+}$, two proteins modified at the same site (His 33) were prepared. Based on the redox potentials of the two ruthenium reagents, intramolecular electron transfer is expected to occur with the same rate, but in opposite directions. In this poster our recent results on these modified proteins will be presented and its implication discussed.

Through Bond and Through Space Limits of the Long Distance Electron Transfer Problem

David N. Beratan and J.N. Onuchic

Department of Chemistry

California Institute of Technology

Pasadena, CA 91125

A model is presented for the long distance electron transfer problem in which the through bond and through space contributions to the tunneling matrix element can be extracted as particular limits of the interaction energy. The relative importance of through bond and through space coupling is discussed. A first model for a donor interacting with a protein might be a "trap" interacting with a periodic potential. In this simple model the decay of the donor state with distance is found. Its decay includes both through bond and through space parts. We find the limits of the problem in which either the through bond or through space pathway dominates. The characteristic distance decay of the interaction energy is different in the two cases. In real systems the through bond pathway is considerably more favorably given the same tunneling distance. In proteins a combination of through bond and through space interactions probably contributes to the donor acceptor interaction.

The Coupling Between Electron Transfer Reactions and Protein Conformational Transitions in Biological Energy Transduction

Bo Cartling

Department of Biophysics

Arrhenius Laboratory

Stockholm University

S-106 91 Stockholm

Sweden

A fundamental mechanism of biological energy transduction is the energetic coupling between electron transfer reactions and strongly endergonic reactions, such as phosphorylation reactions in oxidative and photosynthetic ATP-synthesis. This energetic coupling cannot be accounted for by the electronic - vibrational coupling included in theories of biological electron transfer since excited vibrational energy is dissipated. By partitioning the nuclear degrees of freedom of an electron transfer enzyme into vibrational and conformational types and including electronic - conformational coupling in addition to electronic - vibrational coupling, however, energy transduction can be explained.

We present a stochastic description of protein conformational dynamics and electronic - conformational coupling. It is shown that electron transfer has to be restricted with respect to conformational states to prevent dissipative processes, and the underlying molecular mechanisms of this are discussed. Actual calculations demonstrate different types of transient kinetics, determined by the relative rates of electronic and conformational transitions. The support by available experimental information of the concepts and mechanisms introduced is discussed and further experiments are proposed.

Structural Differences Between Rate States of
Carboxyhemoglobin Compounds

M. Chance[1], B. Chance[1], L. Powers[2], L. Parkhurst[3]

C. Kumar[1] and Y.H. Chou[1]

1-Dept. of Biochemistry and Biophysics

University of Pennsylvania and Inst. for Structural and Functional Studies

Philadelphia, PA

2-AT&T Bell Laboratories

Murray Hill, NJ

3-Dept. of Chemistry

University of Nebraska

Lincoln, NE

The two state model of hemoglobin has been popular for many years but the description of the structural differences between the two states has been established only for the oxy to deoxy transition.

X-ray absorption spectroscopy provides a method of precisely determining the interatomic distances of iron and its ligands in heme proteins. We have studied the CO complexes of various hemoglobins in order to detect structural differences between the R and T states represented by the different compounds. Our results are consistent with an earlier study by Chance et al (Biochem. 22, 3820 (1983)), which showed that when myoglobin-CO is photolyzed, the resultant geminate state trapped at $4°K$ has an expanded iron-pyrrole nitrogen distance (Fe-Np) and a slightly expanded Fe-CO distance compared to the R state. Carp HbCO makes a well characterized shift from R to T states when the pH is lowered or upon the addition of organic phosphates (IHP). We found that not only is the Fe-CO bond enlarged upon the switch to the T state, but the distribution of the bond population is broader. A comparison of Uriches HbCO and Leghemoglobin CO compounds, which also represent R and T state, generally follows the same trend.

Far Ultraviolet Resonance Raman Spectroscopy
of Proteins and Protein Components

Robert A. Copeland and Thomas G. Spiro

Department of Chemistry

Princeton University

Princeton, NJ 08544

Recent advances in laser technology offer an opportunity to exploit the resonance Raman effect using far ultraviolet excitation wavelengths (down to 200nm). Detailed vibrational information can now be obtained on ultraviolet chromophores within protein selectively. We have concentrated attention on the aromatic amino acids, and amide chromophores. Data are presented for tyrosine, tryptophan, and phenylalanine at varying excitation wavelength, and under different conditions (H_2O, D_2O, and high pH). Within proteins, environmentally sensitive amide, tyrosine, and tryptophan modes serve as markers of secondary structure, and solvent accessibility of particular protein segments. Data are presented for hemoglobin, cytochrome c, and the abnormal tyrosine residue of chicken ovomucoid domain III.

Don DeVault

University of Illinois

Department of Physiology and Biophysics

524 Burrill Hall

407 South Goodwin Avenue

Urbana, IL 61801

The geometry of the *R. viridis* reaction center as revealed by the x-ray studies of Deisenhofer et al (1) may be adapted to enhancing the coupling of specific porphyrin vibrations to electron transfer. The orientations of the two BChl (acceptors from BChl$_2$) and the two BPhe (acceptors from BHcl) are edge-on toward their donors while the donors, incn]YZk `whe two BPhe as donors to Q (quinone) are broadside toward their acceptors. This should enhance coupling of the out-of-plane flat-plate type of vibrations of the donor porphyrin to the electron transfer. The equilibrium position of this vibration is a function of the charge on the donor in an electric field perpendicular to the donor plane. This electric field can be that of the transferring electron if the donor is left in a charged state and in this case the electron is directed broadside from the donor. Otherwise the field can come from fixed electric charges in the environment.

The non-heme iron atom is well placed to fill the role of the fixed charge as all four donors which will be neutral when the electron leaves face broadside to the iron atom. Its positive charge would also attract the electron to the general direction of Q which is near by.

The relative orientations of the porphyrins toward each other has been independently measured by Tiede et al (2). They point out that the orientations contradict the suggestions, based on orbital overlap, that parallel would be best.

(1) Deisenhofer, Epp, Miki, Huber and Michel, J. Mol. Biol. *180*, 385-398 (1984).

(2) Tied, Choquet and Breton, Biophys. J. *47*, 443- 447 (1985).

The Transition State for the Change in Hemoglobin Quaternary Structure is T-Like

F.A. Ferrone, A.J. Martino and S. Basak
Department of Physics
Drexel University
Philadelphia, PA
and M. Colette and M. Brunori
University of Rome
Rome, Italy

Using modulated excitation of the carboxy derivative, and monitoring the structural change in the Soret band, we have measured the rate of change between the oxy (R) and deoxy (T) quarternary structure for hemoglobin A and hemoglobin Kansas with three CO ligands bound. Taken with other available data, the data from HbA suggests that the transition state is T like, i.e. the $R \to T$ rate changes much less than the R-T equilibrium changes as ligands are added. This may arise from the control of the transition state by a barrier unconnected with stability of either state, as proposed by Baldwin and Chothia. Alternatively it may be due to the intersection between the potential wells describing the R and the T states being very close to the latter along the reaction coordinate. Because Hb Kansas has a destabilized T state, measurement of its structural kinetics provides a test to distinguish between the two hypotheses. We find that the Kansas mutation changes the equilibrium constant (relative to HbA) by much more than it alters the $R \to T$ rate, in contradiction to the conjecture of Baldwin and Chothia. Thus is appears that the transition state is governed by the very same factor which affect the relative stabilities of the R and T structures.

Electron Transfer in Reaction Centers with Various Quinones Functioning as Q_A

M.R. Gunner, N. Woodbury, W.W. Parson, P.L. Dutton

Department of Biochemistry and Biophysics
University of Pennsylvania
Philadelphia, PA 19104

Department of Biochemistry
University of Washington
Seattle, WA 98195

The native ubiquinone-10 that is tightly bound at the Q_A site of the reaction center of the photosynthetic bacteria *Rps. sphaeroides* can be removed and function can be reconstituted with a variety of other quinones. This is a very good system for studying the respective roles of protein and redox component in determining the rates and free energy of electron transfer between the tightly bound redox sites that are imbedded in the protein. The kinetics of quinone involved internal electron transfers are found to be dependent on what quinone is acting as Q_A.

1) The rate of reduction of Q_A by BPh^- is only weakly quinone dependent.

2) There are two pathways for charge recombination:

a) Direct electron transfer from Q_A^- to $(BChl)_2^+$. For all quinones tested, the rate of electron transfer by this route is found to be only very weakly dependent on the quinone or on the temperature. This suggests that the temperature independence of the rate in the native protein is not due to the exact matching of the energy levels between charge separated state and the ground state.

b) Equilibration of $(BChl)_2^+$ Q_A^- with a thermally accessible higher energy intermediate which then decays to the ground state. This path is substantially faster than the direct route for low midpoint quinones, but it slows with decreasing temperature or decreasing energy of the $(BChl)_2^+$ Q_A^- state.

3) The variation in the kinetics of the "thermal" route for $(BChl)_2^+ \, Q_A^-$ decay correlate extremely well with the changes in the free energy of this state as measured by the delayed fluorescence. Use of this technique shows that the relative $E_{1/2}$ values for Q/Q^- of quinones in the Q_A site are necessarily the same as the relative $E_{1/2}$ measured in solution; this gives a more active role to the protein in determining the *in situ* energetics of the electron transfer reactions; the implications of this effects are important not only to kinetic and thermodynamic studies, but also for structural probes of the site.

Molecular Dynamics Simulation of Ligand Photodissociation from Hemoglobin

Eric R. Henry and William A. Eaton

Laboratory of Chemical Physics

NIADDK, NIH

Bethesda, MD 20205

and

Michael Levitt

Department of Chemical Physics

Wiezmann Institute of Science

Rehovot, Israel

We have simulated the photodissociation of carbon monoxide from the alpha subunit of hemoglobin using the technique of molecular dynamics. Photodissociation was induced by interrupting the trajectory, deleting the iron-ligand bond from the potential function, changing the heme potential function to that of the deoxyheme conformation, and restarting the simulation. Heme potential functions were parameterized to reproduce the energies and forces for the displacement of the iron from the porphyrin plane found in quantum mechanical calculations. Simulations were also performed for an isolated heme-imidazole-CO complexes in a vacuum, in order to assess the effect of the protein on the rate and extent of the iron out-of-plane motion after CO dissociation.

The half-time for this motion was found to be 50-150 femtoseconds for both the isolated heme complex and the complete alpha subunit. This result supports the recent interpretation of a 350 femtosecond spectral relaxation seen in optical absorption studies of photolyzed heme complexes and hemoglobin as arising from the iron motion out of the heme plane.

The Bacteriorhodopsin Puzzle

Leslie A. Kuhn and John S. Leigh, Jr.
Department of Biochemistry and Biophysics
University of Pennsylvania
Philadelphia, PA 19104

Bacteriorhodopsin is an integral membrane protein found in the purple membrane of *halobacterium halobium*. Since electron microscopy data (Henderson and Unwin, 1975) established that the membrane- imbedded part of bacteriorhodopsin consists of seven closely packed α-helices there has been much interest in learning what parts of the amino acid sequence these helices correspond to and how the helices interact. An arrangement of amino acid residues in the seven helices which is consistent with electron and neutron scattering data and general postulates of protein structure is proposed. Fourier transforms of hydrophobicity indexed over the protein sequence suggests possible modes of interaction between helices.

Predicting Membrane Protein Structure

Leslie A. Kuhn and John S. Leigh, Jr.

Department of Biochemistry and Biophysics

University Pennsylvania

Philadelphia, PA 19104

The transmembrane segments of a membrane protein may be predicted from a membrane propensity profile of its amino acid sequence, in which the amino acids are represented by their frequencies of occurrence in a number of putative transmembrane segments. When this sequence has been smoothed by a running average function, the transmembrane segments appear as extended, positive peaks. In an application of this technique to a pool of ten previously studied membrane proteins, the predicted intra- and extra- membrane structures agreed 93.6% on a residue-by-residue basis with the previously suggested structures. This algorithm has been applied to predict the transmembrane segments in subunits I, II, III, IV, V, VIa, VII, VIIIa, and VIIIb of bovine cytochrome c oxidase and has also been used as a tool for studying structural homology between different species cytochrome oxidase subunit I.

Proposal to Detect and Characterize Coherent Vibrational Energy Transport in Alpha-Helical Protein Segments

R.S. Knox

University of Rochester

We propose an experiment designed to create and later detect a localized packet of vibrational energy ("soliton") which has moved through a protein alpha helix. The central idea is to produce a hot electronic excited state, couple part of the energy to vibrational modes of the helix, and probe for a hot ground state of a remote chromophore. The design uses visible and near- UV pump and probe light so that the beginning and end points of the soliton's flight can be determined. The feasibility of the experiment, as well as some of the theoretical problems is poses, and plans being made to perform it, will be discussed. Subpicosecond techniques will be required because we expect coherent motion of the packet over only small distances (up to a few tens of nm). Its speed of propagation is limited to about 1.7 nm/ps (1).

(1) J.M. Hyman, et al., Physica 3D, 23-44 (1981); A.C. Scott, Phys. Rev. A26, 578-595 (1982).

Site Selection Spectroscopy of Proteins

H. Koloczek, K.G. Paul and J.M. Vanderkooi

Department of Biochemistry and Biophysics

University of Pennsylvania

Philadelphia, PA

and

Department of Physiological Chemistry

University of Umea

Umea, Sweden

Optical spectra of biological molecules are structureless under normal measuring conditions. However, using site selection conditions, i.e., narrow laser band excitation and low temperatures (4.2K), resolved spectra are obtained which allow for optical interrogation of single substates of molecules. Quasi-line emission spectra are obtained from the mesoporphyrin derivatives of horse radish peroxidase A and C, Leg-hemoglobin and myoglobin, and iron-free cytochrome c. Excitation into the higher vibrational bands leads to multiple 0-0 emission lines. The distribution of the emission spectra differed with the different proteins, and varied with the addition of substrate, indicating the porphyrin. The extent of the inhomogeneous broadening was about 200 cm^{-1} (5.7 kcal/mole). These findings will be discussed in terms of protein dynamics. (Supported by National Science Foundation PCM 84-0844 and Swedish Medical Research Council 03-6522.)

Segmental Motion at the N-Terminal Structural Domain in Epidermal Growth Factor

K.H. Mayo

Department of Chemistry

Temple University

Philadelphia, PA 19122

The downfield aromatic region of a 500 MHz proton NMR spectrum of mouse epidermal growth factor is shown below. Several of the resonances assigned to the Tyr-10, Try-13 and His-22 residues in mouse epidermal growth factor (mEGF) seem to have one or more associated "minor" resonances (labelled 'B' and 'C') (Mayo, 1984). The term "minor" defines those resonances in the NMR spectrum whose integrated intensities are less than one proton each. Based primarily on the observation of saturation transfer between these minor resonances and the Tyr-10, tyr-13 and His-22 resonances, respectively, a process of slow chemical exchange on a 500 MHz NMR time scale must be occurring between these minor resonances and their respective major resonance counterparts (labelled 'A'). Selective saturation transfer and spin-lattice relaxation (T_1) experiments allow estimation of the exchange rate constants which range from 10 s^{-1} to 700 s^{-1} at 303K. The effect of temperature on the exchange kinetics gives the transition enthalpies ($\sim$ 5 kcal/mole) and entropies ($\sim$ 2 $\times 10^{-1}$ kcal/mole °K) of the exchange process and allows insight into the most probable exchange mechanism. The unequal resonance intensities of the exchanging resonances rule-out 180° "flip" motions about the $C_\beta - C_\alpha$ bond of the tyrosines as the cause for exchange. Rotation about the $C_\alpha - C_\beta$ tyrosine bond also seems unlikely. Since all three residues (i.e. Tyr-10, Tyr- 13 and His-22) are conformed at the N-terminal structural domain of the protein (Mayo, 1984) and the kinetics and thermodynamics of the exchange process are very similar for each of these residues, a common exchange mechanism is, therefore, highly probable. Fluctuations between slightly different conformational states of a segment of the N terminus seem best to account for the data.

Reference: K.H. Mayo, Biochemistry *23*, 4485-4493 (1984).

Metastable Photoproducts from Carbon Monoxy Myoglobin and Hemoglobin

D.L. Rousseau

AT&T Bell Laboratories

Murray Hill, NJ 07974

The photodissociation of carbon monoxy hemoglobin and myoglobin has been studies by low temperature resonance Raman scattering to determine the relaxation processes of the photoproducts. In myoglobin the photoproduct generated at very low temperatures (1.6 and 4.2°K) has an expanded heme core when compared to the deoxy preparation. No other differences were found. At higher temperatures the heme relaxes to take on its deoxy conformation. In the hemoglobin photoproduct, over a wide temperature range many vibrational modes have frequencies which differ from their values in deoxy hemoglobin. The differences in behavior between hemoglobin and myoglobin account for the differences in their biological function. In hemoglobin the state of ligand binding is communicated to the protein through interactions between the heme and the surrounding amino acids thereby triggering the cooperative transition. Myoglobin is noncooperative so the heme-protein interactions seen in hemoglobin are absent.

Charge Recombination Kinetics of cyt c_{558}^+ and Q_A^- in Reaction Centers from Rhodopseudomonas viridis

Robert J. Shopes and Colin A. Wraight

University of Illinois

505 S. Goodwin

289 Morrill Hall

Urbana, IL 61801

Reaction centers from the photosynthetic bacteria *Rp. viridis* contain bound cytochromes, including two high potential cytochrome c_{558}. When depleted of secondary quinone, isolated reaction centers perform the following light-induced reaction sequences:

$$ccPQ_A \xrightarrow{h\nu} ccP^+Q_A^- \rightarrow cc^+PQ_A^- \qquad (Rxn.1)$$

We have observed that Q_A^- and cyt c_{558}^+ decay with the same kinetics. The halftime, $t_{1/2}^c$, did not change when the reaction center concentration was varied 40-fold (0.1 - 4 μ M), implying that the cyt c_{558}^+ Q_A^- pair slowly recombines by an intramolecular process. Preliminary studies on the pH, ionic strength and temperature dependence of this charge recombination show some variation with each of these factors. Of most interest, the reaction is temperature dependent from 40- 20°C, but seems to become temperature independent at about 18°C.

A plausible model for this backreaction involves an electron sharing between P and cyt c_{558} coupled with the charge recombination of P^+ and Q_A^-:

$$cc^+PQ_A^- \overset{K_c}{\leftrightarrow} ccP^+Q_A^- \xrightarrow{t_{1/2}^P} ccPQ_A \qquad (Rxn.2)$$

For this scheme, $t_{1/2}^c$ is given by:

$$t_{1/2}^c = t_{1/2}^P(1 + K_c).$$

We have taken $t_{1/2}^P$ form the recombination kinetics of $P^+Q_A^-$ at high E_h when cyt c_{558} is oxidized, and have calculated K_c from the difference between the midpoint potentials of P and cyt c_{558}. However, the redox states of the reaction center in these measurements are not the same as those involved in the recombination of cyt $c_{558}^+ Q_A^-$, but the true values of Rxn. 2 are experimentally inaccessible. Despite this reservation, we find that the pH dependence of the calculated $t_{1/2}^c$ matched the pH dependence of the observed $t_{1/2}^c$, thus supporting the model.

Horseradish Peroxidase Compound II Heme-Linked Ionization Monitored by the Resonance Raman Fe(IV) = O Stretching Vibration

Andrew J. Sitter, Catherine M. Reczek and James Terner
Department of Chemistry
Virginia Commonwealth University
Richmond, VA 23284

Peroxidases are members of a class of heme proteins which undergo oxidations of the heme to states above iron (III) in the intermediate enzymatic states. Upon reaction with peroxides, the brown resting enzyme, which contains an Fe(III) heme, is converted to a green intermediate known as compound I, which is two oxidation equivalents above the resting enzyme. A one electron reduction of compound I produces a red intermediate known as compound II. Compound II is believed to contain an Fe(IV) porphyrin π-radical cation. Compound II is an Fe(IV) heme.

The Fe(IV) = O resonance Raman stretching vibrations found by ^{18}O-induced shifts were identified for horseradish peroxidase compound [1] and ferryl myoglobin [2]. Here we report that Fe(IV) = O stretching frequency for horseradish peroxidase compound II will switch between two values depending on pH, with pK values corresponding to the previously reported compound II heme-linked ionizations of pK = 6.9 for isoenzyme A-2 and pK = 8.5 for isoenzyme C [3]. Similar pH-dependent shifts of the Fe(IV) = O frequency of ferryl myoglobin were not detected above pH 6. The Fe(IV) = O stretching frequencies of compound II of the horseradish peroxidase isoenzymes at pH FE(IV) = O stretching frequency of ferryl myoglobin. Below the transition points the horseradish peroxidase frequencies were found to be 10 cm^{-1} lower. Frequencies of the Fe(IV) = O stretching vibrations of horseradish to deuterium exchange below the transition point but not above. These results were interpreted to be indicative of an alkaline deprotonation of a distal amino acid group, probably histidine, which is hydrogen bonded to the oxyferryl group below the transition point. Deprotonation of this group at pH values above the pK disrupts hydrogen bonding, raising the $Fe(IV) = 0$ stretching frequency, and is proposed to account for the lowering of compound II reactivity at basic pH.

Compound X is formed by the reaction of horseradish peroxidase with chlorite at alkaline pH. Compound X has been proposed to contain an $Fe(IV)$-OCl heme based on [36]Cl incorporation [4]. We have proposed, alternatively, that compound X contains an $Fe(IV) = 0$ group [1]. The high value of the $Fe(IV) = 0$ vibration of compound II above the transition point at alkaline pH appears to be identical in frequency to what appears to be the $Fe(IV) = 0$ vibration of compound X.

[1] J. Terner, A.J. Sitter and C.M. Raczek, Biochim. Biophys. Acta *828*, 73-80 (1985).

[2] A.J. Sitter, C.M. Reczek and J. Terner, Biochim. Biophys. Acta (1985) in press.

[3] A.J. Sitter, C.M. Reczek and J. Terner, J. Biol. Chem. (1985) in press.

[4] S. Shahangian and L.P. Hager, J. Biol. Chem. *257*, 11529-11533 (1982).

Electrostatic Basis for Vectorial Light-Induced Charge Separation in Photobiological systems

Arieh Warshel

Department of Chemistry

University of Southern California

Los Angeles, CA 90089-0482

The energetics and efficiency of light-induced charge separation across membranes are considered on a molecular level. It is shown how the activation energy that controls the efficiency can be evaluated from the dielectric constant and dielectric relaxation of the system. Examples are given by analyzing the action of artificial and biological photosynthetic systems, including bacterial reactions center (1,2).

1) A. Warshel and D.W. Schlosser, Proc. Natl. Acad. Sci. *78*, 5564 (1982).

2) A. Warshel, Israel J. Chem. *21*, 341 (1981).

Dynamic Features in the Mössbauer Spectra of Heme Proteins

W.W. Wise, G.C. Wagner and P.G. Debrunner

Physics Department

University of Illinois

1110 W. Green St.

Urbana, IL 61801

We report Mössbauer studies of the mean square displacements $\langle x^2(T) \rangle$, the mean square velocities, $\langle v^2(T) \rangle$, and the quadruple splitting $\Delta E_Q(T)$ of heme proteins that probe the motions at the iron active center. Accurate measurements of the recoilless fraction $f(T)$, the isomer shift $\delta(T)$ and the quadrupole splitting $\Delta E_Q(T)$ on ^{57}Fe-enriched oxy- and carbon monoxy-myoglobin (MbO$_2$ and MbCO) in frozen aqueous solution have led to the following conclusions: (i) A harmonic model $\langle x^2 \rangle = \hbar/(6m) \sum (\beta_i^2/\omega_i)$ ctanh $(\hbar\omega_i/2k_BT)$ represents $\langle x^2 \rangle$ quite well up to 180K with a single mode, $\omega \sim 25 cm^{-1}$. The frequency of this vibration is in the range of the delocalized low-frequency modes predicted for globular proteins [1]. The larger contribution of this mode in MbCO than in MbO$_2$, $\beta_i^2/3 \simeq 0.29$ vs. ~ 0.13, is consistent with a stronger coupling to the iron in MbCO. (ii) Assuming that the temperature dependence of $\delta(T)$ is entirely due to the second order Doppler shift a harmonic model for $\langle v^2(T) \rangle$ fits the data well up to 250K with a single mode, $\omega \sim 220 cm^{-1}$ for MbO$_2$ and $\omega \sim 190 cm^{-1}$ for MbCO. The frequencies of these vibrations are similar to those reported for the iron-histidine stretching mode in Mb [2]. A combined fit of $\langle x^2(T) \rangle$ and $\langle v^2(T) \rangle$ for MbCO with the 512 cm^{-1} Fe-CO stretch [3], a low and an intermediate frequency mode, changes the values of ω_1 and β_i very little, suggesting a bimodal frequency spectrum. (iii) For both MbO$_2$ and MbCO we find empirically that an equation of the form $\Delta E_Q(T) = \Delta E_Q^{(0)} + \Delta E_Q^{(2)}$ ctanh$(\hbar\omega/2kT)$ represents the data better than a 2-state Boltzmann average. We therefore propose that the temperature dependence of ΔE_Q is of dynamic origin. The characteristic frequencies, $\omega \sim 25 - 100 cm^{-1}$, are consistent with a delocalized normal mode that incorporates oscillations of the Fe-axial ligand structure.

Supported by Grants GM15406 and NSF DMB82-09616.

[1] M. Go et al, Proc. Natl. Acad. Sci. USA *80*, 3696 (1983).

[2] T. Kitigawa et al. FEBS Lett. *104*, 376 (1979).

[3] M. Tsubaki et al, Biochem. *21*, 1132 (1982).

Participants

Guillermo D. Alegria
c/o Dr. Peter Les Dutton
University of Pennsylvania
Department of Biochemistry/Biophysics
37th & Hamilton Walk
Philadelphia, PA 19104

Ms. Denise M. Alexander
Cornell University
Department of Physics
Clark Hall
Ithaca, NY 14853

Dr. Phillip W. Anderson
Princeton University
Physics Department
Jadwin Hall
Princeton, NJ 08544

Dr. Robert Austin
Princeton University
Department of Physics
Joseph Henry Laboratories
Jadwin Hall
P.O. Box 708
Princeton, NJ 08544

Dr. Gwyn Ballard
The Rockefeller University
1230 York Avenue
New York, NY 10021

Dr. Terence W. Barrett
Code Air 310P
Naval Air Systems Com. Hq.
Washington, DC 20361

Bethany Bechtel
c/o Dr. Peter Les Dutton
University of Pennsylvania
Department of Biochemistry/Biophysics
37th & Hamilton Walk
Philadelphia, PA 19104

Dr. Rolf Bechtold
Rutgers University
Department of Chemistry
P.O. Box 939
Piscataway, NJ 08854

Dr. David N. Beratan
Caltech

Department of Chemistry
164-30
Pasadena, CA 91125

Dr. Marilyn F. Bishop
Drexel University
Department of Physics
Philadelphia, PA 19104

Dr. Kent Blasie
University of Pennsylvania
New Chemistry Building
Department of Chemistry
3301 Spruce Street
Philadelphia, PA 19104

Dr. Ephraim Buhks
B.F. Goodrich
Research & Development Ctr.
9921 Brecksville Road
Breckville, OH 44141

Ms. Barbara Boliano
c/o Dr. Helen Davies
University of Pennsylvania
Microbio. Med.
255 JohnsonPav/G2
37th & Hamilton Walk
Philadelphia, PA 19104

Dr. Burt V. Bronk
Clemson University
Department of Physics and Microbiol.
Clemson, SC 29631

Dr. Grant Bunker
University City Science Center
Inst. for Structural & Functional Studies
3401 Market Street Room 320

Dr. Eric Canal
Rockefeller University
1230 York Avenue
New York, NY 10021

Dr. Bo Cartling
Department of Biophysics
Arrhenius Laboratory
Stockholm University
S-106 91
Stockholm, Sweden

Dr. Marvin Cassman
National Institutes of Health

National Institute of General Medical Sciences
Westwood Building
5333 Westbard Avenue
Bethesda, MD 20205

Dr. Paul Champion
Northeastern University
Department of Physics
3600 Hungtington Avenue
Boston, MA 02115

Dr. Britton Chance
University of Pennsylvania
Biochemistry/Biophysics
Richards Bldg./G4
37th & Hamilton Walk
Philadelphia, PA 19104

Mr. Mark R. Chance
AT&T Bell Labs
600 Mountain Avenue
Murray Hill, NJ 07974

Dr. Yuan-Chin Ching
AT&T Bell Labs
600 Mountain Avenue
Murray Hill, NJ 07974

Dr. Greg Cole
Dupont Corporation
Wilmington, DE 19898

Dr. Robert A. Copeland
Princeton University
Department of Chemistry
Princeton, NJ 08544

Dr. Helen Davies
University of Pennsylvania
Microbio. Med.
255 Johnson Pav/G2
37th & Hamilton Walk
Philadelphia, PA 19104

Dr. Peter G. Debrunner
Department of Physics
University of Illinois
1110 W.Green Street
Urbana, IL 61801

Dr. Don DeVault
Department of Physiol./Biophys.
University Illinois
407 S. Goodwin Avenue
Urbana, IL 61801

Dr. Len Dissado
Department of Physics
Chelsea College
Pulton Place
London, SW65PR, U.K.

Dr. C.D. Durfor
GTE Laboratories
40 Sylvan Road
Waltham, MA 02254

Dr. Peter Les Dutton
University of Pennsylvania
Department of Biochemistry & Biophysics
37th & Hamilton Walk
Philadelphia, PA 19104

Dr. William A. Eaton
National Institutes of Health
Laboratory of Chemical Physics
Bldg. 2, Room B1-04
Bethesda, MD 20205

Dr. S. Walter Englander
Department of Biochemistry
University of Pennsylvania
Medical School
Philadelphia, PA 19104

Dr. George Feher
Department of Physics, b-109
University of California-San Diego
La Jolla, CA 92037

Dr. Jehuda Feitelson
Princeton University
Department of Chemistry
Princeton, NJ 08544

Dr. Frank Ferrone & Dr. Simon Basak
Huan Xiang Zhou, Michael Cho, Anthony Martin
(Students)
Department of Physics
Drexel University
Philadelphia, PA 19104

Dr. Seghart Fischer
Department of Physics
Technical University of Munich
E-8046
Garsching, FDR

Dr. Michelle Franzel
Princeton University
Princeton, NJ 08544

Dr. Hans Frauenfelder
University of Illinois
Department of Physics
Urbana-Champaign, IL 61807

Dr. Joel M. Friedman
AT&T Bell Labs.
Room 1A159
600 Mountain Avenue
Murray Hill, NJ 07974

Dr. Edwin Gabbidon
University of Pennsylvania
Department of Biochem./Biophysics
Richards/G5
37th & Hamilton Walk
Philadelphia, PA 19104

Dr. Debra A. Giammona
M.I.T.
Harvard University
Room 2-204
Cambridge, MA 02139

Dr. Kathleen Giangiacomo
c/o Dr. Peter Les Dutton
University of Pennsylvania
Department of Biochem./Biophysics
37th & Hamilton Walk
Philadelphia, PA 19104

Dr. Vitallii I. Goldanskii
Dr. Yuri Krypyanskii
USSR Academy of Sciences
117977 Ulitza Kosygina
4, Moscow, USSR

Ms. Marilyn Gunner
c/o Dr. Peter Les Dutton
University of Pennsylvania
Department of Biochem./Biophysics
37th & Hamilton Walk
Philadelphia, PA 19104

Dr. John J. Hopfield
Crellin Laboratory
Cal-Tech 164-30
Pasadena, CA 91125

Dr. Robert Goldstein
Stanford University
Department of Cell Biology
Stanford, CA 94305

Dr. Robin M. Hochstrasser

University of Pennsylvania
Department of Chemistry
231 S. 34th Street
Philadelphia, PA 19104

Dr. Lucia E. Hancock
Lehigh University
Box 399
Bethlehem, PA 18015

Dr. Yoshio Hattori
Institute of Atomic Energy
Kyoto University
Uji, Kyoto 611
Japan

Dr. Eric Henry
National Institutes of Health
Lab of Chemistry and Physics
Building 2, Room B1-04
Bethesda, MA 20205

Dr. Barry Honig
Columbia University
Department of Biochemistry and Molecular Biophysics
630 West 168th Street
New York, NY 10032

Toshiko Ichiye
Harvard University
Departmento of Chemistry
12 Oxford Street
Cambridge, MA 02138

Dr. Steven S. Isied
Department of Chemistry
Rutgers University
New Brunswick, NJ 08854

Steven M. Janes
University of Pennsylvania
Department of Chemistry
34th & Spruce Streets
Philadelphia, PA 19104

Carey K. Johnson
University of Pennsylvania
Department of Chemistry/D5
34th & Spruce Streets
Philadelphia, PA 191040

Dr. Joshua Jortner
Tel Aviv University
Department of Chemistry

514

Reamat Aviv 69.978
Israel

Dr. Syed M. Khalid
University City Sciences Center
Institute for Structural & Functional Studies
3401 Market Street, Room 320
Philadelphia, PA 19104

Dr. David Kleinfeld
AT&T Bell Labs
600 Mountain Avenue
Room 1C-535
Murray Hill, NJ 07974

Dr. Robert Knox
Rochester University
Department of Physics & Astronomy
Rochester, NY 14627

Dr. Henry Koloczik
University of Pennsylvania
School of Medicine/G3
Biochem./Biophysics
Philadelphia, PA 19104

Dr. Richard Korszun
University City Sciences Center
Institute for Structural & Functional Studies
3401 Market Street, Room 320
Philadelphia, PA 19104

Dr. Gregory Kovacs
Xerox Research Center
2660 Speakman Drive
Mississauga, Ontario L5K 2L1
Canada

Ms. Kasha Kozinski
University of Pennsylvania
Department of Biochem./Biophysics
Richards/G5
37th & Hamilton Walk
Philadelphia, PA 19104

Dr. James A. Krumhansl
Cornell University
Department of Physics
Ithaca, NY 14853

Dr. Leslie Kuhn
University of Pennsylvania
Department of Biochem./Biophysics
37th & Hamilton Walk
Philadelphia, PA 19104

Dr. Angelo A. Lamola
AT&T Bell Labs
600 Mountain Avenue
Room 1C-417
Murray Hill, NJ 07974

Dr. George Lang
Penn State University
Department of Physics
Davey Building
University Park, PA 16802

Dr. E. Margoliash
Northwestern University
Department of Biochemistry
2153 Sheridan Road
Evanston, IL 60201

Dr. Deven Mayo
Department of Chemistry
Temple University
13th & Norris Streets
Philadelphia, PA 19122

Mr. Steven Meinhardt
University of Pennsylvania
School of Medicine/G4
Department of Biochem./Biophysics
505 Goddard Labs
Philadelphia, PA 19104

Dr. Helmut Michel
Max-Planck-Institut fur Biochemie
8033 Martinsried Bei
Munchen
West Germany

Dr. Chris Moser
c/o Dr. Peter Les Dutton
University of Pennsylvania
Department of Biochem./Biophysics
37th & Hamilton Walk
Philadelphia, PA 19104

Dr. M.E. Michel-Beyerle
Sonderforschungsbereich 143
Institut for Phys. & Theor. Chemie
Technische Universitat Munchen
Lichtenbergstrasse 4
8046 Garching Bei Munchen

Dr. Thomas A. Moore &
Dr. Devens Gust
Department of Chemistry
Arizona State University

Tempe, AZ 85287

Dr. Ali Naquri
University of Pennsylvania
Department of Biochem./Biophysics
Richards/G5
37th & Hamilton Walk
Philadelphia, PA 19104

Dr. Amar Nath
Drexel University
Department of Chemistry
Philadelphia, PA 19104

Dr. James R. Norris
Argonne National Laboratories
Chemistry Division D-200
Argonne, IL 60439

Dr. Thomas M. Nordlund
Rochester University
Department of Physics & Astronomy
Rochester, NY 14627

Tomoko Ohnishi
University of Pennsylvania
School of Medicine
Goddard Labs, 505
37th & Hamilton Walk
Philadelphia, PA 19104

Dr. Mary Jo Ondrechen
Department of Chemistry
Northeastern University
Boston, MA 02115

Dr. Jose Nelson-Onochic
Caltech
164-30
Pasadena, CA 91125

Dr. Lester Packer
University of California
Lawrence Berkeley Lab
Membrane Bioenergetics Group
Berkeley, CA 94720

Dr. Fritz Parak
University of Munster
Institute of Physical Chemistry
4400 Munster, den
Munster, FRG

Dr. Martin Pope
New York University

4 Washington Place, Room 811
New York, NY 10003

Dr. Linda Powers
AT&T Bell Labs, Room 1D-467
600 Mountain Avenue
Murray Hill, NJ 07974

Dr. Zoran D. Popovic
Xerox Corporation
Palo Alto Research Ctr.
3333 Cayote Hill Road
Palo Alto, CA 94304

Ms. Alice R.W. Presley
Drexel University
Department of Physics
Philadelphia, PA 19104

Dr. Catherine M. Reczek
Virginia Commonwealth University
Department of Chemistry
0001 West Main Street
Oliver Hall
Richmond, VA 23284-0001

Dr. Dagmar Ringe
c/o Dr. Gregory A. Petsko
Massachusetts Institute of Technology
Department of Chemistry
Room 18-025
Cambridge, MA 02139

Dr. Heinrich Roder
University of Pennsylvania
School of Medicine/G3
Biochemistry & Biophysics
Philadelphia, PA 19104

Dr. Gerd Rosenbaum
University City Science Center
Institute for Structural & Functional Studies
3401 Market Street, Room 320
Philadelphia, PA 19104

Dr. Denis L. Rousseau
AT&T Bell Labs
600 Mountain Avenue
Murray Hill, NJ 07974

Dr. Massimo Sassaroli
AT&T Bell Labs, Room 10-435
600 Mountain Avenue
Murray Hill, NJ 07974

516

Dr. J. Robert Schrieffer
Department of Physics
University of California
Santa Barbara, CA 93106

Mr. Jonathan Schug
University City Science Center
Institute for Structural & Functional Studies
3401 Market Place, Room 320
Philadelphia, PA 19104

Mr. Robert Shopes
University of Illinois
505 S. Goodwin, Rm. 289
Urbana, IL 61801

Mr. Andrew J. Sitter
Virginia Commonwealth University
Department of Chemistry
1001 West Main Street
Richmond, VA 23284

Dr. Thomas G. Spiro
Department of Chemistry
Princeton University
Princeton, NJ 08544

Dr. Harvey J. Stapleton
Department of Physics
University of Illinois
Urbana, IL 61807

Dr. Edward A. Stern
Department of Physics
University of Washington
Seattle, WA 98195

Dr. Daniel L. Stein
Department of Physics
Princeton University
P.O. Box 708
Princeton, NJ 08544

Dr. Paul Todd
University City Science Center
Bioprocessing & Pharmaceutical ResearchCenter
3401 Market Street, Room 220
Philadelphia, PA 19104

Dr. Jane Vanderkooi
University of Pennsylvania
School of Medicine
Biochem./Biophysics, G3
Philadelphia, PA 19104

Dr. Kurt Warncke
c/o Dr. Peter Les Dutton
University of Pennsylvania
Department of Biochem./Biophysics
37th & Hamilton Walk
Philadelphia, PA 19104

Dr. Arieh Warshel
Department of Chemistry
University of Southern California
Los Angleles, CA 90089

Dr. Watt W. Webb
Cornell University
Applied Physics
Clark Hall
Ithaca, NY 14853

Dr. John J. Wendoloshi
Dupone Experimental Station
Bldg. 328/B15
Wilmington, DE 19898

Dr. Peter G. Wolynes
School of Chem. Sci.
University of Illinois
Urbana, IL 61801

Staff

Sarah B. Congdon, Program Director/ISFS
Sandra Kim, Programmer/Lab Assist.
Ellen Kim, Bookkeeper/Secretary
Rita Kushner, Secretary
University City Science Center
Institute for Structural & Functional Studies
3401 Market Street, Room 320
Philadelphia, PA 19104

Judith R. Flanagan, Program Coordinator
Suzanne White, Public Relations
University City Science Center
3624 Market Street
Philadelphia, PA 19104

Chilton Alter, Elec. Tech.
University of Pennsylvania
Philadelphia, PA 19104

Index

DATE DUE

MAY 2 0 1988		
OCT 1 3 1988		
JUL 2 1 1989		
JAN 03 1995		